Mycorrhizal Technology in Agriculture

From Genes to Bioproducts

Edited by S. Gianinazzi, H. Schüepp,
J.M. Barea and K. Haselwandter

Springer Basel AG

Editors:

Silvio Gianinazzi
CMSE-INRA/Université de Bourgogne
Dijon, France

Hannes Schüepp
Federal Research Station
Wädenswil, Switzerland

José Miguel Barea
Estación Experimental del Zaidín, CSIC
Granada, España

Kurt Haselwandter
Universität Innsbruck
Innsbruck, Austria

Library of Congress Cataloging-in-Publication Data
Mycorrhizal technology in agriculture : from genes to bioproducts / edited by S. Gianinazzi ... [et al.], editors.
p. cm.
Includes bibliographical references.
ISBN 978-3-0348-9444-9 ISBN 978-3-0348-8117-3 (eBook)
DOI 10.1007/978-3-0348-8117-3
1. Mycorrhizas in agriculture. 2. Mycorrhizal fungi. I. Gianinazzi, S.
SB106.M83 M935 2002
631.4'6--dc21 2002074514

Deutsche Bibliothek Cataloging-in-Publication Data
Mycorrhizal technology in agriculture : from genes to bioproducts / ed. by S. Gianinazzi .... - Basel ; Boston ; Berlin : Birkhäuser, 2002
ISBN 978-3-0348-9444-9

ISBN 978-3-0348-9444-9

Originally published by Birkhäuser Verlag, Basel, Switzerland in 2002
Softcover reprint of the hardcover 1st edition 2002

Printed on acid-free paper produced from chlorine-free pulp. TCF ∞
Cover design: Micha Lotrovsky, Therwil, Schweiz
Cover illustration: Six month old vine uninoculated (NM) or inoculated (M) with Endorize® at plantation in an unsterilized soil in a French vineyard. Inserts represents roots with mycorrhizal mycelium colonizing the soil, and identification of the mycorrhizal fungus in plant roots by PCR. Pictures provided by Biorize Ltd (Dijon, F.) and INRA (Dijon, F). Conception by D. van Tuinen

ISBN 978-3-0348-9444-9

9 8 7 6 5 4 3 2 1 www.birkhasuer-science.com

Contents

List of contributors

Henning von Alten, Institute of Plant Diseases, Universität Hannover, Herrenhäuser Str. 2, 30419 Hannover, Germany; e-mail: von-alten@mbox.ipp.uni-hannover.de

David Atkinson, Vice Principal (Research & Education), SAC, West Mains Road, Edinburgh EH9 3JG, UK; e-mail: d.atkinson@ed.sac.ac.uk

Luciano Avio, Centro di Studio per la Microbiologia del Suolo, C. N. R., Via del Borghetto 80, 56124 Pisa, Italy; e-mail: lucavio@agr.unipi.it

Concepción Azcón-Aguilar, Estación Experimental del Zaidín, CSIC, Profesor Albareda 1, 18008 Granada, España; e-mail: cazcon@eez.csic.es

Rosario Azcón, Estación Experimental del Zaidín, CSIC, Profesor Albareda 1, 18008 Granada, España; e-mail: razcon@eez.csic.es

John Baddeley, SAC, Craibstone Estate, Aberdeen, AB21 9YA, UK

Berta Bago, Estación Experimental del Zaidín, CSIC, Profesor Albareda 1, 18008 Granada, España; e-mail: bbago@eez.csic.es

José Miguel Barea, Estación Experimental del Zaidín, CSIC, Profesor Albareda 1, 18008 Granada, España; e-mail: jmbarea@eez.csic.es

Guillaume Bécard, Equipe de Mycologie Végétale, UMR 5546 CNRS/Université Paul Sabatier, Pôle de Technologie Végétale, 31326 Castanet Tolosan, France

Graziella Berta, Dipartimento di Scienze e Tecnologie Avanzate, Università del Piemonte Orientale, Corso Borsalino 54, 15100 Alessandria, Italy; e-mail: berta@mfn.unipmn.it

Valeria Bianciotto, Centro di Studio sulla Micologia del Terreno, CNR, Dipartimento di Biologia Vegetale, Università di Torino, Viale Mattioli 25, 10125 Torino, Italy; e-mail: bianciotto@csmt.to.cnr.it

Bachar Blal, Biorize, 8 Rue sainte Anne, 21000 Dijon, France; e-mail: bachar.blal@epoisses.inra.fr

Paola Bonfante, Centro di Studio sulla Micologia del Terreno, CNR, Dipartimento di Biologia Vegetale, Università di Torino, Viale Mattioli 25, 10125 Torino, Italy; e-mail: p.bonfante@csmt.to.cnr.it

Hermann Bothe, Botanical Institute, The University of Cologne, Gyrhofstr.15, 50923 Köln, Germany; e-mail: Hermann.Bothe@uni-koeln.de

Laurent Brechenmacher, UMR INRA/Université de Bourgogne, BBCE-IPM, CMSE-INRA, BP 86510, 21065 Dijon Cedex, France; e-mail: brechenmacher@epoisses.inra.fr

Catherine Calantzis, Université de Genève, Laboratoire de Bioénergétique, Chemin des Embrouchis 10, 1254 Jussy/Genève, Switzerland; e-mail: catherine.calantzis@bioen.unige.ch

Cinta Calvet, Instituto de Recerca i Tecnologia Agroalimentàries, Departament de Protecció Vegetal, Ctra. de Cabrils s/n, 08348 Cabrils (Barcelona), España; e-mail: cinta.calvet@irta.es

Amelia Camprubí, Instituto de Recerca i Tecnologia Agroalimentàries, Departament de Protecció Vegetal, Ctra. de Cabrils s/n, 08348 Cabrils (Barcelona), España; e-mail: amelia.camprubi@irta.es

Alan Cassells, Department of Plant Science, National University of Ireland, Cork College Road, Cork, Ireland; e-mail: a.cassells@ucc.ie

Justin Clapp, International Institute of Biotechnology, 1/13 Innovation Building 1000, Sittingbourne Research Centre, Sittingbourne, Kent ME9 8HL, UK

Christelle Cordier, UMR INRA/Université de Bourgogne, BBCE-IPM, CMSE-INRA, BP 86510, 21065 Dijon Cedex, France; e-mail: cordier@dijon.inra.fr

Amanda Craven-Griffiths, Research School of Biosciences, University of Kent, Canterbury, Kent CT2 6NJ, UK; e-mail: caavenamanda@praintl.com

John Dodd, International Institute of Biotechnology, 1/13 Innovation Building 1000, Sittingbourne Research Centre, Sittingbourne, Kent ME9 8HL, UK; e-mail: jcdodd@iibiotech.fsbusiness.co.uk

Eliane Dumas-Gaudot, UMR INRA/Université de Bourgogne, BBCE-IPM, CMSE-INRA, BP 86510, 21065 Dijon Cedex, France; e-mail: dumas@epoisses.inra.fr

Victoria Estaun, Instituto de Recerca i Tecnologia Agroalimentàries, Departament de Protecció Vegetal, Ctra. de Cabrils s/n, 08348 Cabrils (Barcelona), España; e-mail: victoria.estaun@irta.es

Falko Feldmann, Institut für Pflanzenkultur, Solkau 2, 29465 Schnega, Germany; e-mail: info@pflanzenkultur.de

Nuria Ferrol, Departamento de Microbiología del Suelo y Sistemas Simbióticos, Estación Experimental del Zaidín (CSIC), Profesor Albareda 1, 18008 Granada, España; e-mail: nferrol@eez.csic.es

Philipp Franken, Max-Planck Institut für terrestrische Mikrobiologie, Karl von Frisch Strasse, 35043 Marburg, Germany; e-mail: frankenp@mailer.uni-marburg.de

Anna Fusconi, Dipartimento di Biologia Vegetale, Università di Torino, Viale Mattioli 25, 10125 Torino, Italy; e-mail: fusconi@bioveg. unito.it

Silvio Gianinazzi, Unité Mixte de Recherche INRA/Université de Bourgogne Biochimie, Biologie Cellulaire et Ecologie des Interactions Plantes/Micro-organismes, CMSE-INRA, BP 86510, 21065 Dijon cedex, France; e-mail: silvio@epoisses.inra.fr

Vivienne Gianinazzi-Pearson, Unité Mixte de Recherche INRA/Université de Bourgogne Biochimie, Biologie Cellulaire et Ecologie des Interactions Plantes/Micro-organismes, CMSE-INRA, BP 86510, 21065 Dijon cedex, France; e-mail: Gianina@epoisses.inra.fr

Manuela Giovannetti, Dipartimento di Chimica e Biotecnologie Agrarie, Università di Pisa, Via del Borghetto 80, 56124 Pisa, Italy; e-mail: mgiova@agr.unipi.it

Preboste Goicoechea, Departamento de Fisiologia Vegetal, Universidad de Navarra, Irunlarrea s/n 31008 Pamplona, España

Armelle Gollotte, UMR INRA/Université de Bourgogne, BBCE-IPM, CMSE/INRA, BP 86510, 21065 Dijon Cedex, France; e-mail: gollotte@epoisses.inra.fr

James Green, SAC, West Mains Road, Edinburgh, EH9 3JG, UK

Carolin Grotkass, Institut für Pflanzenkultur, Solkau 2, 29465 Schnega, Germany; e-mail: info@pflanzenkultur.de

Milan Gryndler, Institute of Microbiology ASCR, Czech Academy of Sciences, Vídenská 1083, 14220 Prague 4, Czech Republic; e-mail: gryndler@biomed.cas.cz

Lucy A. Harrier, Biotechnology Department, Crop and Plant Science Division, Scottish Agricultural College, West Mains Road, Edinburgh, Scotland; e-mail: l.harrier@ed.sac.ac.uk

Kurt Haselwandter, Institut für Mikrobiologie, Universität Innsbruck, Technikerstrasse 25, 6020 Innsbruck, Austria; e-mail: kurt.haselwandter@uibk.ac.at

Ulrich Hildebrandt, Botanical Institute, The University of Cologne, Gyrhofstr.15, 50923 Köln, Germany; e-mail: ulrich.hildebrandt@uni-koeln.de

John E. Hooker, School of Applied Sciences, University of Glamorgan, Pontypridd, Mid-Glamorgan, CF37 1UD, UK; e-mail: jhooker@glam.ac.uk

Maria del Carmen Jaizme-Vega, Instituto Canario de Investigaciones Agrarias, Departamento de Protección Vegetal, Apdo. 60, 38200 La Laguna (Tenerife), Canary Islands, España; e-mail: mcjaizme@icia.rcanaria.es

Peter Jeffries, Research School of Biosciences, University of Kent, Canterbury, Kent CT2 6NJ, UK; e-mail: p.jeffries@ukc.ac.uk

Erik Joner, Centre de Pédologie Biologique, CNRS, 54501 Vandoeuvre-les-Nancy, France; e-mail: joner@cpb.cnrs-nancy.fr

Philippe Lemanceau, CMSE-INRA, UMR INRA/Université de Bourgogne BBCE-IPM, 21065 Dijon, France; e-mail: lemanceau@dijon.inra.fr

Yosi Levy, Department of Citriculture, Institute of Horticulture, Gilat Experiment Station, ARO, Mobile Post Negev 85280, Israel

Corinne Leyval, Centre de Pédologie Biologique, CNRS, 54501 Vandoeuvre-les-Nancy, France; e-mail: leyval@cpb.cnrs-nancy.fr

Steve Millam, Gene Expression Unit, Scottish Crop Research Institute, Invergowrie, Dundee, Tayside, Scotland, UK; e-mail: smilla@scri.sari.ac.uk

Silvia Perotto, Dipartimento di Biologia Vegetale, Università degli Studi di Torino, Viale Mattioli 25, 10125 Torino, Italy; e-mail: perotto@csmt.to.cnr.it

Maria José Pozo, Departamento de Microbiología del Suelo y Sistemas Simbióticos, Estación Experimental del Zaidín, CSIC, Profesor Albareda 1, 18008 Granada, España; e-mail: mjpozo@eez.csic.es

David Read, FRS, Department of Animal & Plant Sciences, University of Sheffield, Sheffield, S10 2TN, UK; e-mail: d.read@sheffield.ac.uk

Juan Manuel Ruíz-Lozano, Estación Experimental del Zaidín, CSIC, Profesor Albareda 1, 18008 Granada, España; e-mail: juanmanuel.ruiz@eez.csic.es

Manuel Sánchez-Díaz, Departamento de Fisiología Vegetal, Universidad de Navarra, Irunlarrea s/n, 31008, Pamplona, España; e-mail: msanchez@unav.es

Cristiana Sbrana, Dipartimento di Chimica e Biotecnologie Agrarie, Università di Pisa, Via del Borghetto 80, 56124 Pisa, Italy; e-mail: sbranac@agr.unipi.it

Andrea Schubert, Dip. Colture Arboree, Università di Torino, Via Leonardo da Vinci 44, 10095 Grugliasco (TO) Italy; e-mail: schubert@agraria.unito.it

Hannes Schüepp, Federal Research Station, 8820 Wädenswil, Switzerland; e-mail: hannes.schuepp@bluewin.ch

Sophie Slezack-Deschaumes, Laboratoire de Phytobiologie Cellulaire, UFR Sciences Vie, Université de Bourgogne, BP 47870, 21078 France

Reto J. Strasser, Université de Genève, Laboratoire de Bioénergétique, Chemin des Embrouchis 10, 1254 Jussy/Genève, Switzerland; e-mail: reto.strasser@bioen.unige.ch

Merope Tsimilli-Michael, Cyprus Ministry of Education and Culture, 1434 Nicosia, Cyprus; e-mail: tsimicha@spidernet.com.cy

Diederik van Tuinen, UMR INRA/Université de Bourgogne, BBCE-IPM, CMSE-INRA, BP 86510, 21065 Dijon Cedex, France; e-mail: vantuinen@epoisses.inra.fr

Katarzyna Turnau, Institute of Botany of the Jagiellonian University, ul. Lubicz 46, 31-512 Kraków, Poland; e-mail: ubturnau@kinga.cyf-kr.edu.pl

Coral del Val Muñoz, Estación Experimental del Zaidín, CSIC, Profesor Albareda 1, 18008 Granada, España

Mauritz Vestberg, Agricultural Research Centre of Finland – MTT, Laukaa Research and Elite Plant Station, Antinniementie 1, 41330 Vihtavuori, Finland; e-mail: mauritz.vestberg@mtt.fi

Miroslav Vosatka, Institute of Botany, Academy of Sciences of the Czech Republic, 252 43 Pruhonice, Czech Republic; e-mail: vosatka@ibot.cas.cz

Christine Watson, Land Management Department, Environment Division, SAC, Craibstone Estate, Aberdeen AB21 9YA, UK; e-mail: c.watson@ab.sac.ac.uk

Stephanie Weidmann, UMR INRA/Université de Bourgogne BBCE-IPM, CMSE-INRA, BP 86510, 21065 Dijon Cedex, France

Preface

Societal expectations in relation to human health today require the production of high quality food and conservation of the environment. Concerns about undesirable side-effects of high-input agriculture have highlighted the contribution made to plant health by soil microorganisms such as arbuscular mycorrhizal fungi (AMF). The ubiquitous symbiosis formed by these fungi with plant roots (arbuscular mycorrhiza) plays an essential role in plant nutrient acquisition/development and in plant/soil protection against environmental stresses.

Substantial progress has been made during the last 25 years in the understanding of arbuscular mycorrhiza (AM) symbioses. Knowledge acquired about this universal fungal/root association underlines the interest of mycorrhizal research for understanding not only basic biological processes like the development of compatibility and synergism between organisms, but also the impact of symbiotic relationships on the acquisition of individual fitness in ecosystems and more generally on the role of symbioses in the evolution of life. Compatibility and mutual development evolved between mycorrhizal partners during land colonization by plants, when the fungal symbiont played a key role in plant adaptation to new environments. Since then, plants and AMF have co-evolved in continuous interaction with their abiotic and biotic environment and have developed a wide range of coordinated mechanisms which have considerably favoured the production of primary biomass. As underlined in this book, knowledge of such mechanisms is of primary importance for promoting sustainable practices in plant production systems as well as in conservation and ecosystem restoration schemes.

Most plant families are able to form mycorrhiza and the AM association is the most common mycorrhiza type involved in agricultural systems. Given the effects of AMF inoculation on plant growth and health, it is generally accepted that appropriate management of this symbiosis should permit reduction of agrochemical inputs, and thus provide for sustainable and low-input plant productivity. Maximum benefits will only be obtained from inoculation with efficient AMF and a careful selection of compatible host/fungus/soil combinations. AM biotechnology is feasible for many crop production systems and the recent development of AMF-specific molecular probes provides tools for monitoring these microsymbionts in soil and roots. However, inoculum production techniques need to be improved for the proper application of AMF in commercial plant production systems, and further research into mechanisms involved in AM development and function is essential to acquire the scientific background for successful exploitation of this symbiosis in agriculture.

Recent developments since the two previous publications from the European network COST (Coopération Scientifique et Technique) on AM research [1, 2] have provided new insight into the biology and function of this plant-fungal interaction. In this context, the aims of this new publication from the COST Action 8.38 (Managing arbuscular mycorrhizal fungi for improving soil quality and plant health in agriculture) are to up-date on new information and to pinpoint future trends with regard to:

- extending knowledge on the genetics, physiology and cell programmes controlling the AM symbiosis,
- understanding the impact of AM on plant/agroecosystem dynamics through the improvement of plant fitness and soil quality,
- analysing the possibilities of technology transfer into commercial practice,
- making plant breeders aware of the potential of AM in selection programmes.

Because mycorrhiza research requires a multi-disciplinary approach, the strategy followed in this book combines scientific expertise and information from a range of well-known authors in the field. The main objective is to provide readers with pre-competitive scientific and practical knowledge essential for successfully using AMF to improve plant health and fitness in the production of high quality food and in the conservation of natural resources. The contents of the different chapters cover the role of AMF in sustainable crop production, ways to improve the biological balance of microorganisms in the mycorrhizosphere, genetic, physiological, cellular and molecular bases of AM functioning, and technologies for inoculum production. The diversity of the topics that are considered reflects the multidisciplinary nature of AM research while, at the same time, assembling knowledge to perceive the AM phenomenon as a whole.

The first part of the book concerns 'population biology' and considers that above-ground plant development is influenced by below-ground microbial activity. AMF represent a major component of the microbial community in the soil, and act as both a reservoir of nutrients and as a conduit for exchange of materials between the soil and the host plant. The fungi are thus an important element for plant productivity. Considerable diversity exists both within and across the currently recognized concepts of AMF species. An understanding of the range of this diversity is essential to establish predictive models of fungal dynamics and their effect on soil quality and plant responses.

Since AMF can act as stress alleviators and phytostimulators, a substantial part of this publication covers the topic 'AMF and plant health'. A variety of biotic and abiotic factors can affect plant health. AMF, by contributing to a more balanced mineral nutrition of plants, by modifying root development and by increasing plant resistance against soil born pathogens, can significantly improve plant health.

AMF also affect rhizosphere colonization by other soil microbes, thereby developing the so-called mycorrhizosphere. Such interactions can influence mycorrhiza formation and functioning and they are a key issue regarding plant fitness and soil quality. These topics are dealt with in several chapters.

Another key topic treated in this book concerns the genetic and cell programmes modulating symbiosis development and function. AMF can alter patterns of gene expression, cell organisation and organ development of host plants, but advances in knowledge about these have been hampered mainly due to a lack of appropriate methodology and to AMF being obligate biotrophs. The conceptual framework for research in this domain has become mature and modern molecular techniques are now adapted for analyses of this unique biological system.

Finally, a group of chapters covers progress made in mycorrhiza technology. The use of AMF in plant biotechnology differs from that of other beneficial soil microorganisms because the fungi involved are obligate symbionts and therefore recalcitrant to pure culture. Thus specific procedures are required to culture and handle them, and specific tools have to be developed and provided to inoculum producers.

In conclusion, the contents of this book are not limited to basic knowledge on the genetics, physiology and ecology of AM. The chapters (i) analyse more deeply the impact of this symbiosis on agroecosystem dynamics, (ii) widely discuss possibilities of technology transfer into commercial practices, (iii) emphasise problems concerning the quality of inoculum production and its proper use and, (iv) present examples of successful introduction of AMF into plant production systems. As a general goal, the book demonstrates that AM symbioses are an essential component to sustain soil quality, plant health and productivity.

This book is geared towards post graduate students, teachers and researchers in the field, and more generally to all professionals wishing to promote the use of biological tools in plant production, soil restoration, land management and, more widely, in bioremediation and sustainable development.

The editors thank all the colleagues who agreed to review the different chapters of the book for their invaluable help.

1. Impact of Arbuscular Mycorrhizas on Sustainable Agriculture and Natural Ecosystems, Birkhäuser Verlag, ISBN 3-7643-5000-8, 1994
2. Mycorrhizas in integrated systems from genes to plant development. European Commission, ISBN 92-827-5676-9, 1996

The Editors:

Silvio Gianinazzi, INRA/CNRS Dijon, France
Hannes Schüepp, FAW Wädenswil, Switzerland
José Miguel Barea, EEZ/CSIC, Granada, Spain
Kurt Haselwandter, University of Innsbruck, Austria

Mycorrhizal Technology in Agriculture
ed. by S. Gianinazzi, H. Schüepp, J.M. Barea and K. Haselwandter

The rhizosphere of mycorrhizal plants

J.M. Barea[1], M. Gryndler[2], P. Lemanceau[3], H. Schüepp[4] and R. Azcón[1]

[1] *Dpto. Microbiología del Suelo y Sistemas Simbióticos, Estación Experimental del Zaidín, Profesor Albareda, E-18008-Granada, Spain*
[2] *Institut of Microbiology ASCR, Czech Academy of Sciences, Vídenská 1083, 14220-Prague, Czech Republic*
[3] *CMSE-INRA, UMR INRA/Université de Bourgogne BBCE-IPM, 21065 Dijon cedex, France*
[4] *Swiss Federal Research Station, CH-8820 Wädenswil, Switzerland*

Keywords: Plant health, soil quality, root exudates, mycorrhizosphere, N_2-fixing bacteria, phosphate-solubilizing microorganisms, biological control agents, PGPR

Introduction

Providing that appropriate carbon substrates are available, microbial communities are able to develop a range of activities which are crucial in maintaining a biological balance in soil (Bowen and Rovira 1999), a key issue for the sustainability of either natural ecosystems or agroecosystems (Kennedy and Smith 1995). Soil-borne microbes have a particular microhabitat in which to flourish. In particular, they are bound to the surface of soil particles or found in soil aggregates, while others interact specifically with the plant root system (Glick 1995). The root-soil interface is actually a dynamic changing environment, a microcosm where microorganisms, plant roots and soil constituents interact (Lynch 1990; Azcón-Aguilar and Barea 1992; Linderman 1992; Barea 1997, 2000, Kennedy 1998; Bowen and Rovira 1999; Gryndler 2000), to develop what is known as the *rhizosphere* (Hiltner 1904). The rhizosphere, therefore, is the zone of influence of plant roots on the associated microbiota and soil components, and is clearly a different physical, chemical and biological environment from the bulk soil (Bowen and Rovira 1999), where an altered microbial diversity and increased activity and number of microorganisms is characteristic (Kennedy 1998). Actually, the structure and diversity of populations of fluorescent pseudomonads associated with roots were shown to differ significantly from those of soil populations. Rhizosphere and non-rhizosphere populations could be discriminated on the basis of their ability to use specific organic compounds (Lemanceau et al. 1995; Latour et al. 1996), to mobilize ferric iron (Lemanceau et al. 1988) or to reduce nitrogen oxides (Clays-Josserand et al. 1995).

The supply of photosynthates and decaying plant material to the root-associated microbiota, together with microbial-induced changes in rooting patterns and the supply of available nutrients to plants, as derived from microbial activities, are key issues in rhizosphere formation and functioning. The release of organic material occurs mainly as root exudates, which act as either signals or growth substrates (Werner 1998). Rhizosphere functioning is known to markedly influence plant fitness and soil quality so that, microorganisms associated with plant roots would help the host plant to adapt to stress conditions concerning water and mineral nutrition, and soil-borne plant pathogens (Lynch 1990; Bethlenfalvay and Schüepp 1994).

For the compartmentalization of the rhizosphere in the *broad sense*, Kennedy (1998) suggested that there are three separate but interacting components, namely the rhizosphere, the rhizoplane, and the root itself. The rhizosphere is the zone of soil influenced by roots through the release of substrates that affect microbial activity. The rhizoplane is actually the root surface, but also includes the strongly adhering soil particles. The root itself is part of the system because of certain microorganisms, the *endophytes*, which colonize the root tissues and carry out activities involved in plant growth promotion and plant protection (Kloepper 1994; Chanway 1996; Sturz et al. 2000). Microbial colonization of the rhizoplane and/or the root tissues is known as *root colonization*, while the colonization of the adjacent volume of soil under the influence of the root is known as *rhizosphere colonization* (Kloepper et al. 1991).

Two main groups of microorganisms can be distinguished: saprophytes and symbionts. Both of them comprise detrimental, neutral and beneficial bacteria and fungi. Detrimental microbes include the major plant pathogens, as well as minor parasitic and non-parasitic, deleterious rhizosphere organisms, either bacteria or fungi (Weller and Thomashow 1994; Nehl et al. 1996). Beneficial microorganisms are known to play fundamental roles in agroecosystem and natural ecosystem sustainability, and some of them can be used as inoculants to benefit plant growth and health (Alabouvette et al. 1997; Barea et al. 1997; Cordier et al. 2000). A subset of the total rhizosphere bacterial community, termed *rhizobacteria*, are known to display a specific ability for root colonization (Kloepper 1994, 1996). The beneficial root colonizing rhizosphere bacteria, the so-called *plant growth promoting rhizobacteria* (PGPR), carry out many important ecosystem processes, such as those involved in the biological control of plant pathogens, nutrient cycling and/or seedling establishment (Kloepper et al. 1991; Lugtenberg et al. 1991; Haas et al. 1991; Lemanceau and Alabouvette 1993; O´Gara et al. 1994; Weller and Thomashow 1994; Glick 1995; Broek and Vanderleyden 1995; Bashan and Holguin 1998; Barea 2000). As indicated before, the endophytic microorganisms develop activities involved in plant growth promotion and plant protection (Kloepper 1994; Chanway 1996; Sturz et al. 2000; Sturz and Novak 2000). Even non-symbiotic microorganisms may be endophytes and colonize the root tissues. For example, Duijff et al. (1997) found that the presence of a pseudomonad strain in tomato root cells was shown to induce resistance in this host plant (Duijff et al. 1998) as also found for other plants (Van Loon et al. 1998).

Arbuscular mycorrhizal fungi (AMF) and nitrogen (N_2)-fixing bacteria are particularly important mutualistic symbionts (Barea 1997). As universally accepted, the main mycorrhizal functions include improvement of plant establishment, enhancement of nutrient uptake, protection against cultural and environmental stresses, and the improvement of soil structure (Barea et al. 1997). This is particularly relevant for the arbuscular mycorrhizal (AM) associations.

Because of current public concerns about the side-effects of agrochemicals, attention is being given to research areas concerning biological balance in soil, microbial diversity, microbial dynamics in soil and rhizosphere interactions, and because these processes impact functioning and sustainability of either natural or agro-ecosystems (Bethlenfalvay and Linderman 1992; Barea and Jeffries 1995; Kennedy and Smith 1995), the study of interactions in the rhizosphere of mycorrhizal plants is a topic of current concern, and is the subject of this chapter.

Certainly, plant health and productivity depend on soil quality which, in turn, is dependent on the diversity and effectiveness of its microbiota (Bethlenfalvay and Schüepp 1994). This statement is well illustrated by some soils known to be naturally suppressive to soil-borne diseases such as fusarium-wilts (Alabouvette and Lemanceau 1996), take-all (Weller and Cook 1988), and root-rot of tobacco (Stutz et al. 1986). The natural suppressiveness of these soils has been ascribed to the activity of their microbiota (Cook and Baker 1983).

Mycorrhiza-induced changes in rhizosphere functioning

Mycorrhiza establishment is known to change the mineral nutrient composition, hormonal balance, C allocation patterns, and other aspects of plant physiology (Harley and Smith 1983; Smith et al. 1994; Azcón-Aguilar and Bago 1994). Thus, the AM symbiotic status changes the chemical composition of root exudates while the development of a mycorrhizal soil mycelium introduces physical modifications into the environment surrounding the roots. A typical property of the AMF soil mycelium is to serve as a carbon source to microbial communities, even outside the limit of the rhizosphere, and it results in an important contribution through interactions with components of the microbiota to improve plant growth and health, and soil quality (Bethlenfalvay and Schüepp 1994).

Mycorrhiza-induced changes both quantitatively and qualitatively affect the microbial populations in either the rhizosphere and/or the rhizoplane (Azcón-Aguilar and Barea 1992; Linderman 1992; Barea 1997). Large numbers of bacteria (including actinomycetes) and fungi can be associated with AM fungal structures (Filippi et al. 1998). Many specific cases have been described, for example, a strain of *Paenibacillus* sp. was isolated from the mycorrhizosphere of sorghum (Budi et al. 1999). Particular groups of microorganisms, like those involved in nitrogen transformation, are affected differentially in the rhizosphere of AM-plants (Amora-Lazcano et al. 1998). Interestingly, some rhizobial and pseudomonad bacteria adhere to the hyphae of AMF, which appear to be a vehicle for rhizosphere colonization by these bacteria (Bianciotto et al.

1996). An extreme case of close interactions is that of *Burkholderia* bacteria which have been identified as endosymbionts in AMF of the Gigasporaceae (Bianciotto et al. 2000; Ruiz-Lozano and Bonfante 2000).

In addition to changes in the rhizosphere of a mycorrhizal plant, there are specific modifications in the environment surrounding the mycorrhizal mycelium itself which are responsible for the development of the so-called *mycorrhizosphere* (Linderman 1992). In addition to this term, the soil space affected by extraradical hyphae is also called *mycosphere* (Linderman 1988) or *hyphosphere* as an analogy with the term rhizosphere (Gryndler 2000). Since the AMF mycelium releases energy-rich organic compounds, increased growth and activity of saprophytes is expected to occur in the mycorrhizosphere. However, the enrichment of this particular environment by organic compounds is probably much lower than that of the rhizosphere which corresponds to lower counts of bacteria in mycospheric soil, compared to those in the rhizosphere (Andrade et al. 1997). Bacteria isolated from the surface of hyphae of AMF extracted from the soil were mostly Gram-negative while both Gram-positive and Gram-negative bacteria were detected in bulk soil and rhizosphere (Vosátka 1996). An indirect cause of the 'mycorrhizosphere effect' (Barea 1997) derives from the recognized role of AMF colonization in changing a wide range of morphological parameters in developing root systems (Atkinson et al. 1994; Berta et al. 1995), with greater root branching being the most commonly described effect. Undoubtedly these changes must affect establishment and activity of microorganisms in the rhizosphere/mycorrhizosphere environment.

Early information reviewed by Puppi et al. (1994) and Barea (1997) indicated that the establishment of inoculated or naturally occurring microorganisms in the rhizosphere was affected by AM development. More recent information supports previous statements and both beneficial and detrimental effects of AMF on the establishment of bacterial inoculants have been found (Christensen and Jakobsen 1993; Ravnskov et al. 1999). For example, AM establishment may selectively influence the persistence of bacterial inoculants and resident soil bacteria (Andrade et al. 1998a). In particular, AM fungal inoculation improved the establishment of both inoculated and indigenous phosphate-solubilizing rhizobacteria (Toro et al. 1997; Barea et al. 2001).

It is noteworthy that microbial populations established in the rhizosphere can affect mycorrhiza formation (Germida and Walley 1996). Deleterious rhizosphere bacteria have been found to interfere with mycorrhiza formation and/or functioning (Nehl et al. 1996), and mycoparasitic relationships involving AMF have been described (Jeffries 1997).

With respect to beneficial microbial effects on mycorrhiza formation early reports reviewed by Azcón-Aguilar and Barea (1992) showed that the rate of spread of both arbuscular and ectomycorrhizal fungi on the root system of an appropriate host was stimulated by rhizosphere microorganisms. As an example, the classical study by Bowen and Theodorou (1979) described a range of situations showing either beneficial, neutral or detrimental effects of bacterial treatments on the colonization of the roots of *Pinus radiata* by different species of ectomycorrhizal fungi. Undoubtedly the result depends on the microbial strain and the mycorrhizal fungus involved (Garbaye and Bowen 1987; Chanway et al. 1991). More recent reviews discuss the ecological significance and the potential use of the so-called mycorrhiza helper bacteria (MHB) to stimulate mycelial growth of AMF or enhance mycorrhizal formation (Garbaye 1994; Azcón-Aguilar and Barea 1995; Frey-Klett et al. 1997; Barea 1997). It appears that changes in the root exudation patterns, and in the hormonal balance of the plant, are involved in the establishment and development of the mycorrhizal symbiosis. Soil microorganisms are able to produce compounds that increase root cell permeability, thereby enhancing the rates of root exudation. This, in turn, would stimulate AM fungal mycelia in the rhizosphere and facilitate root penetration by the fungus. Plant hormones, as produced by soil microorganisms, are known to affect AM fungal establishment on root cortex (Azcón-Aguilar and Barea 1992, 1995; Barea 1997, 2000).

A detailed study of 46 bacterial isolates obtained from the bulk soil and hyphae of three different soils (each involving arable and non-cultivated samples) revealed that isolates from hyphae did not show a substantial stimulatory effect on growth of hyphae of *Glomus claroideum in vitro* but produced more uniform effects on hyphal growth than the isolates obtained from bulk soil (Gryndler et al. 2000). A larger number of stimulatory isolates was obtained from tilled soil samples but the bacteria showing the strongest stimulation of hyphal growth were isolated from non-cultivated soils. The results indicate that the effect on AMF of the bacteria population may vary in different soils.

Interactions involved in nutrient cycling and plant growth promotion

Numerous soil microorganisms are able to change the bioavailability of mineral plant nutrients. This could account for the improvement of plant growth by means of synergistic interactions of such microorganisms with AMF. It has also been suggested that selectivity ('specificity') is involved in these interactions (Azcón 1989). From the perspective of sustainability, the re-establishment of nutrient cycles after any process of soil degradation is of interest, as is the understanding of the microbial interactions responsible for the subsequent management of such natural resources, either for a low-input agricultural technology (Bethlenfalvay and Linderman 1992; Gianinazzi and Schüepp 1994; Jeffries and Barea 2000) or for the re-establishment of the natural vegetation in a degraded area (Miller and Jastrow 1994; Barea and Jeffries 1995).

The interactions of AMF and rhizobial N-fixing symbiotic bacteria greatly benefit the biological N inputs to soil-plant systems (Barea et al. 1997). However, to achieve this beneficial effect 'specific' bacteria/AMF combinations must first be selected (Azcón et al. 1991; Ruíz-Lozano and Azcón 1993). The role of AMF in improving nodulation and N_2 fixation is now universally recognized and the use of the isotope ^{15}N has made it possible to ascertain and quantify the amount of N which actually is fixed in a particular situation, and the contribution of the AM symbiosis to the processes involved. Many studies have been addressed to elucidate the physiological and biochemical basis of AMF × *Rhizobium* interactions to improve legume productivity (Barea et al. 1992). It is accepted that the main AM effect in enhancing *Rhizobium* activity is through a generalized stimulation of host nutrition, but more localized effects may occur at the root or nodule level. Interactions can also take place either at the pre-colonization stages, when both microorganisms interact as rhizosphere inhabitants, or during the development of the tripartite symbiosis (Azcón-Aguilar and Barea 1992; Barea et al. 1997). The influence of host genotype in these interaction has also been argued (Monzón and Azcón 1996).

Studies by Goicoechea et al. (1995; 1996; 1997; 1998) support the involvement of plant hormones, polyamines, proline and photosynthetic parameters in the interactions between AM- and the rhizobial symbiosis to help legumes to overcome drought stress. Similarly AMF can help rhizobial symbiosis under salinity stress conditions (Azcón and El-Atrash 1996). The effect of AMF inoculation to alleviate drought stress-induced nodule senescense has been described recently (Ruiz-Lozano et al. 2001).

With regard to multimicrobial interactions, Requena et al. (1997) tested AMF, *Rhizobium* spp. and PGPR isolated from a representative area of a desertification-threatened semi-arid ecosystem in the south-east of Spain. Microbial isolates were characterized and screened for effectiveness in soil microcosms. *Anthyllis cytisoides* L. a mycorrhiza-dependent pioneer legume, dominant in the target mediterranean ecosystem, was the test plant. Several microbial cultures from existing collections were also included in the screening process. Overall, *Glomus coronatum*, native to the field site, was more effective than the exotic *Glomus intraradices* in co-inoculation treatments. In general, the results support the importance of physiological and genetic adaptation of microbes to the environment, thus local isolates must be used. Many microbial combinations were effective in improving either plant development, nutrient uptake, N_2-fixation (^{15}N) or root system quality. Selective and specific functional compatibility relationships among the microbial inoculants, with respect to plant response, were observed. Since some PGPR may improve nodulation by *Rhizobium* spp. (Halverson and Handelsman 1991; Staley et al. 1992; Azcón 1993), certain PGPR-*Rhizobium* interactions could be relevant to mycorrhizosphere interactions. *Rhizobium* spp. have been also suggested to be rhizosphere colonizing bacteria in non-legume hosts (Schloter et al. 1997; Galleguillos et al. 2000).

The rhizobacteria of the genus *Azospirillum* are known to benefit plant development and yield under appropriate conditions (Okon 1994; Costacurta and Vanderleyden 1995; Bashan 1999). Interactions between AMF and *Azospirillum* have been reviewed by Volpin and Kapulnik (1994) and the main conclusions are that these bacteria, by influencing the morphology, geometry and physiology of the root system, could enhance mycorrrhizal formation and response and that AMF may improve *Azospirillum* establishment. It has been suggested that ^{15}N methods should be used to ascertain whether N_2-fixation is involved in the interaction with AMF (Barea et al. 1992 and 1997).

In vitro experiments showed that many soil microrganisms are able to solubilize phosphate ions from sparingly soluble inorganic and organic P compounds (Whitelaw 2000). Synergistic microbial interactions involving these microorganisms and AMF occur (Barea et al. 1997) and the use of phosphate solubilizing rhizobacteria (PSR) is giving new impetus to the research. For example, the interactive effect of PSR and AMF on plant use of soil P sources of low bioavalibility (endogenous or added as rock phosphate) was evaluated using a soil microcosm system which integrated ^{32}P isotopic dilution approaches (Toro et al. 1997). The rhizobacteria behaved as MHB promoting AM establishment by either the indigenous or inoculated AM endophytes. Because the bacteria did not change root weight, length or specific root length, they probably acted by improving the pre-colonization stages of AM formation. The dual inoculation treatment significantly increased biomass and N and P accumulation in plant tissues. Regardless of the rhizobacterial strain and of whether or not rock phosphate had been added, AMF-inoculated plants displayed lower specific activity ($^{32}P/^{31}P$) than their comparable controls, suggesting that the plants were using P sources otherwise unavailable. It therefore appears that these rhizosphere/mycorrhizosphere interactions contributed to the biogeochemical P cycling thus promoting a sustainable nutrient supply to plants.

Microbiologically-treated agrowastes, involving *Aspergillus niger,* improved rock phosphate availability by mycorrhizal nodulated alfalfa (Rodriguez et al. 1998), thus displaying a sustainable approach based on biotechnological inputs involving mycorrhizal-microbe-rhizosphere interactions.

In relation to current biosafety-related concerns a series of experiments were carried out to compare the effects on AM formation and function of a wild type (WT) *Rhizobium meliloti* strain with those of its genetically modified (GM) derivative. This GM *Rhizobium* was developed to improve the nodulation competitiveness of the WT strain (Sanjuan and Olivares 1991). It was found that the GM rhizobial strain did not interfere with any processes related to mycorrhiza formation by the representative AMF *Glomus mosseae*. The parameters tested include spore germination, mycelial growth from the mycorrhizal propagules and AMF 'entry point' formation on the developing root system of the host plant *Medicago sativa.* Indeed, the GM *Rhizobium* increased the number of AMF colonization units and the nutrient acquisition ability in mycorrhizal plants, with respect to the WT rhizobial strain (Tobar et al. 1996). The mycorrhizal development and quality of nodulation increased with time, and co-incided with increased biomass and nutrient (N, P) uptake in the host plant. The establishment of the symbiotic interactions also induced changes in root morphology, in particular, the degree of branching increased and the number of lateral roots was higher in mycorrhizal plants inoculated with the GM *Rhizobium* strain (Barea et al. 1996).

In another experiment, a soil microcosm system was used to evaluate the interactive effects of some multifunctional microbial inoculation treatments and rock phosphate application on N and P acquisition by alfalfa plants. The microbial inocula consisted of the WT *Rhizobium meliloti* strain, its GM derivative, which had an enhanced competitiveness, the AMF *Glomus mosseae* and a PSR. Inoculated microorganisms established in the root tissues and/or in the rhizosphere soil of alfalfa plants. Measurements of the $^{15}N/^{14}N$ ratio in plant shoots indicate an enhancement of the N_2 fixation rates in *Rhizobium*-inoculated AM-plants, over that achieved by the same *Rhizobium* strain in non-mycorrhizal plants. Regardless of the *Rhizobium* strain and of whether or not rock phosphate was added, AM-inoculated alfalfa showed a lower specific activity ($^{32}P/^{31}P$) than did their comparable non-mycorrhizal controls, suggesting that the plant was using P sources otherwise unavailable (Toro et al. 1998).

These and other examples (Kim et al. 1998; Gryndler and Hrselova 1998; Belimov at al. 1999; Vosatka and Gryndler 1999, 2000) showed that co-inoculation with particular selected bacteria has frequently been shown to improve plant growth and nutrient uptake in mycorrhizal plants. However, in many cases it is difficult to associate certain physiological effects of arbuscular mycorrhiza exclusively with the altered physiology of mineral nutrition or water uptake. Changed patterns of allocation of photoassimilated carbon (photosynthates) and their utilization could also be involved. As previously stated, the rhizosphere may be seen as an interface between the root and soil environment, in which the exchange of nutrients (including carbon), energy and information occurs. Since the flux of carbon rich compounds into the rhizosphere is significant and might reach several dozens or even hundreds of milligrams per gram root per day (Toal et al. 2000), the question is how AMF commonly present in the rhizosphere affect this flux. AMF, as ubiquitous inhabitants of rhizospheres/mycorrhizospheres in arable and natural soils, may affect the carbon cycling in the ecosystems if they are able to change the rhizospheric carbon fluxes. Very little is known about the interrelation between global carbon cycle and AM symbiosis (Fitter et al. 2000). However, if the fungi regulate the soil carbon fluxes, it could be also important at the global ecosystem scale.

For example, if the community of AMF in a prairie soil was eliminated (by the application of a fungicide), the total soil microbial biomass and substrate induced respiration increased while the fungal biomass decreased (Smith et al. 2000). This was followed by changes in populations of fungal feeding and predatory nematodes and most probably also for other groups of soil organisms. This represents a demonstration of the effect on soil carbon fluxes mainly by the presence/absence of AMF. If these fungi are present in the soil, a great part of the root allocated photoassimilated carbon may be transported by hyphae into the bulk of soil, and released into the environment. Here it would be utilized in other ways and by other soil organisms than if it was exuded in the proximity of the root surface where biomass has accumulated as a result of the rhizosphere effect (Jakobsen and Rosendhal 1990).

The immobilization of organic carbon in the roots and the mycelial biomass is an other way by which AMF can affect soil carbon dynamics. In mycorrhizal plants, a significant part of photoassimilated carbon is transported into the mycelium of the symbiotic fungus. Since the turnover of organic carbon immobilized in the AMF mycelium is likely to be faster when compared to that immobilized in the root tissues, the allocation of the carbon into the hyphae instead of in the root tissues may result in a faster average turnover of organic carbon in the soil and an increased carbon mineralization (Fitter et al. 2000). However, these hypotheses cannot be widely applied since further extensive research is needed to fully understand the behavior of different pools of biologically immobilized carbon and to understand the effects of arbuscular mycorrhizas at an ecosystem scale (Norby and Jackson 2000). Combinations of different microorganisms (*Glomus mosseae* BEG12 with *Agrobacterium radiobacter* K1026 and *Glomus mosseae* BEG12 with *Pseudomonas fluorescens* C7) applied to vitroplants of strawberry clearly showed to promote plant growth (Cordier et al. 1999).

Interactions for the biological control of root pathogens

Arbuscular mycorrhizal associations have been shown to reduce damage caused by soil-borne plant pathogens. However, the enhancement of root resistance/tolerance exerted by the AM symbiosis is not shown by all AMF, it cannot be applied for all pathogens, and it is not expressed in all substrates or environmental conditions (Linderman 1994, 2000; Azcón-Aguilar and Barea 1996).

The mechanisms that have been suggested to explain the protective action of AM symbiosis are discussed by Pozo et al. (in this book). The particular relevance, however, of microbial changes in the mycorrhizosphere for biocontrol will be discussed here.

Because AM formation induces changes in host physiology, these can modify root exudation patterns (Azcón-Aguilar and Bago 1994; Smith et al. 1994), and consequently, may cause qualitative and/or quantitative alterations in microbial populations in the rhizosphere. There is strong evidence that microbial shifts occur in the mycorrhizosphere, and that the resulting microbial equilibria could influence the growth and health of the plants. Although this effect has not been specifically evaluated as a mechanism for AM associated biological control, there are indications that such a mechanism may operate (Azcón-Aguilar and Barea 1992, 1996; Linderman 1994, 2000).

Changes in the populations of soil microorganisms induced by AM formation, may lead to stimulation of certain components of the resident microbiota which may be antagonistic to root pathogens. Meyer and Linderman (1986) found that the number of sporangia and zoospores formed by cultures of *Phytophthora cinnamomi* was reduced by the application of extracts of rhizosphere soil from AM plants. Secilia and Bagyaraj (1987) isolated more pathogen-antagonistic actinomycetes from the rhizosphere of AM plants than from that of the corresponding non-mycorrhizal controls, an effect which also appears to depend on the AMF involved. Recent studies have corroborated these findings and demonstrated that such an effect is dependent on the AMF involved as well as the substrate and host plant (Linderman 1994; Azcón-Aguilar and Barea 1996). The topic obviously merits further attention in order to elucidate how microbiota-mediated changes are involved in biological control by AM associations.

It is accepted that the prophylactic ability of AMF could be exploited in association with other rhizosphere microorganisms, antagonistic against root pathogens, that are being used as biological control agents (Linderman 1994, 2000; Nemec 1997; Barea et al. 1998). To give a better protection than that performed by each organism separately, the organisms associated must be compatible with each other and express additive and even synergistic modes of action (Alabouvette and Lemanceau 1997). This was shown for the combination of specific strains of fluorescent pseudomonads and non-pathogenic *Fusarium oxysporum* (Lemanceau and Alabouvette 1991; Lemanceau et al. 1992). More recently, a *Paenibacillus* sp. strain has proved not only to be compatible with AM development in tomato roots but also to act in synergy with AMF to increase the level of resistance against soil-borne pathogenic fungi, particularly *Phytophthora parasitica* (Budi et al. 1999). Biological control of soil-borne diseases is known to result (i) from the reduction of the saprophytic growth of the pathogens and then of the frequency of the root infections (microbial antagonism) and (ii) from the stimulation of the defense reactions of the host-plants (induced systemic resistance). The molecular bases of the microbial antagonism and of the induced systemic resistance achieved by selected rhizobacteria are being investigated (O´Gara et al. 1994; Kloepper 1994; Cook et al. 1995; Thomashow and Weller 1995; Chen et al. 1996; Van Loon et al. 1998). Obviously, AMF and PGPR may have several molecular mechanisms in common regarding biocontrol and these could be investigated following common approaches.

Several studies have suggested that microbial antagonists of fungal pathogens, either fungi or PGPR, do not exert an antimicrobial effect against AMF (Calvet et al. 1993; Barea et al. 1998; Edwards et al. 1998; Vazquez et al. 2000). It was recently shown that specific strains of AMF and fluorescent Pseudomonads, that positively influence plant growth, can live together within the same root cell (Gianinazzi-Pearson, Arnould, Loison, Lemanceau, unpublished data). Microbial antagonists of fungal pathogens may even improve the development of the mycosymbiont and facilitate AM formation (Linderman 1994; Barea et al. 1998; Edwards et al. 1998; Budi et al. 2000). The interaction of a PGPR and AMF affected reproduction of a root-knot nematode (Siddiqui and Mahmood 1998). Studies are currently being carried out to test the beneficial effect of the combinations of glomalean fungi and bacterial strains (fluorescent Pseudomonads and *Paenibacillus* sp.) on plant-growth and plant-health of tomato in field conditions (Escande, Gianinazzi, Blal, Lemanceau, unpublished data).

Interactions improving soil quality

The interactions between AMF and rhizobial strains to improve revegetation processes for desertified ecosystems based on the use of shrub legumes belonging to the natural succession have received considerable attention in the last decade (Herrera et al. 1993). In this context, a number of experiments have been carried out aimed at assessing the long-term benefits of inoculation with these two types of plant symbionts not only on the establishment of target legume species but also on changes induced by the symbiotically tailored seedlings in key physico-chemical soil properties affecting soil quality (Requena et al. 2001). This microbial management practice is important because, as a result of the degradation/desertification processes, disturbance of natural plant communities is often accompanied or preceded by loss of physico-chemical and biological soil properties, such as soil structure, plant nutrient availability, organic matter content, microbial activity, etc. (Barea and Jeffries 1995; Schreiner et al. 1997).

A representative area within a desertified semi-arid ecosystem in the Sierra de los Filabres, Almeria (southeast Spain), was chosen for field studies on this topic. The existing natural vegetation was a degraded shrubland where *Anthyllis cytisoides*, a drought-tolerant legume able to form symbioses with both rhizobial and AM microsymbionts, was the dominant species (Requena et al. 1997).

The experimental variables tested in transplanted seedlings were survival rates, growth, N- fixation, and N-transfer from N-fixing to non-fixing species associated in the natural succession. Improvements in soil quality in terms of N content, levels of organic matter, and hydrostable soil aggregates were also analysed in the rhizosphere of the target mycorrhizal and nodulated plants. In addition, it was determined whether these changes were accompanied by an increase in the inoculum potential of AMF. A long-term improvement in the physico-chemical properties was evident in the soil around the *Anthyllis* plants inoculated with an AM fungal inoculum. The benefit includes an increased N content and a higher amount of organic matter and soil aggregation. It can be assumed that the increase in N content in the rhizosphere of the legume can be accounted for by an improvement in nodulation and N-fixation capacity resulting from inoculation with AMF (Barea et al. 1992). The improvement of soil aggregation contributes to maintain good water infiltration rates, good tilth and adequate aeration for plant growth thus improving soil quality (Wright and Upadhyaya 1998). The important role of the soil mycelium of AM fungal mycelium in the formation of water stable soil aggregates is well-documented (Andrade et al. 1995 and 1998b; Bethlenfalvay et al. 1999), and the involvement of glomalin, a glycoprotein produced by the external hyphae of AMF, has been demonstrated (Wright and Upadhyaya 1998). Glomalin has been suggested to contribute to the hydrophobicity of soil particles and also, because its glue-like hydrophobic nature, to participate in the initiation of soil aggregates.

Inoculation with native AMF also benefited plant growth, N fixation and N-transfer. Improved N status of non-leguminous plants grown in association with legumes has previously been described for agricultural crops (Azcón-Aguilar et al. 1979), but this was the first demonstration of this phenomenon for natural plant communities in a semi-arid ecosystem. It was also evidence for the important role of the mycotrophic shrub legumes as a source of AM fungal inoculum for the surrounding area and in improving N nutrition for non-N-fixing vegetation. The results support the general conclusion that the introduction of target indigenous species of plants, associated with a managed community of microbial symbionts, is a successful biotechnological tool to aid the recovery of desertified ecosystems, suggesting that this represents the initial steps in the restoration of a self-sustaining ecosystem.

Conclusion

Mycorrhiza formation changes several aspects of plant physiology and some nutritional and physical properties of the rhizosphere, i. e. the zone of influence of plant roots on microbial populations and other soil constituents. These effects modify the colonization patterns of the root/mycorrhiza environments (mycorrhizosphere) by soil micro-organisms. The rhizosphere of mycorrhizal plants, in practice the mycorrhizosphere, harbors a great array of microbial activities responsible for several key ecosystem processes. The reviewed information on current developments concerning the microbial interactions between AMF and other members of rhizosphere microbiota demonstrated that: (i) soil microorganisms affect mycorrhiza formation; (ii) mycorrhiza establishment changes size and diversity of microbial population in the mycorrhizosphere; (iii) many interactions benefit nutrient cycling and plant growth; (iv) certain interactions co-operate for the biological control of soil-borne plant pathogens; and (v) some other interactions improve soil quality. In summary, microbial interactions in the rhizosphere of mycorrhizal plants improve plant fitness and soil quality, critical issues for sustainable agricultural developments and ecosystem functioning.

References

Alabouvette C, Lemanceau P (1997) Joint action of microbials for disease control. In Hall F R, Menn J J (eds) Biopesticides: use and delivery. The Humana Press inc, Totowa, NJ, US, pp 117-135

Alabouvette C, Schippers B, Lemanceau P, Bakker P A H M (1997) Biological control of fusarium-wilts: towards development of commercial product. In: Boland G J, Kuykendall L D (eds) Plant microbe interactions and biological control. Marcel Dekker, Inc., New York, US, pp 15-36

Amora-Lazcano E, Vázquez M M, Azcón R (1998) Response of nitrogen-transforming microorganisms to arbuscular mycorrhizal fungi. Biol Fertil Soil 27:65-70

Andrade G, Azcón R, Bethlenfalvay G J (1995a) A Rhizobacterium modifies plant and soil responses to the mycorrhizal fungus *Glomus mosseae.* Appl Soil Ecol 2:195-202

Andrade G, Linderman R G, Bethlenfalvay G J (1998b) Bacterial associations with the mycorrhizosphere and hyphosphere of the arbuscular mycorrhizal fungus *Glomus mosseae.* Plant Soil 202:79-87

Andrade G, Mihara K L, Linderman R G, Bethlenfalvay G J (1997) Bacteria from rhizosphere and hyphosphere soils of different arbuscular mycorrhizal fungi. Plant Soil 192:71-79

Andrade G, Mihara K L, Linderman R G, Bethlenfalvay G J (1998) Soil aggregation status and rhizobacteria in the mycorrhizosphere. Plant Soil 202:89-96

Atkinson S, Berta G, Hooker J E (1994) Impact of mycorrhizal colonisation on root architecture, root longevity and the formation of growth regulators. In: Gianinazzi S, Schüepp H (eds) Impact of arbuscular mycorrhizas on sustainable agriculture and natural ecosystems. ALS, Birkhäuser Verlag, Basel, Switzerland, pp 47-60

Azcón R (1989) Selective interaction between free-living rhizosphere bacteria and vesicular-arbuscular mycorrhizal fungi. Soil Biol Biochem 21:639-644

Azcón R (1993) Growth and nutrition of nodulated mycorrhizal and non-mycorrhizal *Hedysarum coronarium* as a results of treatments with fractions from a plant growth-promoting rhizobacteria. Soil Biol Biochem 25:1037-1042

Azcón R, Barea J M (1997) Mycorrhizal dependency of a representative plant species in mediterranean shrublands (*Lavandula spica* L) as a key factor to its use for revegetation strategies in desertification-threatened areas. Appl Soil Ecol 7:83-92

Azcón R, El-Atrash F (1997) Influence of arbuscular mycorrhizae and phosphorus fertilization on growth, nodulation and N_2 fixation (^{15}N) in *Medicago sativa* at four salinity levels. Biol Fert Soil 24:81-86

Azcón R, Rubio R, Barea J M (1991) Selective interactions between different species of mycorrhizal fungi and *Rhizobium meliloti* strains, and their effects on growth, N_2-fixation (^{15}N) and nutrition of *Medicago sativa L.* New Phytol 117:399-404

Azcón-Aguilar C, Azcón R, Barea J M (1979) Endomycorrhizal fungi and *Rhizobium* as biological fertilizers for *Medicago sativa* in normal cultivation. Nature 27:235-237

Azcón-Aguilar C, Bago B (1994) Physiological characteristics of the host plant promoting an undisturbed functioning of the mycorrhizal symbiosis. In: Gianinazzi S, Schüepp H (eds) Impact of arbuscular mycorrhizas on sustainable agriculture and natural ecosystems. ALS, Birkhäuser Verlag, Basel, Switzerland, pp 47-60

Azcón-Aguilar C, Barea J M (1992) Interactions between mycorrhizal fungi and other rhizosphere microorganisms. In: Allen M J (ed) Mycorrhizal functioning. an integrative plant-fungal process. Routledge, Chapman & Hall Inc., New York, pp 163-198

Azcón-Aguilar C, Barea J M (1995) Saprophytic growth of arbuscular-mycorrhizal fungi. In: Hock B, Varma A (eds) Mycorrhiza structure function, molecular biology and biotechnology. Springer-Verlag, Heidelberg, pp 391-407

Azcón-Aguilar C, Barea J M (1996) Arbuscular mycorrhizas and biological control of soil-borne plant pathogens. An overview of the mechanisms involved. Mycorrhiza 6:457-464

Barea JM (1997) Mycorrhiza/bacteria interactions on plant growth promotion. In: Ogoshi A, Kobayashi L, Homma Y, Kodama F, Kondon N, Akino S (eds) Plant growth-promoting rhizobacteria, present status and future prospects. OECD, Paris, pp 150-158

Barea J M (2000) Rhizosphere and mycorrhiza of field crops. In: Toutant J P, Balazs E, Galante E, Lynch J M, Schepers J S, Werner D, Werry P A (eds) Biological resource management: connecting science and policy (OECD). INRA, Editions and Springer, pp 110-125

Barea J M, Andrade G, Bianciotto V, Dowling D, Lohrke S, Bonfante P, O'Gara F, Azcón-Aguilar C (1998) Impact on arbuscular mycorrhiza formation of *Pseudomonas* strains used as inoculants for the biocontrol of soil-borne plant fungal pathogens. Appl Environ Microbiol 64:2304-2307

Barea J M, Azcón R, Azcón-Aguilar C (1992) Vesicular-arbuscular mycorrhizal fungi in nitrogen-fixing systems. In: Norris J R, Read D J, Varma A K (eds) Methods in microbiology. Academic Press, London, pp 391-416

Barea J M, Azcón-Aguilar C, Azcón R (1997) Interactions between mycorrhizal fungi and rhizosphere microorganisms within the context of sustainable soil-plant systems. In: Gange A C, Brown V K (eds) Multitrophic interactions in terrestrial systems. Blackwell Science, Oxford, pp 65-77

Barea J M, Jeffries P (1995) Arbuscular mycorrhizas in sustainable soil plant systems. In: Hock B, Varma A (eds) Mycorrhiza structure function, molecular biology and biotechnology. Springer-Verlag, Heidelberg, pp 521-559

Barea J M, Tobar R M, Azcón-Aguilar C (1996) Effect of a genetically-modified *Rhizobium meliloti* inoculant on the development of arbuscular mycorrhizas, root morphology, nutrient uptake and biomass accumulation in *Medicago sativa* L. New Phytol 134:361-369

Barea J M, Toro M, Orozco M O, Campos E, Azcón R (2001) The application of isotopic (^{32}P and ^{15}N)

dilution techniques to evaluate the interactive effect of phosphate-solubilizing rhizobacteria, mycorrhizal fungi and *Rhizobium* to improve the agronomic efficiency of rock phosphate for legume crops. Nutrient Cycling Agroecosys (In press)

Bashan Y (1999) Interactions of *Azospirillum* spp. in soils: a review. Biol Fertil Soils 29:246-256

Bashan Y, Holguin G (1998) Proposal for the division of plant growth-promoting rhizobacteria into two classifications: biocontrol-PGPB (plant growth-promoting bacteria) and PGPB. Soil Biol Biochem 30:1225-1228

Belimov A A, Serebrennikova N V, Stepanok V V (1999) Interaction of associative bacteria and an endomycorrhizal fungus with barley upon dual inoculation. Microbiology 68:104-108

Berta G, Trotta A, Fusconi A, Hooker J E, Munro M, Arkinson D, Giovannetti M, Morini S, Fortuna P, Tisserant B, Gianinazzi-Pearson V, Gianinazzi S (1995) Arbuscular mycorrhizal induced changes to plant growth and root system morphology in *Prunus cerasifera*. Tree Physiol 15:281-293

Bethlenfalvay G J, Cantrell I C, Mihara K L, Schreiner R P (1999) Relationships between soil aggregation and mycorrhizae as influenced by soil biota and nitrogen nutrition. Biol Fertil Soils 28:356-363

Bethlenfalvay G J, Linderman R G (eds) (1992) Mycorrhizae in Sustainable Agriculture. ASA Special publication No. 54, Madison, Wisconsin

Bethlenfalvay G J, Schüepp H (1994) Arbuscular mycorrhizas and agrosystem stability. In: Gianinazzi S, Schüepp H (eds) Impact of arbuscular mycorrhizas on sustainable agriculture and natural ecosystems. Birkhäuser, Basel, pp 117-131

Bianciotto V, Lumini E, Lanfranco L, Minerdi D, Bonfante P, Perotto S (2000) Detection and identification of bacterial endosymbionts in arbuscular mycorrhizal fungi belonging to the family Gigasporaceae. Appl Environ Microbiol 66:4503-4509

Bowen G D, Rovira A D (1999) The rhizosphere and its management to improve plant growth. Adv Agron 66:1-102

Bowen G D, Theodorou C (1979) Interactions between bacterial and ectomycorrhizal fungi. Soil Biol Biochem 11:119-126

Broek A V, Vanderleyden J (1995) Genetics of the *Azospirillum*-plant root association. Crit Reviews Plant Sci 14:445-466

Budi S W, Van Tuinen D, Martinotti G, Gianinazzi S (1999) Isolation from *Sorghum bicolor* mycorrhizosphere of a bacterium compatible with arbuscular mycorriza development and antagonistic towards soilborne fungal pathogens. Appl. Environ. Microbiol. 65:5148-5150

Calvet C, Pera J, Barea J M (1993) Growth response of marigold (*Tagetes erecta* L) to inoculation with *Glomus mosseae*, *Trichoderma aureoviride* and *Phythium ultimum* in a peat-perlite mixture. Plant Soil 148:1-6

Caron M (1989) Potential use of mycorrhizae in control of soilborne diseases. Can J Plant Pathol 11:177-179

Christensen H, Jacobsen I (1993) Reduction of bacterial growth by a vesicular-arbuscular mycorrhizal fungus in the rhizosphere of cucumber (*Cucunis sativus* L). Biol Fertil Soils 15:253-258

Cook R J, Baker K F (1983) The nature and practice of biological control of plant pathogens. American Phytopathological Society, St. Paul p 539

Cook R J, Thomashow L S, Weller D M, Fujimoto D, Mazzola M, Bangera G, Kim D S (1995) Molecular mechanisms of defense by rhizobacteria againsts root disease. Proc Nalt Acad Sci USA 92:4197-4201

Chanway C P (1996) Endophytes: they´re not just fungi!. Can J Bot 74:321-322

Chanway C P, Radley R A, Holl F B (1991) Inoculation of conifer seed with plant growth promoting *Bacillus* strains causes increased seedling emergence and biomass. Soil Biol Biochem 23:575-580

Chen J, Jacobson L M, Handelsman J, Goodman R M (1996) Compatiility of systemic acquired resistance and microbial biocontrol for suppression of plant disease in a laboratory assay. Mol Ecol 5:73-80

Clays-Josserand A, Lemanceau P, Philippot L, Lensi R (1995) Influence of two plant species (flax and tomato) on the distribution of nitrogen dissimilative abilities within fluorescent *Pseudomonas* spp. Appl. Environ. Microbiol. 61:1745-1749

Cordier C, Lemoine M C, Lemanceau P, Gianinazzi-Pearson V, Gianinazzi S (1999) The beneficial rhizosphere: a necessary strategy for microplant production. Acta Horticulturae 530:259-265

Duijff B J, Gianinazzi-Pearson V, Lemanceau P (1997) Involvement of the outermembrane lipopolysaccharides in the endophytic root colonization of tomato roots by biocontrol

Pseudomonas fluorescens strain WCS417r. New Phytol. 135:325-334
Duijff B J, air D, Olivain C, Alabouvette C, Lemanceau P (1998) Implication of systemic induced resistance in the suppression of fusarium wilt of tomato by *Pseudomonas fluorescens* WCS417r and nonpathogenic *Fusarium oxysporum* Fo47. Eur J Plant Pathol 104:903-910
Edwards S G, Young J P W, Fitter A H (1998) Interactions between Pseudomonas fluorescens biocontrol agents and *Glomus mosseae*, an arbuscular mycorrhizal fungus, within the rhizosphere. FEMS Microbiol Letters 116:297-303
Filippi C, Bagnoli G, Citernesi A S, Giovannetti M (1998) Ultrastructural spatial distribution of bacteria associated with sporocarps of *Glomus mosseae*. Symbiosis 24:1-12
Fitter A H, Heinemeyer A, Staddon P L (2000) The impact of elevated CO_2 and global climate change on arbuscular mycorrhizas: a mycocentric approach. New Phytol 147:179-187
Frey-Klett P, Garbaye J, Berge O, Heulin T (1997) Metabolic and genotypic fingerprinting of fluorescent *Pseudomonads* associated with the Douglas Fir-*Laccaria biocolor* mycorrhizosphere. Appl Environ Microbiol 63:1852-1860
Garbaye J (1994) Helper bacteria: A new dimension to the mycorrhizal symbiosis. New Phytol 128:197-210
Garbaye J, Bowen G D (1987) Effect of different microflora on the success of ectomycorrhizal inoculation of *Pinus radiata*. Can J Fores Res 17:941-943
Germida J J, Walley F L (1996) Plant growth-promoting rhizobacteria alter rooting patterns and arbuscular mycorrhizal fungi colonization of field-grown spring wheat. Biol Fertil Soils 23:113-120
Gianinazzi S, Schüepp H (eds) (1994) Impact of Arbuscular Mycorrhizas on Sustainable Agriculture and Natural Ecosystems. ALS, Birkhäuser Verlag, Basel
Glick B R (1995) The enhancement of plant growth by free-living bacteria. Can J Microbiol 41:109-117
Goicoechea N, Antolín M C, Strnad M, Sánchez-Díaz M (1996) Root cytokinins, acid phosphatase and nodule activity in drought-stressed mycorrhizal or nitrogen-fixing alfalfa plants. J Experiment Bot 47:683-686
Goicoechea N, Antolín M C, Sánchez-Díaz M (1997) Influence of arbuscular mycorrhizae and *Rhizobium* on nutrient content and water relations in drought stressed alfalfa. Plant Soil 192:261-268
Goicoechea N, Doléza K, Antolín M C, Strnad M, Sánchez-Díaz M (1995) Influence of mycorrhizae and *Rhizobium* on cytokinin content in drought-stressed alfalfa. J Experiment Bot 46:1543-1549
Goicoechea N, Szalai G, Antolín M C, Sánchez-Díaz M, Paldi E (1998) Influence of arbuscular mycorrhizae and *Rhizobium* on free polyamines and proline levels in water-stressed alfalfa. J Plant Physiol 153:706-711
Gryndler M, Hrselova H (1998) Effect of diazotrophic bacteria isolated from a mycelium of arbuscular mycorrhizal fungi on colonization of maize roots by *Glomus fistulosum*. Biol Plant 41:617-621
Gryndler M, Hrselová H, Stríteská D (2000) Effect of soil bacteria on growth of hyphae of the arbuscular mycorrhizal (AM) fungus *Glomus claroideum*. Folia Microbiologica 45: (In press)
Haas D, Keel C, Laville J, Maurhofer M, Oberliansli T, Schnider U, Voisard C, Wüthrich B, Defago G. (1991) Secondary metabolites of *Pseudomonas fluorescens* strain CHA0 involved in the suppresion of root diseases. In: Hennecke H, Verma D P S (eds) Advances in molecular genetics of plant-microbe interactions. Kluwer Academic publisher, Dordrecht, pp 450-456
Halverson L J, Handelsman J (1991) Enhacement of soybean nodulation by *Bacillus ceresus* UW95 in the field and in a growth chambers. Appl Environ Microbiol 57:2767-2770
Harley J L, Smith S E (eds) (1983) Mycorrhizal symbiosis. Academic Press, New York
Herrera M A, Salamanca C P, Barea J M (1993) Inoculation of woody legumes with selected arbuscular mycorrhizal fungi and rhizobia to recover desertified mediterranean ecosystems. Appl Environ Microbiol 59:129-133
Hiltner L (1904) Über neuere Erfahrungen und Probleme auf dem Gebiet der Bodenbakteriologie und unter besonderer Berücksichtigung der Gründüngung und Brache. Arb Dtsch Landwirtsch Ges 98:59-78
Jakobsen I, Rosendahl L (1990) Carbon flow into soil and external hyphae from roots of mycorrhizal cucumber plants. New Phytol 115:77-83
Jeffries P (1997) Mycoparasitism. In: Wicklow Södertröm (eds) The mycota IV environmental and microbial relationships. Springer-Verlag, Berlin, Heidelberg, pp 149-164
Jeffries P, Barea J M, Dodd J C (2000) Restoration of environmental diversity by effective ecosymbiont monitoring ('REDEEM'). In: Balabanis P, Peter D, Ghazi A, Tsogas M (eds) Mediterranean

desertification research results and policy implications. Eruropean Communities, Luxenburgo, pp 487-495
Kennedy A C (1998) The rhizosphere and spermosphere In: Sylvia D M, Fuhrmann J J, Hartel P G, Zuberer D A (eds) Principles and applications of soil microbiology. Prentice Hall, Upper Saddle River, New Jersey, pp 389-407
Kennedy A C, Smith K L (1995) Soil microbial diversity and the sustainability of agricultural soils. Plant Soil 170:75-86
Kim K Y, Jordan D, McDonald G A (1998) Effect of phosphate-solubilizing bacteria and vesicular-arbuscular mycorrhizae on tomato growth and soil microbial activity. Biol Fertil Soils 26:79-87
Kloepper J W (1994) Plant growth-promoting rhizobacteria (other systems) In: Okon Y (ed) *Azospirillum*/plant associations. CRC Press, Boca Raton, pp 111-118
Kloepper J W (1996) Host specificity in microbe-microbe interactions. BioScience 46:406-409
Kloepper J W, Zablotowick R M, Tipping E M, Lifshitz R (1991) Plant growth promotion mediated by bacterial rhizosphere colonizers. In: Keister D L, Cregan P B (eds) The rhizosphere and plant growth. Kluwer Academic Publishers, Dordrecht, pp 315-326
Latour X, Corberand T, Laguerre G, Allard F, Lemanceau P (1996) The composition of fluorescent pseudomonad population associated with roots is influenced by plant and soil type. Appl. Environ. Microbiol. 62:2449-2556
Lemanceau P, Alabouvette C (1991) Biological control of fusarium diseases by fluorescent *Pseudomonas* and non-pathogenic *Fusarium*. Crop Protect. 10:279-286
Lemanceau P, Alabouvette C (1993). Suppression of fusarium-wilts by fluorescent pseudomonads: mechanisms and applications. Biocontrol Science and Technology 3:219-234
Lemanceau P, Samson R, Alabouvette C (1988). Recherches sur la résistance des sols aux maladies. XV. Comparaison des populations de *Pseudomonas* fluorescents dans un sol résistant et un sol sensible aux fusarioses vasculaires. Agronomie 8:243-249
Lemanceau P, Bakker P A H M, De Kogel W J, Alabouvette C, Schippers B (1992) Effect of pseudobactin 358 production by *Pseudomonas putida* WCS358 on suppression of fusarium wilt of carnations by nonpathogenic *Fusarium oxysporum* Fo47. Appl. Environ. Microbiol. 58:2978-2982.
Lemanceau P, Corberand T, Gardan L, Latour X, Laguerre G, Boeufgras J M, Alabouvette C (1995) Effect of two plant species flax (*Linum usitatissinum* L.) and tomato (*Lycopersicon esculentum* Mill.) on the diversity of soilborne populations of fluorescent pseudomonads. Appl. Environ. Microbiol. 61: 1004-1012
Linderman R G (1992) Vesicular-arbuscular mycorrhizae and soil microbial interactions. In: Bethlenfalvay G J, Linderman R G (eds) Mycorrhizae in sustainable agriculture. ASA Spec. Publ., Madison, Wisconsin, pp 45-70
Linderman R G (1994) Role of VAM fungi in biocontrol. In: Pfleger F L, Linderman R G (eds) Mycorrhizae and plant health. APS Press, St Paul, pp 1-26
Lugtenberg B J J, Weger de L A, Bennett J W (1991) Microbial stimulation of plant growth and protection from disease. Current Opinion Microbiol 2:457-464
Lynch J M (ed) (1990) The rhizosphere. John Wiley, New York
Miller R M, Jastrow J D (1994) Vesicular-arbuscular mycorrhizae and biogeochemical cycling. In: Pfleger F L, Linderman R G (eds) Mycorhizae and plant health. APS Press, St. Paul, MN, pp 189-212
Monzón A, Azcón R (1996) Relevance of mycorrhizal fungal origin and host plant genotype to inducing growth and nutrient uptake in *Medicago* species. Agric Ecosyst Environm 60:9-15
Nehl D B, Allen S J, Brown J F (1996) Deleterious rhizosphere bacteria: An integrating perspective. Appl Soil Ecol 5:1-20
Nemec S (1997) Longevity of microbial biocontrol agents in a planting mix amended with *Glomus intraradices*. Biocontrol Sci Technol 7:183-192
Norby R J, Jackson R B, (2000) Root dynamics and global change: seeking an ecosystem perspective. New Phytol 147:3-12
O´Gara F, Dowling D N, Boesten B (eds) (1994) Molecular Ecology of Rhizosphere Microorganisms. VCH, Weinheim, Germany
Okon Y (eds) (1994) *Azospirillum*/plant associations. CRC Press, Boca Raton
Puppi G, Azcón R, Höflich G (1994) Management of positive interactions of arbuscular mycorrhizal fungi with essential groups of soil microorganisms. In: Gianinazzi S, Schüepp H (eds) Impact of arbuscular mycorrhizas on sustainable agriculture and natural ecosystems. ALS, Birkhäuser Verlag, Basel Switzerland, pp 201-215

Ravnskov S, Nybroe O, Jakobsen I (1999) Influence of an arbuscular mycorrhizal fungus on *Pseudomonas fluorescens* DF57 in rhizosphere and hyphosphere soil. New Phytol 142:113-122
Requena N, Jimenez I, Toro M, Barea J M (1997) Interactions between plant-growth- promoting rhizobacteria (PGPR), arbuscular mycorrhizal fungi and *Rhizobium* spp. in the rhizosphere of *Anthyllis cytisoides*, a model legume for revegetation in mediterranean semi-arid ecosystems. New Phytol 136:667-677
Requena N, Perez-Solis E, Azcón-Aguilar C, Jeffries P, Barea J M (2001) Management of indigenous plant-microbe symbioses aids restoration of desertified. Appl Environ Microbiol 67:000-000
Rodríguez R, Vassilev N, Azcón R (1998) Increases in growth and nutrient uptake of alfalfa grown in soil amended with microbially-treated sugar beet waste. Appl Soil Ecol 330:1-7
Ruiz-Lozano J M, Azcón R (1993) Specificity and functional compatibility of VA mycorrhizal endophytes in association with *Bradyrhizobium* strains in *Cicer arietinum*. Symbiosis 15:217-226
Ruiz-Lozano J M, Bonfante P (2000) A *Burkholderia* strain living inside the arbuscular mycorrhizal fungus *Gigaspora margarita* possesses the *vacB* gene, which is involved in host cell colonization by bacteria. Microbial Ecol 39:137-144
Ruiz-Lozano J M, Collados C, Barea J M, Azcón R (2001) Arbuscular mycorrhizal symbiosis can alleviate drought-induced nodule senescence in soybean plants. New Phytol 15:493-502
Sanjuan J, Olivares J (1991) Multicopy plasmids carrying the *Klebsiella pneumoniae* nifA gene enhance *Rhizobium meliloti* nodulation competitiveness on alfalfa. Mol Plant Micr Interac 4:365-369
Schloter M, Wiehe W, Assmus B, Steindl H, Becke H, Höflich G, Hartmann A (1997) Root colonization of different plants by plant-growth-promoting *Rhizobium leguminosarum* bv. trifolii R39 studied with monospecific polyclonal antisera. Appl Environ Microbiol 63:2038-2046
Schreiner R P, Mihara KL, McDaniel H, Bethlenfalvay G J (1997) Mycorrhizal fungi influence plant and soil functions and interactions. Plant Soil 188:199-209
Siddiqui Z A, Mahmood I (1998) Effect of a plant growth promoting bacterium, an AM fungus and soil types on the morphometrics and reproduction of Meloidogyne javanica on tomato. Appl Soil Ecol 8:77-84
Smith S E, Gianinazzi-Pearson V, Koide R, Cairney J W G (1994) Nutrient transport in mycorrhizas: structure, physiology and consequences for efficiency of the symbiosis. In: Robson A D, Abbott L K, Malajczuk N (eds) Management of mycorrhizas in agriculture, horticulture and forestry. Kluwer Academic Publishers, Dordrecht, pp 103-113
Smith M D, Hartnett D C, Rice C W (2000) Effects of long-term fungicide applications on microbial properties in tallgrass prairie soil. Soil Biol Biochem 32:935-946
Staley T W, Lawrence E G, Nance E L (1992) Influence of a plant growth-promoting pseudomonad and vesicular-arbuscular mycorhizal fungus on alfalfa and birdsfoot trefoil growth and nodulation. Biol Fertil Soils 14:175-180
Sturz A V, Christie B R, Nowak J (2000) Bacterial endophytes: potential role in developing sustainable systems of crop production. Crit Rev Plant Sci 19:1-30
Sturz A V Nowak J (2000) Endophytic communities of rhizobacteria and the strategies required to create yield enhancing associations with crops. Appl Soil Ecol 15:183-190
Stutz E W, Défago G, Kern H (1986) Naturally occuring fluorescent Pseudomonads in suppression of black root rot of tobacco. Phytopathology 76:181-185
Toal M E, Yeomans C, Killham K, Meharg A A (2000) A review of rhizosphere carbon flow modelling. Plant Soil 222:263-281
Tobar R M, Azcón-Aguilar C, Sanjuán J, Barea J M (1996) Impact of a genetically modified *Rhizobium* strain with improved nodulation competitiveness on the early stages of arbuscular mycorrhiza formation. Appl Soil Ecol 4:15-21
Toro M, Azcón R, Barea J M (1997) Improvement of arbuscular mycorrhizal development by inoculation with phosphate-solubilizing rhizobacteria to improve rock phosphate bioavailability (^{32}P) and nutrient cycling. Appl Environ Microbiol 63:4408-4412
Toro M, Azcón R, Barea J M (1998) The use of isotopic dilution techniques to evaluate the interactive effects of *Rhizobium* genotype, mycorrhizal fungi, phosphate-solubizing rhizobacteria and rock phosphate on nitrogen and phosphorus acquisition by *Medicago sativa*. New Phytol 138:265-273
Van Loon L C, Bakker P A H M, Pieterse C M J (1998) Systemic resistance induced by rhizosphere bacteria. Annu. Rev. Phytopathol. 36:453-483
Vázquez M M, Cesar S, Azcón R, Barea J M (2000) Interactions between arbuscular mycorrhizal fungi and other microbial inoculants (*Azospirillum, Pseudomonas, Trichoderma*) and their effects

on microbial population and enzyme activities in the rhizosphere of maize plants. Appl Soil Ecol 15:261-272

Volpin H, Kapulnik Y (1994) Interaction of *Azospirillum* with beneficial soil microorganisms. In: Okon Y (ed) *Azospirillum*/plant associations. CRC Press, Boca Raton, pp 111-118

Vosátka M (1996) Soil bacteria - a component of plant, soil and arbuscular mycorrhizal fungal interactions. In: Azcon-Aguilar C, Barea J M, (eds) Mycorrhizas in integrated systems - from genes to plant development. European Commission Report EUR 16728, Brussels, Luxembourg, pp 613-618

Vosátka M, Gryndler M (1999) Treatment with culture fractions from *Pseudomonas putida* modifies the development of *Glomus fistulosum* mycorrhiza and the response of potato and maize plants to inoculation. Appl Soil Ecol 11:245-251

Vosátka M, Gryndler M (2000) Response of micropropagated potatoes transplanted to peat media to post-vitro inoculation with arbuscular mycorrhizal fungi and soil bacteria. Appl Soil Ecol 15:145-152

Weller D M, Howie W J, Cook RJ (1998) Relationship between in vitro inhibition of *Geaumannomyces graminis* var. tritici and suppression of take-all of wheat by fluoresent pseudomonads. Phytopathology 78:1094-1100

Weller D M, Thomashow L S (1994) Current challenges in introducing beneficial microorganisms into the rhizosphere. In: O´Gara F, Dowling D N, Boesten B, Weinheim V C H (eds) Molecular ecology of rhizosphere microorganisms biotechnology and the release of GMOs. Germany, pp 1-18

Werner D (1998) Organic signals between plants and microorganisms. In: Pinton R, Varanini Z, Nannipieri P (eds) The rhizosphere: biochemistry and organic substances at the soil-plant interfaces. Marcel Dekker Inc., New York

Whitelaw M A (2000) Growth promotion of plants inoculated with phosphate-solubilizing fungi. Adv Agron 69:99-151

Wright S F, Upadhyaya A (1998) A survey of soils for aggregate stability and glomalin, a glycoprotein produced by hyphae of arbuscular mycorrhizal fungi. Plant Soil 198:97-107

Mycorrhizal Technology in Agriculture
ed. by S. Gianinazzi, H. Schüepp, J.M. Barea and K. Haselwandter

Arbuscular mycorrhizal fungi and soil bacteria: from cellular investigations to biotechnological perspectives

V. Bianciotto[1], S. Perotto[1], J.M. Ruiz-Lozano[2] and P. Bonfante[1]

[1] *Centro Studio sulla Micologia del Terreno and Dipartimento di Biologia Vegetale, Viale Mattioli 25, I-10125 Torino, Italy*
[2] *Departamento de Microbiologia del Suelo y Sistemas Simbioticos, Estación Experimental del Zaidín, Profesor Albareda 1, E-18008 Granada, Spain*

Keywords: Rhizosphere, beneficial microorganisms, mycorrhizal fungi, PGPRs, saprotrophs, symbionts

Introduction

The rhizosphere is a dynamic environment in which bacteria, viruses, fungi, and microfauna, develop, interact and take advantage of organic matter released by the root (Weller and Thomashow 1994). A substantial consequence of this richness, in comparison with the bulk soil, is an intense microbial activity that results in changes in root development and growth of the whole plant. Two main groups of microorganisms are stimulated to grow in the rhizosphere: saprotrophs and symbionts, both comprised of bacteria and fungi that can be detrimental, neutral or beneficial to plants. In this article, we will focus on two categories of beneficial microorganisms of the rhizosphere, the plant-growth-promoting rhizobacteria (PGPRs) and the mycorrhizal fungi.

PGPRs include bacteria with different life styles that are active in many key processes, such as biological control of plant pathogens, nutrient cycling and seedling establishment. Some PGPRs (e.g. pseudomonads and *Azospirillum*) accomplish their whole life cycle in the rhizosphere soil and interact only loosely with the plant, whereas others, like rhizobia, have both a free-living and a complex symbiotic stage inside the root tissues. PGPRs can influence plant growth directly, by releasing a variety of compounds ranging from mineral nutrients to phytohormones, or indirectly, by protecting the plant against pathogens through the production of antimicrobial compounds (Défago and Keel 1995; Keel and Défago 1997).

Mycorrhizal fungi colonize the root tissues and establish with their host a mutualistic association. Mycorrhiza is an essential feature of the biology and ecology of most terrestrial plants, and influences their growth, water and nutrient absorption, and protects them from root diseases (Smith and Read 1997). By far the most common type of mycorrhizal association found in land plants is formed by AMF. These Zygomycetes reside in the rhizosphere as hyphae, spores and other propagules, and occupy the rhizoplane during their interaction with the root (Bianciotto and Bonfante 1999).

The importance of mycorrhizal fungi in the rhizosphere and their beneficial effects on plants have been documented widely (see references in Smith and Read 1997). More recently, the mutual interactions between AMF and other components of the microbial communities in the rhizosphere have also been investigated (Linderman 1992; Azcón-Aguilar and Barea 1992; Garbaye 1994; Barea 1997). Consequently, it is now known that AMF can influence microbial populations directly in the rhizosphere, or indirectly by modifying host physiology and patterns of root exudation. Moreover, many experiments demonstrate that rhizobacteria stimulate mycorrhizal fungal growth (Mayo et al. 1986; Azcón-Aguilar et al. 1986; Azcón 1987; Linderman and Pauliz 1990; Requena et al. 1999) and mycorrhiza formation and development (Azcón-Aguilar and Barea 1992; Barea 1997). Garbaye (1994), for example, has elegantly shown how certain bacterial strains can increase a root's ability to establish an ectomycorrhizal symbiosis, and has proposed a new bacterial category to describe this effect: the mycorrhization helper bacteria (MHB). However, despite the extensive research to elucidate the mechanisms responsible for the stimulatory effect of soil microbes on AMF (Azcón-Aguilar and Barea 1992) it is still unknown how microbes might signal the fungus and elicit observed increases in mycelial development (Requena 1998). Several mechanisms have been proposed to explain these effects, including the production of vitamins, amino acids, phytohormones and/or cell wall hydrolytic enzymes. Some of these may directly influence the germination and growth rate of fungal structures, whereas others may act on root development and susceptibility to infection.

The outcome of these associations between AMF and PGPRs can be synergisms that can have profoundly beneficial effects on plant health (Azcón-Aguilar and Barea 1992; Hodge 2000). For example, dual inoculation with a PGPR (*Pseudomonas putida*) and an AMF induced additional growth enhancement of subterranean clover compared to single inoculation (Meyer and Lindermann 1986). Similar data were obtained for tomato plants when biocontrol *Pseudomonas* strains producing antifungal compounds were used in combination with the AMF *Glomus mosseae* (Barea et al. 1998). A *Paenibacillus* strain isolated from the rhizosphere of *Sorghum bicolor*, besides having antagonistic activity towards soilborne fungal pathogens, also stimulated mycorrhiza formation (Budi et al. 1999). In other experiments, colonization and activity of *Glomus mosseae* were unaltered in the presence of three biocontrol *P. fluorescens* CHA0 strains, although the presence of *Glomus mosseae* increased the total population of *P. fluorescens* (Edwards et al. 1998).

Despite our understanding of the molecular dialogue between plants and rhizobacteria or AMF (Gianinazzi-Pearson 1996; Harrison 1997, 1999) is growing rapidly, information concerning the molecular cross talk between rhizobacteria and AMF is practically lacking. In fact, the molecular mechanisms underlying the cell-to-cell communication between AMF and other soil microorganism are totally unknown. Recently, Requena et al. (1999) studied changes in the gene expression of the AMF *Glomus mosseae* in response to the rhizobacterium *Bacillus subtilis* by using differential RNA display. They found that this bacterial strain, which can increase growth of the AMF mycelium, specifically induces in the fungus the expression of *GmFOX2*, a highly conserved gene encoding a multifunctional protein involved in peroxisomal beta-oxidation.

Cellular and molecular bases of the physical association between rhizobacteria and AMF

Microorganisms in natural environments live predominantly attached to solid surfaces. These sessile microbial communities, comprising either single or multiple species, are commonly referred to as biofilms (Costerton et al. 1995). The formation of bacterial biofilms on the plant root surface may be very significant, since biological activity of some PGPRs depends on their ability to colonise the rhizosphere and to remain associated with the root (Lugtenberg and Dekkers 1999).

While numerous experiments have demonstrated the role of several surface components in the physical interactions between rhizosphere bacteria and the plant root during biofilm development, there is little information on cellular interactions between PGPRs and AMF. Bacterial attachment generally proceeds through two consecutive steps. Appendages such as pili, fimbriae and flagella are involved in the initial attachment of bacteria to solid surfaces, and their role has been demonstrated for a number of pathogenic and beneficial plant/microbe interactions (Vandebroek and Vanderleyden 1995).

During the second step, extracellular carbohydrate polymers are responsible for the firm anchoring of bacteria to the plant surface. For example, cellulose fibrils have been identified at the attachment site of *Rhizobium leguminosarum* (Smit et al. 1992). In addition to a role in adhesion, extracellular polysaccharides (EPS) form the matrix that embeds bacterial colonies in biofilms (Costerton et al. 1995).

The formation of bacterial biofilms around the hyphae of mycorrhizal fungi has been demonstrated for ectomycorrhizal (Sen et al. 1996) and AMF (Bianciotto et al. 1996; Filippi et al. 1998). Several rhizobacteria described as good root colonizers and biocontrol agents are able to attach to the hyphal surface, suggesting that the mechanisms involved may be quite similar for the two systems. For example, Bianciotto et al. (1996) investigated by electron and laser scanning confocal microscopy whether PGPRs, some of which are used as biocontrol agents, adhere to AM structures through a direct cell-to-cell interaction. PGPRs strains already identified as strong root colonizers, such as *Pseudomonas fluorescens* strain WCS365, could form a thick coat on the AM fungal hyphae (Fig. 1 A, C) and on auxiliary cells (Fig. 1 B, D), as do strains

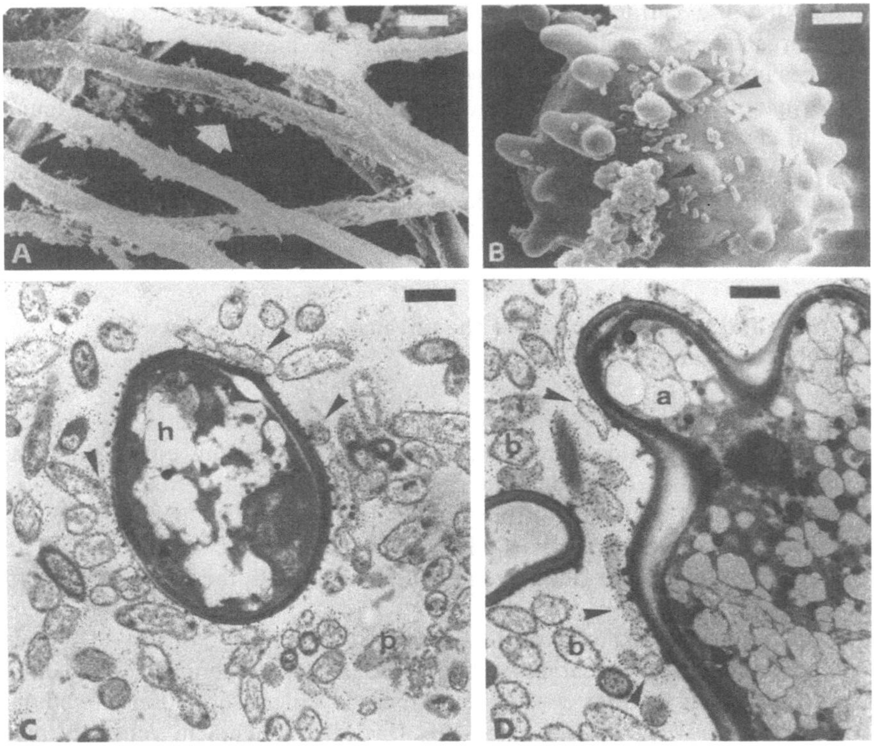

Figure 1. Scanning and transmission electron micrographs of AM fungal structures formed by *Gigaspora margarita*, coated with bacterial cells of *Pseudomonas fluorescens* strain WCS365. (A) Attachment of bacterial cells to the surface of fungal hyphae. The hyphae appear completely coated by the bacteria (arrow). Bar = 12,5 µm. (B) Bacterial cells (arrows) attached to an auxiliary cell formed *in vitro*. Bar = 4 µm. (C) Hypha and (D) auxiliary cell, surrounded by numerous bacterial cells (b) that show a labelling after treatement with an antiserum that recognizes this strain. Only some of the bacteria are in direct contact with the fungal surface (arrowheads). (a) auxiliary cell, (h) hypha. In (C) and (D) bar = 1,3 µm.

of *Rhizobium leguminosarum*. Extracellular material of bacterial origin was visible around attached bacteria and may mediate fungal/bacterial interactions (Fig. 2).

To understand the role of different surface components in the attachment of PGPRs to AM fungal structures and to mycorrhizal roots, we have tested mutants of *Azospirillum brasilense* and *R. leguminosarum* impaired in the production of extracellular polysaccharides (EPS). The data obtained from experiments using these mutants (Bianciotto et al. 2001a) demonstrate that, for both bacterial species, EPS play a crucial role in the anchoring of bacteria and in the formation of biofilms on the root and the AMF. Impairment in EPS production was mirrored in fact by a strong decrease in the number of bacterial cells on both surfaces.

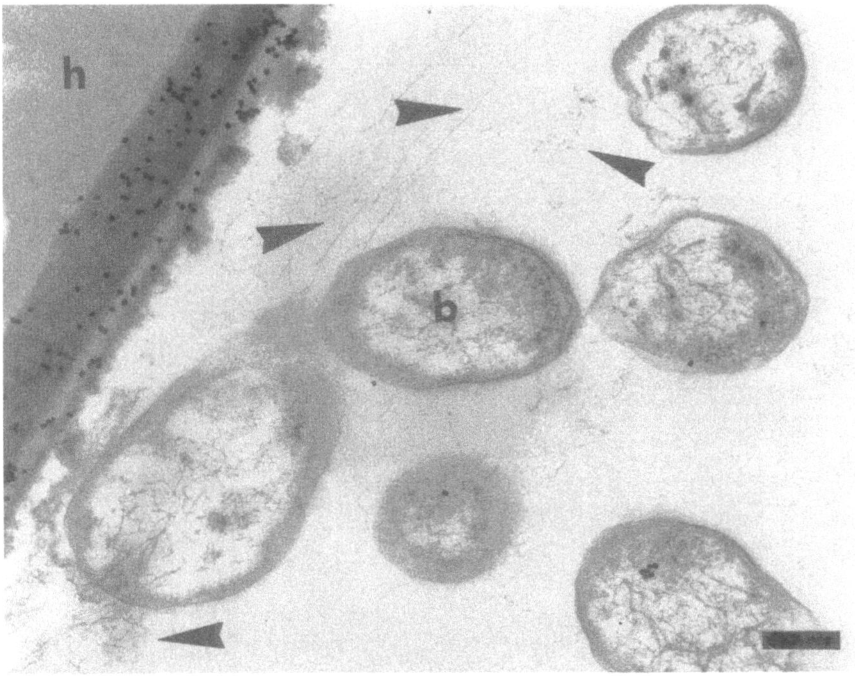

Figure 2. Ultrastructural study of the attachment of *Pseudomonas fluorescens* strain WCS365 to AM fungal structures. Extracellular material (arrows) is found between individual bacterial cells (b) in the biofilm surrounding an hypha (h). The fungal wall is labelled with the WGA-gold complex specific for chitin. Bar = 0,2 μm.

Further support for the importance of EPS in biofilm formation has emerged from investigation of mutants of *P. fluorescens* with increased production of acidic EPS (Bianciotto et al. 2001b).

In Figure 3A, the non-mucoid wild type strain *P. fluorescens* (CHA0) adhered very little on a hypha of *Gigaspora margarita*, whereas the EPS over-producer, mucoid strains (CHA211 and CHA213M) formed a dense and patchy bacterial layer on the fungal structures (Fig. 3B, C). The observation that mucoid *P. fluorescens* mutants were able to form a more conspicuous bacterial layer compared with the non-mucoid wild type indicates that extracellular acidic EPS is involved, at least *in vitro*, in the association of bacteria with solid substrates such as roots and AM surfaces.

Morover, it is likely that the anchoring of rhizobacteria to AM fungal structures may have ecological significance. Like roots, mycorrhizal fungi can release exudates to create a niche relatively rich in organic compounds compared to the bulk soil, the hyphosphere (Marschner 1995), where a specialised microbial community can establish (Andrade et al. 1997; Frey et al. 1997; Timonen et al. 1998). Colonisation and adhesion of bacterial cells on the

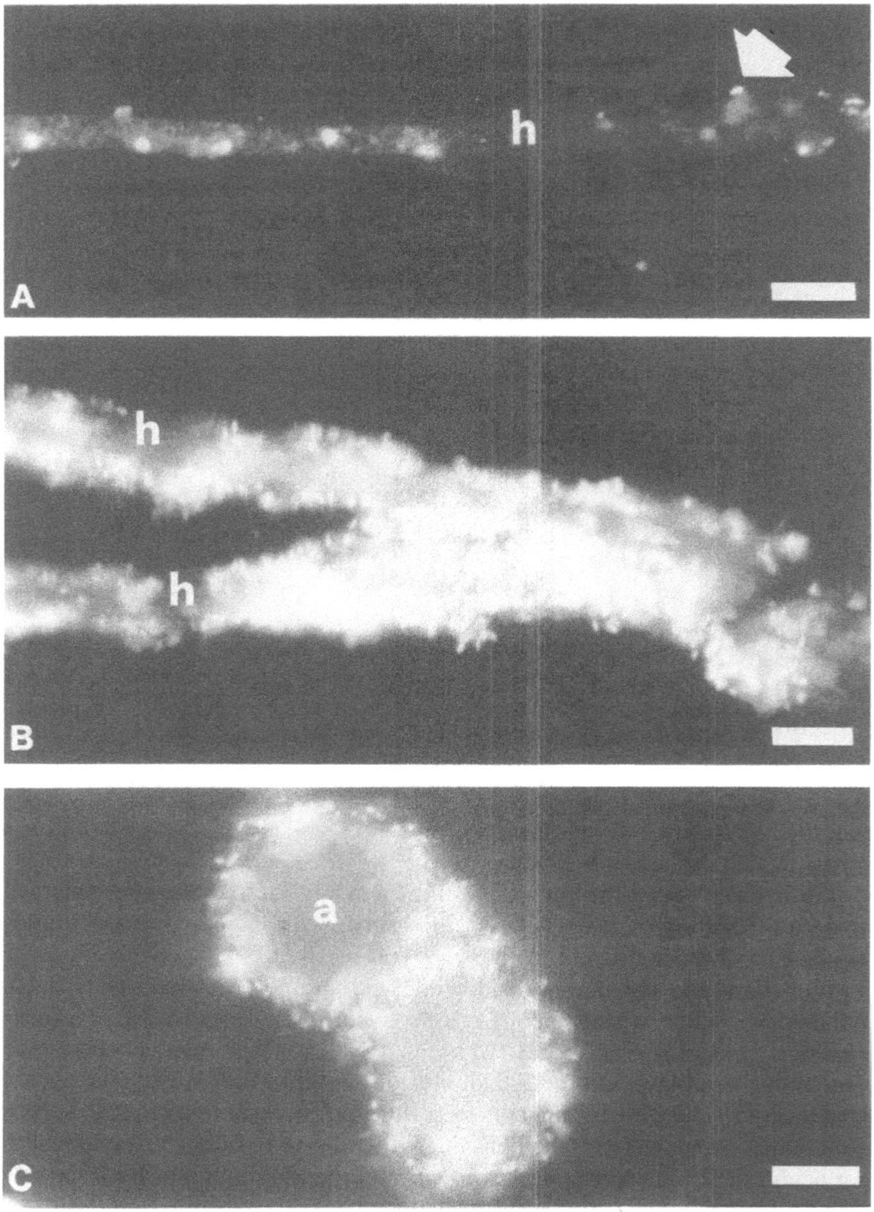

Figure 3. Confocal microscopy images of the association of wild-type *P. fluorescens* strain CHA0 and mucoid strains to mycelium and auxiliary cells of a *Gigaspora margarita* spore germinated *in vitro*. (A) The non-mucoid strain CHA0 shows very little ability to associate with the mycelium grown *in vitro*. Only few bacteria are attached to the hypha (arrow). (B) The mucoid strain CHA211 forms a dense bacterial layer on the hyphae (h). (C) Bacterial cells of the mucoid strain CHA213 cover the auxiliary cells surface. Bars = 10,5 μm

hyphal surface is likely to play an important role in the establishment of the hyphosphere community, and our results strongly suggest that EPS are involved in this process.

A general role of EPS is to protect bacterial cells against desiccation and harsh environments (Ophir and Gutnick 1994). Increased EPS production in the mucoid derivatives of *P. fluorescens* CHA0 may thus have biotechnological applications. Future research should elucidate whether survival of *P. fluorescens* mutants in the soil may benefit from their increased mucoidity. Moreover, the wild type strain CHA0 displays biocontrol activity, and data obtained *in vitro* suggest that the antifungal compounds responsible for biocontrol activity remain entrapped in the EPS of mucoid strains, thus reaching a high local concentration (Schnider et al. 1997). It is important to elucidate whether this feature of mucoid colonies may lead to increased plant protection, the thicker bacterial biofilm forming a shield rich in bioactive compounds around the root, or whether a limited diffusion of these compounds may hamper biocontrol activity.

Endosymbiotic bacteria of AMF

Besides the physical interactions occurring between soil bacteria and AM fungal structures, a peculiar feature of some isolates of AMF is to host intracellular structures very similar to bacteria, called Bacteria-like Organisms (BLOs) and first described in the 1970s (Mosse 1970; see Scannerini and Bonfante 1991 for a review). Further investigation on these BLOs, including the demonstration of their prokaryotic nature, was hampered for long time because of the inability to grow them *in vitro*. A combination of morphological observations and molecular analyses have allowed us to identify BLOs in the isolate *Gigaspora margarita* BEG 34 as true bacteria belonging to the genus *Burkholderia*, and to start unravelling their symbiotic relationship with AMF (Bianciotto et al. 1996). Recently, *in situ* hybridisation with specific probes has confirmed the topological position of these endobacteria (Bianciotto et al. 2000).

The detection of *Burkholderia* in the spores and hyphae of the BEG 34 isolate, originally from a New Zealand soil, has raised the question of whether the same bacteria can be found in isolates from different geographic areas and in other Glomales. To address this question, many isolates of different origin from the three AM fungal families (Glomaceae, Gigasporaceae and Acaulosporaceae) were investigated by confocal microscopy and PCR amplification to detect the presence and nature of endosymbiotic bacteria. Bacterial DNA can be amplified from all these families (Bianciotto et al. 1996; Hosny et al. 1999), but more detailed investigations of Glomaceae are hampered by the low number of intracellular bacteria and by the strong microbial contamination of the spore surface (Bianciotto, unpublished observations). We focussed thus our attention on the Gigasporaceae, which comprise the genera

Gigaspora and *Scutellospora*, and eleven fungal isolates collected from different geographic areas and belonging to six different species were analysed (Bianciotto et al. 2000). With the exception of the four isolates of *Gigaspora rosea*, bacteria could be visualized in the cytoplasm of all fungal isolates, and their DNA could be amplified with universal bacterial primers. We demonstrated by sequencing that at least three different fungal species in the two genera *Gigaspora* and *Scutellospora* harbour endosymbiotic bacteria in their cytoplasm that are closely related to each other and belong to the genus *Burkholderia.*

Intracellular symbioses raise fascinating questions about the acquisition of the endosymbionts, their transmission and the evolution of partner adaptations (Margulis and Chapman 1998). Nothing is known about the molecular mechanisms that allow entry of endobacteria into the AMF, and about the control of colonization of the AM cytoplasm. Perotto and Bonfante (1997) have suggested the possibility that a rhizospheric bacterial strain once acquired the possibility to actively invade the fungal cytoplasm. Colonization determinants common in pathogenic and symbiotic bacteria have been described by Galan and Collmer 1999. Another important colonization determinant is *vacB*, originally described in *Shigella flexneri* and enteroinvasive *Escherichia coli* (EIEC) as a chromosomal gene required for the expression of virulence genes (Tobe et al. 1992). Altough VacB was first defined as a virulence factor, it is currently known to be an exoribonuclease RNase R involved in post-transcriptional processing of mRNAs. It modulates the ability of bacteria to adhere and to penetrate cells, and later spread for full virulence expression (Cheng et al. 1998). A *vacB*-like gene has been isolated and characterized (Ruiz-Lozano and Bonfante 2000) from a genomic library developed from *Gigaspora margarita* spores and also representative of the *Burkholderia* genome (van Buuren et al. 1999). The design of specific primers on the *vacB* nucleotide sequence demonstrated that the endosymbiotic *Burkholderia* strain possesses the molecular determinant required for the colonization of a eukaryotic cell. The authors hypothesized that the *vacB* gene was part of a genetic region acquired by a rhizospheric *Burkholderia* strain that together with other genes, enabled this strain to establish a symbiotic interaction with the AMF *Gigaspora margarita*. Supporting their hypothesis, a corresponding DNA fragment could be amplified from some rhizospheric *Burkholderia* isolates (Ruiz-Lozano and Bonfante 2000).

What can be the metabolic role of the endosymbiotic *Burkholderia* strain?

Despite the huge number of endobacteria found in the cytoplasm of AMF (about 250.000 in a single spore of *Gigaspora margarita*), their functional significance is not clear. AMF mostly assist their host plant by providing phosphate (Harrison 1999). To investigate whether *Burkholderia* can somehow influence phosphorus metabolism in AM, degenerated oligonucleotide primers were designed, based on conserved regions of a gene coding for a subunit of a bacterial phosphate transporter. These primers amplified a DNA fragment from the genomic library of *Gigaspora margarita*, that was used as probe and led to the isolation and characterization of the complete bacterial operon coding for a high-affinity phosphate transporter (Ruiz-Lozano and Bonfante 1999). The organization of the *Burkholderia* operon (e.g. gene order, direction of transcription) is the same as in *E. coli* and similar to many other bacteria, thus confirming the notion that this type of transporter is highly conserved. To demonstrate that the operon is contained in the genome of the intracellular *Burkholderia*, and to exclude the possibility that it may derive from surface spore contaminants, the same procedure already used for the *vacB* gene was followed. Specific primers were designed and used in PCR on DNA extracted from carefully surface sterilized spores. The primers successfully amplify fragments of the expected size in spores of *Gigaspora margarita* and *S. persica*, which contain related endobacteria (Bianciotto et al. 2000), whereas no amplification occurred on DNA from the related species *Gigaspora rosea*, which is devoid of intracellular bacteria.

Since some free-living *Burkholderia* are known to fix nitrogen (Gillis et al. 1995), a strategy was developed to investigate whether *nif* genes are present also in the genome of endosymbiotic *Burkholderia*. Preliminary investigations based on the use of a *nifDK* probe from *Azospirillum brasilense* have led to the identification of several positive clones during the screening of the genomic *Gigaspora margarita* library. The putative protein encoded by one of these clones showed a very high degree of sequence similarity (>90%) with the NifD protein from different nitrogen fixing microorganisms. In addition, RT-PCR experiments performed with specific primers on the mRNA extracted from germinating spores of *Gigaspora margarita* indicate that the transcript is present at this developmental stage (Minerdi et al. 2001). It will be exciting to get the complete sequence of these *nif* genes, to understand whether and when they are transcribed and whether a functional nitrogenase is produced.

The discovery in the genome of endosymbiotic *Burkholderia* of genes involved in some important metabolic functions opens of course a number of intriguing questions. The presence for example of a bacterial phosphate transporter system indicates that endobacteria may influence the phosphate flux that takes place between the AMF and the host plant. Since AMF possess a high affinity phosphate transporter that is active in the extraradical mycelium (Harrison and van Buuren 1995), endobacteria may have direct access to this phosphorus source and they could use it for their own metabolism, thus reducing the phosphorus flux to the root. In the same way, the presence of *nif* genes in the genome of the bacteria opens the possibility for N_2 fixation in the ensemble *Burkholderia*/AMF and/or *Burkholderia*/AMF/host plant. Physiological studies will be required to properly address this question. As a first approach, the effects of intracellular bacteria on plant growth and nutrient uptake were investigated by comparing in pot experiments the growth effects of a mycorrhizal strain which contains endobacteria (*Gigaspora margarita*) with those of a taxonomically related species lacking them (*Gigaspora rosea*). The results obtained on lettuce plants showed that *Gigaspora margarita* had a higher symbiotic efficiency in terms of nutrient acquisition and plant growth. Moreover, plants colonized by *Gigaspora margarita* had the highest N and P content when compared with plants colonized by *Gigaspora rosea* or *Glomus versiforme* (Ruiz-Lozano and Bonfante 2001). Even if the experimental approach was not optimal, because two different, though related, species were compared, the results suggest that the occurrence of endobacteria has no negative influence on the phosphorus flux to the root. On the contrary, the endosymbionts may lead indirectly to a positive effect on plant growth and nutrition, at least under poor soil conditions.

Conclusions and biotechnological perspectives

The development of inocula based on the use of plant growth-promoting microorganisms will be key to the future of sustainable agriculture. Investigation of the cellular bases of interactions between AMF and PGPRs is a preliminary step that will allow us to better define the parameters/traits necessary to develop an efficient mixed inoculum. For example, adhesive properties of bacterial microorganisms to the AM fungal structures may be important to make a stable inoculum. Our results suggest that this trait may be improved with the selection of strains with a relatively high production of EPS. The use of more adhesive, mucoid PGPR strains for AMF spore coating may increase the density of bacterial inoculum, decrease the stress provoked by desiccation, and may also help to transport bacteria more effectively to the root surface.

Genomics is a powerful approach to study the potential properties of an organism, and the technology has now progressed to provide relatively rapid results. This technique, applied to the genomic libraries already available for *Gigaspora margarita*, will help to establish the features of the endobacterial population, which has so far remained hidden inside its fungal hosts. The preliminary data obtained indicate the presence of interesting metabolic genes. A better understanding of this fascinating biological system may allow us to exploit this metabolic potential and to transfer these bacterial capabilities to tuned mycorrhizal inocula.

Acknowledgements
We wish to thank Raffaella Balestrini for the transmission electron microscopy images and Maria Teresa Della Beffa for the reference list. The research was supported by the National Council of Research (CNR, Italy) by the European Project IMPACT (Contract BIO-CT96-0027) and by the National Project "Produzione Agricola nella Difesa dell'Ambiente" (PANDA). J.M.R.L. was funded by a E.U. research training grant (Contract BIO4-CT97-5118).

References

Andrade G, Mihara K L, Linderman R G, Bethlenfalvay G J (1997) Bacteria from rhizosphere and hyphosphere soils of different arbuscular-mycorrhizal fungi. Plant and Soil 192:71-79

Azcón R (1987) Germination and hyphal growth of *Glomus mosseae* in vitro: effects of rhizosphere bacteria and cell-free culture media. Soil Biol. Biochem. 19:417-419

Azcón-Aguilar C, Barea J M (1992) Interaction between mycorrhizal fungi and other rhizosphere microorganisms. In: Allen M J (ed) Mycorrhizal functioning: an integrative plant-fungal process. Chapman & Hall, New York pp 163-198

Azcón-Aguilar C, Diaz-Rodriguez R M, Barea J M (1986) Effect of soil micro-organisms on spore germination and growth of the vesicular-arbuscular mycorrhizal fungus *Glomus mosseae*. Trans. Br. Mycol. Soc. 86:337-340

Barea J M (1997) Mycorrhiza/bacteria interactions on plant growth promotion. In: Ogoshi A, Kobayashi K, Homma Y, Kodama F, Kondo N, Akino S (eds) Plant growth-promoting Rhizobacteria: present status and future prospects. OECD Paris pp 150-158

Barea J M, Andrade G, Bianciotto V, Dowling D N, Lohrke S, Bonfante P, O'Gara F, Azcón-Aguilar C (1998) Impact on arbuscular mycorrhiza formation of *Pseudomonas* strains used as inoculants for the biocontrol of soil borne plant fungal pathogens. Appl Environ Microbiol 64:2304-2307

Bianciotto V, Andreotti S, Balestrini R, Bonfante P, Perotto S (2001a) Extracellular polysaccharides are involved in the attachment of *Azospirillum brasilense* and *Rhizobium leguminosarum* to arbuscular mycorrhizal structures. European Journal of Histochemistry 45, in press

Bianciotto V, Andreotti S, Balestrini R, Bonfante P, Perotto S. (2001b) Mucoid mutants of the biocontrol strain *Pseudomonas fluorescens* CHA0 show increased ability in biofilm formation on mycorrhizal and non mycorrhizal carrot roots. Molecular Plant Microbe Interactions 14:255-260

Bianciotto V, Bandi C, Minerdi D, Sironi M, Tichy H V, Bonfante P (1996) An obligately endosymbiotic fungus itself harbors obligately intracellular bacteria. Appl Environ Microbiol 62:3005-3010

Bianciotto V, Bonfante P (1999) Presymbiotic versus symbiotic phase in arbuscular endomycorrhizal fungi. In: Varma A, Hock B (eds) Mycorrhiza, Springer Berlin pp 229-251

Bianciotto V, Lumini E, Lanfranco L, Minerdi D, Bonfante P, Perotto S (2000) Detection and identification of bacterial endosymbionts in arbuscular mycorrhizal fungi belonging to Gigasporaceae. Appl Environ Microbiol 66:4503-4509

Budi S W, van Tuinen D, Martinotti G, Gianinazzi S (1999) Isolation from the *Sorghum bicolor* mycorrhizosphere of a bacterium compatible with arbuscular mycorrhiza development and antagonistic towards soilborne fungal pathogens. Applied Environmental Microbiology 65:5148-5150

Cheng Z F, Zuo Y, Li Z, Rudd K E, Deutscher M (1998) The *vacB* gene required for virulence in *Shigela flexneri* and *Escherichia coli* encodes the exoribonuclease RNase R. J Biol Chem 273:14077-14080

Costerton J W, Lewandowski Z, Caldwell D E, Korber D R, Lappin-Scott H M (1995) Microbial biofilms. Annu Rev Microbiol 49:711-45

Défago G, Keel C (1995) Pseudomonads as biocontrol agents of diseases caused by soilborne pathogens. In: Hokkanen H M T, Lynch J M (eds) Benefits and risks of introducing biocontrol agents. University Press, Cambridge. pp 137-148

Edwards S G, Young J P W, Fitter A H (1998) Interactions between *Pseudomonas fluorescens* biocontrol agents and *Glomus mosseae*, an arbuscular mycorrhizal fungus, within the rhizosphere. FEMS Microbiology Letters 166(2):297-303

Filippi C, Bagnoli G, Citernesi A S, Giovannetti M (1998) Ultrastructural spazial distribution of bacteria associated with sporocarps of *Glomus mosseae*. Symbiosis 24:1-12

Frey P, Frey-Klett P, Garbaye J, Berge O, Haeulin T (1997) Metabolic and genotypic fingerprinting of fluorescent pseudomonads associated with the Douglas Fir-*Laccaria bicolor* mycorrhizosphere. Appl Environ Microbiol 63:1852-1860

Galán J E, Collmer A (1999) Type III secretion machines: Bacterial devices for protein delivery into host cells. Science 284:1322-1328

Garbaye J (1994) Helper bacteria: a new dimension to the mycorrhizal symbiosis. New Phytol 128:197-210

Gianinazzi-Pearson V (1996) Plant Cell Responses to Arbuscular Mycorrhizal Fungi: Getting to the Roots of the Symbiosis. Plant Cell 8:1871-1883

Gillis M, Tran Van T, Bardin R, Goor M P, Hebbar A, Willems P, Segers K, Kerster T, Heulin,

Fernandez M P (1995) Polyphasic taxonomy in the genus *Burkholderia* leading to an emended description of the genus and proposition of *Burkholderia vietnamiensis* sp. nov. for N_2-fixing isolates from rice in Vietnam. Int J Syst Bacteriol 45:274-289

Harrison M J (1997) The arbuscular mycorrhizal symbiosis: an underground association. Trends Plant Sci 2:54-59

Harrison M J (1999) Molecular and cellular aspects of the arbuscular mycorrhizal symbiosis. Annu Rev Plant Physiol Plant Mol Biol 50:361-389

Harrison M J, Van Buuren M L (1995) A phosphate transporter from the mycorrhizal fungus *Glomus versiforme*. Nature 378:626-629

Hodge A (2000) Microbial ecology of arbuscular mycorrhiza. FEMS Microbiology Ecology 32:91-96

Hosny M, van Tuinen D, Jacquin F, Fuller P, Zhao B, Gianinazzi-Pearson V, Franken P (1999) Arbuscular mycorrhizal fungi and bacteria: how to construct prokaryotic DNA-free genomic libraries from the Glomales. FEMS Microbiology Letters 170:425-430

Keel C, Défago G (1997) Interactions between beneficial soil bacteria and root pathogens: mechanism and ecological impact. In: Gange A C, Brown V K (eds) Multitrophic interactions in terrestrial system. Blackwell Sci. London pp 27-46.

Linderman R G (1992) Vesicular-arbuscular mycorrhizae and soil microbial interactions. In: Bethlenfalvay G J, Linderman R G (eds) Mycorrhizae in sustainable agriculture. pp 45-70

Linderman R G, Pauliz T C (1990) Mycorrhiza-rhizobacterial interactions. In: Hornby D, Cook R J, Henis Y, Ko W H, Rovira A D, Schippers B, Scott P R (eds) Biological control of soil-borne plant pathogens. CAB International, Wallingford, UK. pp 261-283

Lugtenberg B J J, Dekkers L C (1999) What makes *Pseudomonas* bacteria rhizosphere competent? Environ Microbiol 1:9-13

Margulis L, Chapman M J (1998) Endosymbioses: cyclical and permanent in evolution. Trends Microbiol 6:342-345

Marschner H (1995) Mineral nutrition of higher plants, 2nd edition, Academic Press, London

Mayo K, Davis R E, Motta J (1986) Stimulation of germination of spores of Glomus versiforme by spore-associated bacteria. Mycologia 78:426-431

Meyer J R, Lindermann R G (1986) Selective influence on population of rhizosphere or rhizoplane bacteria and actinomycetes by mycorrhizas formed by *Glomus fasciculatum*. Soil Biology and Biochemistry 18:191-196

Minerdi D, Fani R, Gallo R, Boarino A, Bonfante P (2001) Nitrogen fixation genes in an endosymbiotic *Burkholderia* strain. Appl Environ Microbiol (in press)

Mosse B (1970) Honey coloured sessile Endogone spores. Changes in fine structure during spore development. Arch Mikrobiol 74:146-159

Ophir T, Gutnick D L (1994) A role of exopolysaccharides in the protection of microorganisms from desiccation. Appl Environ Microbiol 60:740-745

Perotto S, Bonfante P (1997) Bacterial associations with mycorrhizal fungi: close and distant friends in the rhizosphere. Trends Microbiol 5:496-501

Requena N (1998) Mycorrhizal interactions in the rhizosphere. In: Varma A (ed) Microbes for health, wealth and sustainable environment. Malhorta Pub. House, New Delhi pp 725-738

Requena N, Füller P, Franken P (1999) Molecular characterization of *GmFOX2*, an evolutionarily highly conserved gene from the mycorrhizal fungus *Glomus mosseae*, down-regulated during interaction with rhizobacteria. Molecular Plant-Microbe Interactions 12:934-942

Ruiz-Lozano J M, and Bonfante P (2001) Intracellular *Burkholderia* strain has no negative effect on the symbiotic efficiency of the arbuscular mycorrhizal fungus *Gigaspora margarita*. Plant Growth Regulation (in press)

Ruiz-Lozano J M, Bonfante P (1999) Identification of a putative P-transporter operon in the genome of a *Burkholderia* strain living inside the arbuscular mycorrhizal fungus *Gigaspora margarita*, J Bacteriol 181:4106-4109

Ruiz-Lozano J M, Bonfante P (2000) Intracellular *Burkholderia* of the arbuscular mycorrhizal fungus *Gigaspora margarita* possesses the *vacB* gene, which is involved in host cell colonization by bacteria. Microbial Ecology 39:137-144

Sarand I, Timonen S, Koivula T, Haahtela K, Sen R, Romantschuk M (1999) Biodegradation of m-toluate by Scots pine, mycorrhizal fungi and the associated bacteria. Journal of Applied Microbiology 86:817-826

Scannerini S, Bonfante P (1991) Bacteria and bacteria like objects in endomycorrhizal fungi (Glomaceae). In: Margulis L, Fester R (eds) Symbiosis as source of evolutionary innovation: speciation and morphogenesis. The MIT Press: Cambridge, MA, USA pp 273-287

Schnider U, Keel C, Blumer C, Michaux P, Défago G, and Haas D (1997) Regulation of mucoidity in the biological control agent *Pseudomonas fluorescens* CHA0 and construction of an exopolysaccharide-overproducing variant. In: Ogoshi A, Kobayashi K, Homma Y, Kodama F, Kondo N, Akino S (eds) Plant growth-promoting Rhizobacteria: present status and future prospects. OECD, Paris pp 349-352

Sen R, Nurmiaho-Lassila E L, Haahtela K, Korhonen T (1996) Attachement of Pseudomonas fluorescens strains to the cell walls of ectomycorrhizal fungi. In: Azcón-Aguilar C, Barea J M (eds) Mycorrhizas in integrated systems. From genes to plant development. COST action 821, U.E pp 661-664

Smit G, Swart S, Lugtenberg B J, Kijne J W (1992) Molecular mechanisms of attachment of *Rhizobium* bacteria to plant roots. Mol Microbiol 6:2897-903

Smith S E, Read D J (1997) Mycorrhizal Symbiosis. Academic Press, San Diego

Timonen S, Jørgensen K S, Haahtela K, Sen R (1998) Bacterial community structure at defined locations of *Pinus sylvestris - Suillus bovinus* and *Pinus sylvestris – Paxillus involutus* mycorrhizospheres in dry pine forest humus and nursery peat. Can J Microbiol 44:499-513

Tobe T, Sasakawa C, Okada N, Honma Y, Yoshikawa M (1992) *vacB*, a novel chromosal gene required for expression of virulence genes on the large plasmid of *Shigela flexneri*. J Bacteriol 174:6359-6367

van Buuren M, Lanfranco L, Longato S, Minerdi D, Harrison M J, Bonfante P (1999) Construction and characterization of genomic libraries of two endomycorrhizal fungi: *Glomus versiforme* and *Gigaspora margarita*. Mycol Research 103:955-960

Vandebroek A, Vanderleyden J (1995) The role of bacterial motility, chemotaxis and attachment in bacteria-plant interactions. Mol Plant Microbe Interact 8:800-810

Weller D M, Thomashow L S (1994) Current challanges in introducing beneficial microorganisms into the rhizosphere. In: O'Gara F, Dowling D N, Boesten B (eds) Molecular ecology of rhizosphere microorganism biotechnology and the release of GMOs. VCH, Weinheim, Germany pp 1-18

Mycorrhizal Technology in Agriculture
ed. by S. Gianinazzi, H. Schüepp, J.M. Barea and K. Haselwandter

Bases of the obligate biotrophy of arbuscular mycorrhizal fungi

B. Bago[1,3] and G. Bécard[2]

[1] *Departamento de Microbiología del Suelo y Sistemas Simbióticos, Estación Experimental del Zaidín (CSIC), Profesor Albareda 1, E-18008 Granada, Spain*
[2] *Equipe de Mycologie Végétale, UMR 5546 CNRS/Université Paul Sabatier, Pôle de Technologie Végétale, F-31326 Castanet Tolosan, France*
[3] *Centro de Investigaciones sobre Desertificación (CSIC/UV/GV), Camí de la Marjal s/n, E-46470-Albal (Valencia), Spain*

Keywords: Asymbiotic and presymbiotic growth, carbon metabolism, fatty acid synthesis, symbiotic growth, root exudates, storage lipid translocation

Introduction

The obligate biotrophic nature of arbuscular mycorrhizal fungi (AMF) has long been accepted (Azcón-Aguilar et al. 1999), although its absolute occurrence under natural conditions has never been proved. Saprophytic capabilities of AMF have been suspected based on observations of prolonged independent growth of these fungi *in vitro* (Hepper 1983; Mosse 1988; Strullu et al. 1997; reviewed by Azcón-Aguilar et al. 1999), but all efforts of culturing these organisms continuously under axenic conditions have failed up to now.

What makes AMF obligate endosymbionts? This is an important question for both fundamental and applied reasons. On the one hand, since the AM symbiosis is an ancestral plant-fungus association (Simon et al. 1993; Taylor et al. 1995), the elucidation of the cellular and molecular mechanisms that underlie the obligate biotrophy of the fungal partner may help to understand other plant-microbe biotrophic relationships. On the other hand, this elucidation is a prerequisite to better manipulating and exploiting AM symbioses: depending on the circumstances, the mycotrophic role of the plant may have to be minimized to improve the cost/benefit relationship and plant productivity, or may have to be artificially enhanced to promote fungal growth and inoculum production. One can hypothesize also that understanding the AMF obligate biotrophy will provide clues to growing these organisms in pure culture. Our failure to cultivate AMF axenically considerably limits progress in basic research in addition to limiting biotechnological strategies for the large scale production of AM fungal inoculum.

Hepper (1987) drew attention to three aspects when addressing the question of AM fungal biotrophy: i) nutrition, since the growth of the fungus may depend on specific nutrients supplied by the host ii) physical aspects, since some growth conditions *in vivo* may be essential for *in vitro* growth; and iii) genetics, since the fungus may have lost part of its genetic material or may have part of its genome repressed. Because of the ancient endosymbiotic origin of AMF, it is conceivable that horizontal genetic transfer occurred and that AMF underwent some genetic regression. However, all cytochemical, biochemical, metabolic and genome investigations carried out the last 25 years provided no indication supporting the latter hypothesis. Rather, they indicated that AMF resemble saprophytes when considering genome size, metabolic capabilities, independent capacities for DNA, RNA and protein synthesis, etc. Recent data, based on a more careful examination of fungal growth stimulation during the presymbiotic stage, and of fungal C metabolism during both free living stages and symbiosis has provided indirect evidence that the fungus must interact closely with its host to fully express its genetic potential.

Changes in AM fungal biology from the asymbiotic to the symbiotic stage

Spore germination – asymbiotic growth

Resting spores of AMF are produced in the rhizosphere by the extraradical mycelium (ERM) during the symbiotic plant-fungus association. Once the life of the host plant is over, the spores represent the ultimate conservation form of the fungus waiting for the establishment of a symbiotic association with a new host. The resting spores of most AMF can germinate spontaneously in water and in the absence of a host root (asymbiotic growth), as long as a number of abiotic conditions are satisfied (e.g. proper range of temperature, water availability and pH). Sometimes germination can be significantly improved with a CO_2-enriched atmosphere and with a pre-incubation treatment under cold temperature. Many AMF also produce intraradical vesicles that can serve as conservation and propagation structures, but the physiology of these vesicles has not been investigated as much.

In any case, AM fungal asymbiotic growth is characterized by a limited development of the germ tube, both in terms of duration (days) and hyphal elongation (several mm or cm depending on the fungus). It has been shown that most of the major metabolic carbon pathways of the fungus are active during this life stage, including the use of exogenous sugars such as glucose and fructose (see below), and that DNA synthesis and nuclear division occurs (Bianciotto and Bonfante 1993; Bécard and Pfeffer 1993). However, this asymbiotic growth ceases rapidly, long before the spore reserves are consumed. Random autolytic processes occur along the germ tube (Bago et al. 1998c), or else the fungus can retract its cytoplasm (Logi et al. 1998) back to the spore, and regerminate later (Koske 1981). Since the germ tubes rather than the resting spores can sense the presence of a plant host (see below), one can hypothesise that AMF have developed mechanisms, as a survival strategy, to minimize energy loss during spore germination. Significant growth and full utilisation of spore reserves seem to be unlocked only in the presence of a host (Bécard and Piché 1989a). This growth switch from the asymbiotic to the presymbiotic stage is the first plant intervention in fungal development. What prevents AMF from actively growing during germination? What are the plant factors that trigger active fungal growth? What are the fungal genetic and metabolic targets of this activation process? Are the plant factors responsible for this presymbiotic growth activation still necessary to control the symbiotic life stage of the fungus? Investigations are in progress to try to answer these challenging questions, since they will provide clues to understand the obligate biotrophic nature of AMF.

Stimulation by root exudates – presymbiotic growth

After spore germination the AM fungus becomes dependent upon the presence of an adequate host for further growth and development (Mosse and Hepper 1975; Elias and Safir 1987; Bécard and Piché 1989a b; Gianinazzi-Pearson et al. 1989; Schreiner and Koide 1993; Tawaraya et al. 1996; Giovannetti et al. 1993 and 1996; Giovannetti and Sbrana 1998). We define this growth activation as presymbiotic because it does not require physical root-fungus contact, but rather some specific diffusible root exudates and/or volatiles. The fungal growth response is generally characterized by a stimulation of hyphal elongation, hyphal branching and sometimes by the production of auxiliary cells. Hyphal branching, which is the equivalent of cell proliferation for a filamentous fungus, is a remarkable response whose intensity increases with root proximity. This intense branching has been described by Powell (1976) as preinfection fan like structures and by Giovannetti et al. (1993, 1996) as the result of the presence of root signals exclusively produced by mycotrophic plants. More recently Nagahashi and Douds (1999) developed an *in vitro* assay, based on the branching response of *Gigaspora gigantea*, to test fractions of root exudates. Buée (2000) used this bioassay to chemically and biologically characterize the root factors responsible for such a hyphal branching response. By investigating root exudates from several plant species, he confirmed the presence of a branching factor only produced by mycotrophic plants.

It was shown with *Gigaspora* species (Bécard and Piché 1989b; Chabot et al. 1992; Poulin et al. 1993; Buée and Bécard, unpublished) that the presymbiotic stage requires the simultaneous presence of root exudates and CO_2. Bécard and Piché (1989b) hypothesized that CO_2 could be fixed by *G. rosea* as an anaplerotic source of carbon. Dark fixation was later confirmed by NMR spectroscopic analyses with *G. intraradices,* as discussed below. Although not involving a *net* C gain, dark fixation of CO_2 may activate fungal metabolism and growth by providing additional input of C to the fungus. Studies on more specific root stimulatory compounds have often used CO_2 incubators adjusted to 2% (Bécard et al. 1992; Chabot et al. 1992; Bécard et al. 1995; Poulin et al.

1993, 1997; Balaji et al. 1995; Douds et al. 1996; Nagahashi and Douds 1999; Buée et al. 2000). A constant CO_2 atmospheric concentration avoids uncontrolled variation in the recycling of the respired CO_2, and variation in the fungal response.

Root stimulatory compounds are expected to be potentially produced by a wide spectrum of plant species. A number of metabolites including various flavonoids, phenolic acids, and polyamines have been tested empirically on germinating spores (Gianinazzi-Pearson et al. 1989; Tsai and Phillips 1991; Nair et al. 1991; Bécard et al. 1992; Chabot et al. 1992; Poulin et al. 1993; Douds et al. 1996; Ghachtouli et al. 1996). Some flavonoids were greatly stimulatory (reviewed by Vierheilig et al. 1998) as well as some polyamines (Ghachtouli et al. 1996). A stereochemistry-activity relationship was found for the flavonoids and this relationship could vary with the fungal species. This feature and the fact that they were active at micromolar concentration fitted their putative role as signals. In addition, flavonoids are known for their estrogenic activity in vertebrates, and a possible presence of estrogen-binding sites has been proposed in AMF (Poulin et al. 1997). Although these chemicals are potentially present in root exudates of many plants (Nair et al. 1991; Tsai and Phillips 1991; Bel-Rhlid et al. 1993), their natural occurrence as essential plant signals for mycorrhizal establishment has been questioned (Bécard et al. 1995). *In vitro* and *in vivo* experiments showed that mycorrhiza formation and normal symbiotic fungal growth and development could be obtained in the complete absence of flavonoids.

Attempts were recently carried out to isolate the active stimulatory molecules from crude root exudates (Nagahashi and Douds 1999; Buée et al. 2000). Some purified fractions of root exudates have been isolated with a strong activity on fungal branching (Nagahashi and Douds 1999) and cell proliferation (Buée et al. 2000). *Gigaspora gigantea* was routinely used in the bioassays but other species (*Glomus* and *Gigaspora*) responded similarly in the presence of the purified exudates. The stimulatory activity was found in root exudates of all the host plants tested (8 species) but not in root exudates of non-hosts (4 species). Previous data showing the absence of stimulatory activity in root exudates of non-mycotrophic plants was strengthened here because concentrated fractions from non-mycotrophic plants could be tested with no risk of artificially concentrating other compounds with inhibitory effects. Preliminary chemical analyses indicated that the responsible substances were low molecular weight, thermo-resistant lipophilic molecules, but were not flavonoids. Their presence could mimic that of an entire root and they were still active at infinitesimal concentration (Buée 2000). The exact chemical nature of these molecules has yet to be determined. When they are supplied very locally, the stimulated hyphal segment responds by producing profuse branches within 5 h. The active root molecules seem to behave like signals, necessary to switch the fungus to a new morphogenic programme. We lack genetic evidence that the presence of these molecules in the rhizosphere is essential for mycorrhizal establishment. If this is the case, we should be able to isolate plant mutants unable to produce these molecules and that exhibit a Myc^{-} phenotype when germinating spores are used as inoculum.

Preliminary investigations have been made to define, at the cellular level, which new functions are triggered in the fungus when stimulated by root exudates. In the presence of root volatiles (including CO_2) and root exudates, germ tubes of *Gigaspora rosea* exhibited higher phosphate uptake and plasmalemma H^+-ATPase activity (Lei et al. 1991). The intracellular pH of this fungal system was higher by 0.2 pH units when growing in the presence of a host root (Jolicoeur et al. 1998). With *Gigaspora gigantea*, Buée et al. (2000) were able to show that in the presence of active semi purified root fractions, the rapid branching response (5 h) of the fungus was correlated to a higher mitotic activity. The intracellular pH of *Gigaspora gigantea* was also higher 24 h after the addition of the branching factor (Buée 2000) confirming, on another fungus and with more purified root exudates, the observation of Jolicoeur et al. (1998). Transcriptional analyses, using differential display (DDRT-PCR) and subtractive hybridisation techniques (SSH) are presently carried out to isolate the fungal genes specifically expressed by the root stimulus (Tamasloukht et al. unpublished). These analyses will allow identification of the very first genes induced by the branching factor. They will provide valuable information to know whether the fungus has been energized by the root factor leading to higher ATPase activity, intracellular pH, membrane transport and nuclear division, or whether the fungus has been more directly activated in its cell cycle.

We have seen that some plant molecules, the chemical nature of which is yet to be determined, can regulate intimate functions of AM fungal cells during the presymbiotic stage at extremely low concentration. We will see below that other fungal functions linked with carbon metabolism, are also regulated by the host plant during the symbiotic stage.

The symbiotic AMF – morphogenic differentiation

During the establishment of the symbiosis, the AM fungus develops within the root structures constituting the intraradical mycelium (IRM) that has biotrophic access to the plant-derived C sources it requires to develop (Smith and Read 1997; Smith and Smith 1997; Bago 2000). As a result of this biotrophic C uptake, fungal growth is invigorated and a profuse hyphal network (the extraradical mycelium, ERM) is produced (Smith and Read 1997; Friese and Allen 1991; Bago et al. 1998a). The ERM develops within the substrate surrounding the host root, and it is very active in acquiring mineral (and perhaps organic) nutrients (George et al. 1995; Smith and Read 1997; Bago 2000; Koide and Kabir 2000; Joner et al. 2001). Similarly to all growing fungal colonies, the AMF colony (comprising both IRM and ERM) undergoes initially an *assimilative* state (Bago et al. 1998a) in which the acquired nutrients are mostly invested in growth. This stage is characterized by the production of young, profusely branched structures putatively related to nutrient uptake (revised by Bago 2000): arbuscules (Bonfante-Fasolo 1984; Alexander et al. 1989; Smith and Read 1997) by the IRM, and of the so-called branched absorbing structures (BAS; Bago et al. 1998b; Declerck et al. 2000; Karandashov et al. 2000) in the external substrate. Upon nutrient starvation, fungi usually enter a *sporulation* phase, in which production of reproductive structures occurs. In most cases (specially when producing long-term resting spores) these structures accumulate storage C compounds. In AMF a sporulation phase also takes place (Bago et al. 1998a), and it is characterised by the asexual production (both intra- and extraradically) of resting spores. AM fungal spores contain mostly storage lipids (Sward 1981; Bonfante et al. 1994; Sancholle et al. 2000), but also glycogen deposits (Bonfante et al. 1994) and numerous (2000 to 9000) nuclei (Cooke et al. 1987; Bécard and Pfeffer 1993). When fully mature these spores are ready to germinate thus re-initiating the AM fungal life cycle.

One of the reasons proposed for the failure of the fungus to complete its life cycle in the absence of symbiosis is the existence of a metabolic failure in C metabolism (reviewed by Azcón-Aguilar et al. 1999). Nevertheless, neither assays of involved enzymatic activities revealed the origin of such a metabolic blockage, nor did any of the numerous C sources assayed in synthetic media induce the fungus to fulfil its life cycle in the absence of symbiosis with a host root (Hepper 1987; Bécard and Piché 1989a; reviewed by Azcón-Aguilar et al. 1999). In recent years the application of a range of experimental techniques including molecular biological methods, NMR spectroscopy, respirometry and *in vivo* microscopy has greatly contributed to our understanding of metabolism in AMF, both under asymbiotic and symbiotic conditions (for reviews see Harrison 1999; Bago et al. 2000). The development of *in vitro* cultures of AMF with host roots (AM monoxenic cultures, Bécard and Fortin 1988; St. Arnaud et al. 1996) also contributed to these advances.

Carbon metabolic pathways in AMF

Assimilative phase

During the assimilative stages, intraradical AM fungal hyphae acquire exogenous hexose very efficiently (Shachar-Hill et al. 1995; Solaiman and Saito 1997). Plant-derived hexose is taken up by the intraradical fungus (Fig. 1) and used via glycolysis and the pentose phosphate pathway and the tricarboxylic acid (TCA) cycle. AMF also transform plant-derived hexose into two typical fungal carbohydrates: trehalose and glycogen (Shachar-Hill et al. 1995).

These carbohydrates may act as short and medium-term C storage compounds, thus maintaining constant the cytosolic glucose concentration (Pfeffer et al. 1999; Bago et al. 2000). Hexose is incorporated by AMF without significant recycling via the mannitol cycle or hexose/triose futile cycling, which contrasts to free living ectomycorrhizal fungi (Martin et al. 1985; Pfeffer et al.

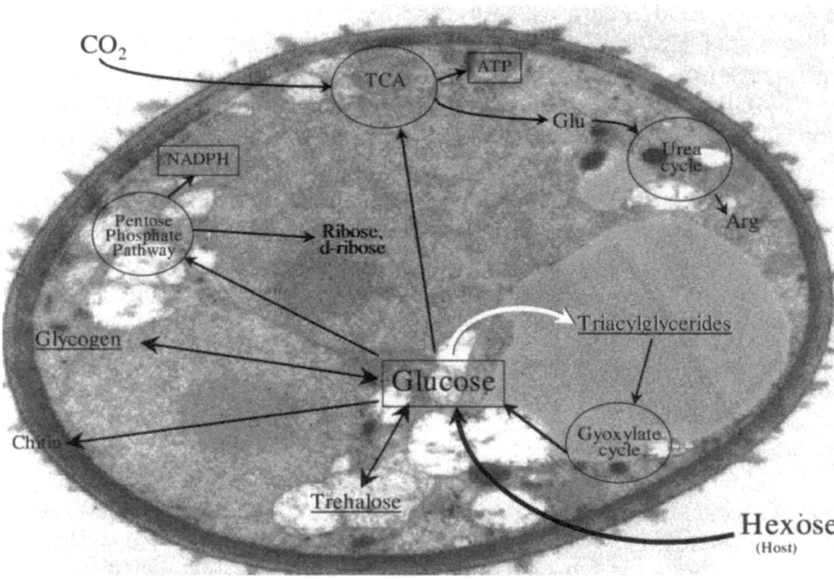

Figure 1. Carbon metabolic pathways known to be active in AMF. The white arrow indicate the metabolic pathway (fatty acid synthesis) suggested to be involved in AM fungal obligate biotrophy.

1999). Fungal trehalose undergoes rapid turnover (Shachar-Hill et al. 1995; Fig. 1), and it seems not to have a role in C export to the extraradical mycelium (Bago et al. 2000). Glycogen is also turned over (Fig. 1), and it accumulates as granular deposits within active arbuscular branches and intraradical hyphae (Bonfante-Fasolo 1984) and in the thinnest branches of extraradical BAS (Bago et al. 1998b). It may be the very first C currency shipped from the intraradical to the extraradical mycelium in early fungal developmental stages (discussed in Bago et al. 2000; Bago et al. unpublished).

There is a third major C storage compound synthesized by the intraradical hyphae during symbiosis: triacylglycerides (Fig. 1; Pfeffer et al. 1999). Triacylglycerol (TAG) accumulates as neutral lipid bodies (oleosomes, Murphy 1991) in arbuscular trunks and intercellular hyphae (Bonfante-Fasolo 1984). From there they are exported towards the extraradical mycelium (Pfeffer et al. 1999). Little or no gluconeogenesis is found in the intraradical fungal colony (Bago et al. 1999) which appears to present a mainly glycolytic metabolic behavior during assimilative stages (Bago et al. 2000).

^{13}C-NMR experiments show that the ERM of AMF is unable to acquire exogenously provided hexose, and that TAG synthesis in the ERM is either absent, or a very minor process during the assimilative phase (Pfeffer et al. 1999). This could imply the lack of the appropriate hexose transporters in extraradical structures, and that *de novo* fatty acid synthesis is not functional

(Bago et al. 2000). Extraradical hyphae seem to be specialised for acquiring soil nutrients, and this is reflected by the fact that some transporters (e.g. P transporters) are expressed in ERM but not in IRM (Harrison and van Buuren 1995). All this results in an interesting bipolar metabolic behavior of intra- and extraradical AM fungal symbiotic hyphae with the assimilative external hyphae exclusively relying on the export of carbon from the IRM. Once exported TAGs arrive in the ERM they are catabolized by fungal lipases and β-oxidation, then converted to glucose through the glyoxylate cycle and gluconeogenesis (Lammers et al. 2001). The ERM *gluconeogenically* active metabolism contrasts with the mainly *glycolytic* status of the intraradical fungal colony, and reflects once more that symbiotic AMF behave as 'metabolic bipoles'.

Sporulation phase

During sporulation intraradical hyphae continue to directly incorporate hexose into trehalose and glycogen (Shachar-Hill et al. 1995; Pfeffer et al. 1999), with TAGs as the main C form exported to the extraradical mycelium (Pfeffer et al. 1999; Bago et al. 1999; Bago et al. 2001). Storage lipids also accumulate intraradically within developing vesicles and spores. Similarly to the assimilative phase, intraradical sporulating hyphae show a mainly glycolytic behaviour (Pfeffer et al. 1999; Bago et al. 1999), i.e., this is maintained throughout their entire life cycle. Extraradical hyphae also maintain their gluconeogenic behavior during sporulation, but in this case storage lipids shipped from the IRM are accumulated within developing spores rather than being used for growth purposes. Resting spores are the turning point in AM fungal life cycle, and they constitute the key reserve of both genetic material and carbon stores from one to another AMF generation. Although glycogen deposits are abundant in AM spores (Bonfante et al. 1994) TAGs are largely the most abundant C store (up to 95%, Beilby 1983; Jabaji-Hare 1988; Bécard et al. 1991; Bonfante et al. 1994). Therefore AMF absolutely depend on the intraradically pre-synthesized TAGs to form new resting spores, i.e. to fulfil their life cycle.

Carbon metabolism in the asymbiotic AMF

The morphogenic and metabolic abilities depicted above reveal AMF as well organized, highly polarized organisms in which the intraradical and extraradical mycelia have distinct metabolic roles. This symbiotic differentiation could be attributed to the differential expression of fungal metabolic genes, according to nutrient availability. As far as we know the genetic pool of AMF is the same for intraradical and extraradical mycelium, all of it coming from nuclei contained in AM fungal spores; what would then be the C metabolic capabilities of asymbiotic germ-tubes?

Germinating spores have a mainly gluconeogenic behavior, depending on the mobilization of lipid stores to sustain growth (Jabaji Hare 1988; Bago et al. 1999). During the asymbiotic stages, observations have been made that are consistent with significant carbon fluxes through β-oxidation and the glyoxylate cycle, as well as through the tricarboxylic acid cycle, non-photosynthetic one-carbon metabolism and the pentose phosphate pathway, as it is also the case in ERM (Bago et al. 1999). However, and similarly to IRM, germ-tubes also undergo glycolysis and are able to take up limited amounts of hexose (Bago et al. 1999). At first glance AM fungal germ-tubes have all C metabolic pathways functional, and therefore they should be expected to develop normally in the absence of symbiosis; nevertheless, they are still unable to complete their life cycle unless they colonize a host root and 'differentiate' into IRM and ERM. How is this possible? There is a notable absence on the above list of metabolic abilities of AM germ-tubes: storage lipid synthesis (or, more precisely, *de novo* fatty acid synthesis), a process that either does not take place or is very minor during asymbiosis. This process seems to be *mostly* carried out by intraradical hyphae (Pfeffer et al. 1999; Bago et al. 1999), and it has been hypothesized to be one reason that prevents AM fungal asymbiotic growth (Bago et al. 2000).

The 'lipid question'

Despite its (at least, quantitative) importance, the study of storage lipid metabolism in AMF has been relatively neglected. Lösel and Cooper (1979) were the first to demonstrate that the lipid component of onion mycorrhizal roots became labeled when plants photosynthesized in the presence of $^{14}CO_2$, or were supplied with exogenous labeled sucrose, acetate or glycerol. These results suggested that AMF metabolize part of the acquired host-derived sugar into lipid, as it was previously hypothesized (Bevege et al. 1975). Gaspar et al. (1997) reported a net increase in total lipids with time (mainly in TAGs) in mycorrhizal roots and AM extraradical hyphae. Such increase was proportional to intraradical colonization development and extraradical hyphae spreading. The authors concluded that lipids in extraradical hyphae were synthesized from host-derived C sources, but they did not define where (intraradically or extraradically) the synthesis of such storage lipids takes place.

The evident importance of storage lipids in AM fungal biology justifies a closer review of what is known on lipid metabolism in these fungi, a subject which will be considered in the following section.

This point was clarified by NMR spectroscopy studies using *G. intraradices* monoxenic cultures (Pfeffer et al. 1999): the results showed that the TAGs of extraradical hyphae had been previously synthesized by the intraradical mycelium and exported. Also, whereas in intraradical hyphae the ^{13}C-labeling pattern is consistent with host-derived hexoses being incorporated (via glycolysis, part of the TCA cycle, fatty acid synthesis and TAG assembling,

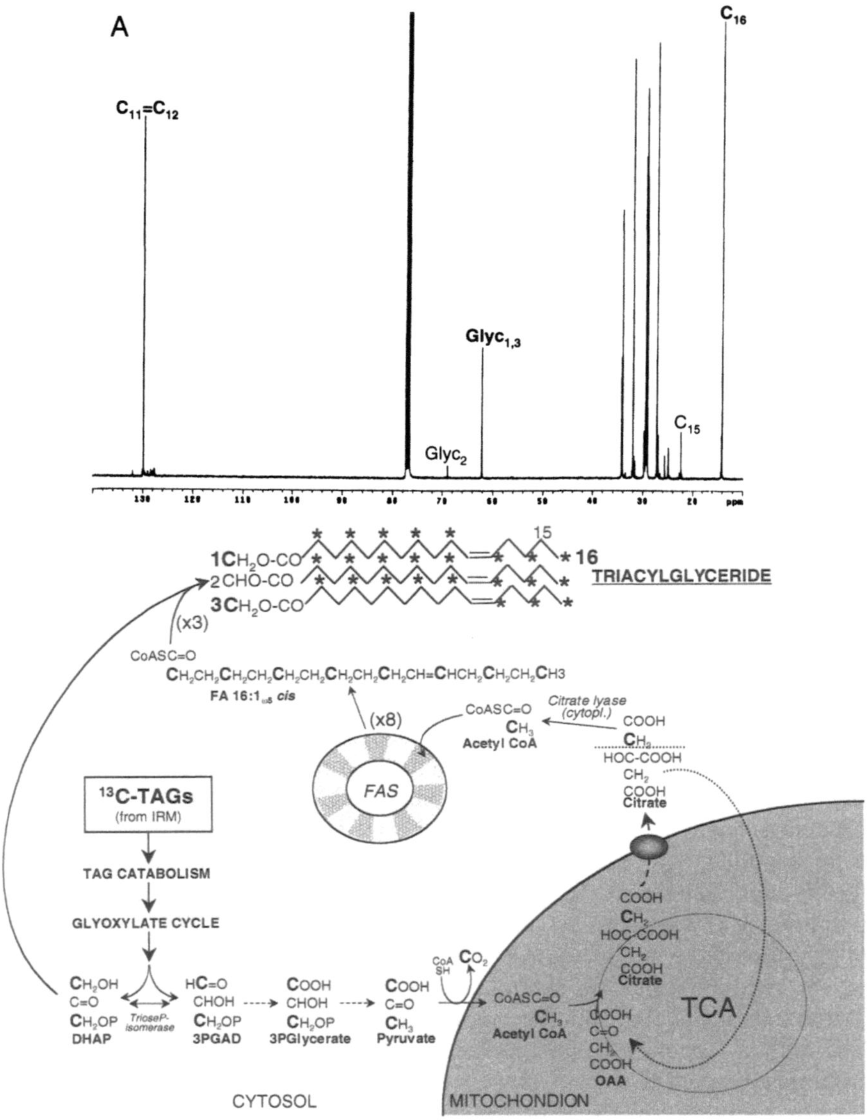

Figure 2 A. The lack of fatty acid synthesis in germinating spores as revealed by ^{13}C-NMR spectroscopy, explanation in the text.

Figure 2A) into TAGs, in the case of both ERM and germinating spores the ^{13}C-label provided as acetate (permeant substrate) did not become incorporated neither in the glyceryl nor the fatty acid moieties of the TAG molecule (Fig. 2B) (Pfeffer et al. 1999; Bago et al. 1999); however, when supplied as glycerol (also permeant) ^{13}C did not incorporate into fatty acids, but appeared

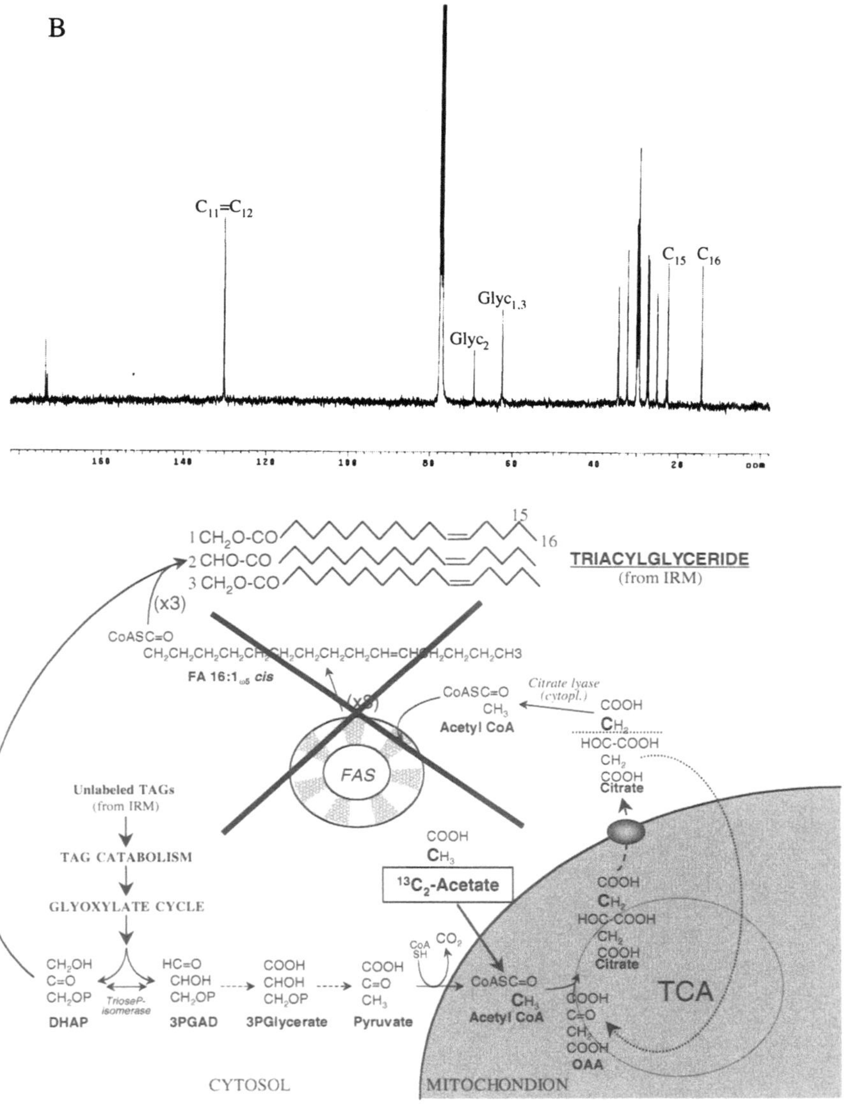

Figure 2 B.

in the glyceryl moieties of TAGs (Bago et al. unpublished results). Therefore, whereas glycerol moieties are synthesized by intraradical and extraradical hyphae, and by asymbiotic germ-tubes, fatty acid synthesis occurs mainly in the intraradical fungal phase. The inability (or insufficient ability) of both the ERM and the asymbiotic fungus to fully synthesize its own storage lipids may have its origin in a biochemical blockage on the fatty acid biosynthetic pathway. The exact metabolic step in which such a failure takes place has not been yet identified, hopefully forthcoming studies combining different techniques such as molecular biology and cytological localization will shed some light on this important issue.

Lipid translocation along the AM fungal colony

As mentioned, the absence of *de novo* fatty acid synthesis by the ERM makes this phase of the fungal colony dependent on the IRM for TAGs needed to sustain growth and sporulation. Therefore, very efficient processes for storage lipid translocation must take place between the IRM and ERM in AM symbioses.

Fungal storage lipids accumulate as spherical droplets of about 1 μm diameter, whose presence has long since been recognised in AMF by colorimetric and electron microscopy techniques (Fig. 3a; Nemec 1981; Bonfante-Fasolo 1984; Cooper 1984). Both intraradical and extraradical fungal structures contain lipid bodies: they are least apparent within young arbuscules (Nemec 1981; Cooper 1984) and 'BAS' (Bago et al. 1998b), and most abundant in AM fungal vesicles and spores (Nemec 1981; Bonfante-Fasolo 1984; Cooper 1984; Bonfante et al. 1994). The use of AM monoxenic cultures combined with multiphoton microscopy and fluorochrome labeling (Fig. 3b, c, d) is revealing new information concerning the *in vivo* translocation of such lipid bodies along

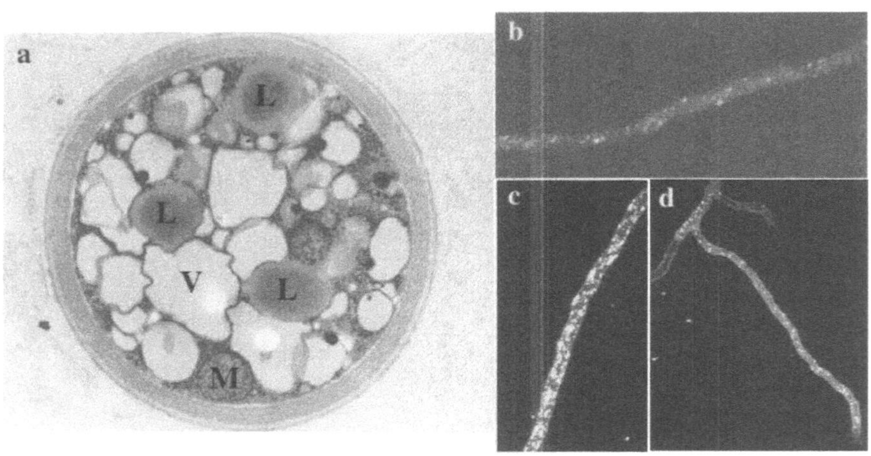

Figure 3. Microscopy of lipid globules in AM fungal hyphae. a, Transverse section of a growing germ tube of *Glomus intraradices* showing its cytoplasmic content and organelles. L, lipid globules; M, mitochondria; V, vacuoles. b to d, multiphoton microscopy of lipid globules travelling along asymbiotic (b) or symbiotic (c, d) hyphae of *G. intraradices*. Note the lower amount of lipid globules under asymbiotic conditions (b), and the depletion in lipid globules in a BAS branch (d).

extraradical hyphae (Bago et al. 2001). Two types of transport seem to be involved in lipid translocation: cytoplasmic streaming and directed translocation, probably involving cytoskeleton elements. A shortage in lipid globules might be at the origin of fungal growth arrest during asymbiosis (Fig. 3b), and in the ERM huge amounts of storage lipids are translocated (Fig. 3c). A depletion in the amount of oleosomes in external hyphal areas where lipid consumption may be expected to be at its highest (apices of growing runner hyphae and ephemeral, putatively nutrient-acquiring structures, i.e. BAS) was also observed (Fig. 3c).

Concluding remarks

The establishment of symbiosis is crucial for the AM fungus to fulfil its life cycle. It seems to be also the only way for it to become 'differentiated' both morphologically (i.e. formation of structures adapted to the intra-/extraradical environment (Bago 2000) and biochemically (i.e. formation of a 'metabolic bipole', Bago et al. 2000); therefore it is possible that such a morphological/biochemical differentiation is at the basis of AM fungal obligate biotrophy. The bases of such a 'differentiation' remain completely unknown, but they should certainly have their origin in the host plant. The root apoplast, where the IRM develops, may provide a particular environment where the fungus is exposed to specific physical or chemical factors, necessary to induce the differentiation programme. Some fungal genes in IRM nuclei may be specifically induced or repressed.

The need for AMF to differentiate their C metabolism in order to be fully functional has been pinpointed. Yet there is still much to do in order to fill gaps in our knowledge of AM C biochemistry and metabolism. Genes encoding the key enzymes governing the metabolic pathways putatively involved in AM fungal differentiation should be localized, and their expression studied. During the presymbiotic stage, when the fungus is activated by some plant factor(s), C metabolism, including lipid synthesis, should be carefully examined to see whether or not it is proportionally activated. If not, the plant factors responsible for the induction of the fungal fatty acid synthesis will have to be sought in the root apoplastic environment. Also, characterization and *in situ* localization of C transporters the fungus uses in each of its life stages would be of great interest. Only the localisation and study of such regulatory processes would allow us to fully understand the metabolic bases of AM symbiosis.

References

Alexander T, Toth R, Meier R, Weber H C (1989) Dynamics of arbuscule development and degeneration in onion, bean and tomato with reference to vesicular-arbuscular mycorrhizae in grasses. Can J Bot 67:2505-2513

Azcón-Aguilar C, Bago B, Barea J M (1999) Saprophitic growth of AMF. In: Varma A, Hock B (eds) Mycorrhiza: structure, function, molecular biology and biotechnology. 2nd Ed. Springer-Verlag, Berlin pp 391-407

Bago B (2000) Putative sites for nutrient uptake in arbuscular mycorrhizal fungi. Plant Soil (in press)

Bago B, Azcón-Aguilar C Piché Y (1998a) Architecture and developmental dynamics of the external mycelium of the arbuscular mycorrhizal fungus *Glomus intraradices* grown under monoxenic conditions. Mycologia 90:52-62

Bago B, Azcón-Aguilar C, Goulet A, Piché Y (1998b) Branched absorbing structures (BAS): a feature of the extraradical mycelium of symbiotic arbuscular mycorrhizal fungi. New Phytol 139:375-388

Bago B, Zipfel W, Williams R C, Chamberland H, Lafontaine J-G, Webb W W, Piché Y (1998c) *In vivo* studies on the nuclear behavior of the arbuscular mycorrhizal fungus *Gigaspora rosea* grown under axenic conditions. Protoplasma 203:1-15

Bago B, Pfeffer P E, Douds D D, Bécard G, Shachar-Hill Y (1999) Carbon metabolism in spores of the arbuscular mycorrhizal fungus *Glomus intraradices* as revealed by nuclear magnetic reonance spectroscopy. Plant Phys 121:263-271

Bago B, Pfeffer P E, Shachar-Hill Y (2000). Carbon metabolism and transport in arbuscular mycorrhizas. Plant Physiol 124:949-957

Bago, B., Zipfel, W., Williams, R C., Jun J, Arreola R, Lammers PJ, Pfeffer, P.E., Shachar-Hill, Y. (2001). Translocation and utilization of fungal storage lipid in the arbuscular mycorrhizal symbiosis. Plant Physiol (In press).

Balaji B, Poulin M J, Vierheilig H, Piché Y (1995) Responses of an arbuscular mycorrhizal fungus, *Gigaspora margarita*, to exudates and volatiles from the Ri T-DNA-transformed roots of nonmycorrhizal and mycorrhizal mutants of *Pisum stivum* L Sparkle. Exp Mycol 19:275-283

Bécard G, Fortin A (1988) Early events of vesicular-arbuscular mycorrhiza formation on Ri T-DNA transformed roots. New Phytol 108:211-218.

Bécard G, Piché Y (1989a) New aspects on the acquisition of biotrophic status by a VAM fungus, *Gigaspora margarita*. New Phytol 112:77-83

Bécard G, Piché Y (1989b) Fungal growth stimulation by CO_2 and root exudates in the vesicular-arbuscular mycorrhizal symbiosis. Appl Environ Microbiol 55:2320-2325

Bécard G, Doner L W, Rolin D B, Douds D D, Pfeffer P E (1991) Identification and quantification of trehalose in vesicular-arbuscular mycorrhizal fungi by *in vivo* ^{13}C NMR and HPLC analyses. New Phytol 118:547-552

Bécard G, Douds D D, Pfeffer P E (1992) Extensive hyphal growth of vesicular-arbuscular mycorrhizal fungi in the presence of CO_2 and flavonols. Appl Environ Microbiol 58:821-825

Bécard G, Pfeffer P E (1993) Status of nuclear division in arbuscular mycorrhizal fungi during *in vitro* development. Protoplasma 174:62-68.

Bécard G, Taylor L P, Douds D D, Pfeffer P E, Doner L W (1995). Flavonoid are not necessary plant signal compounds in arbuscular mycorrhizal symbioses. Mol Plant Microbe Interact 8:252-258

Beilby JP (1983) Effects of inhibitors on early protein, RNA, and lipid synthesis in germinating vesicular-arbuscular mycorrhizal fungal spores of *Glomus caledonium* Can J Bot 29:596-601

Bel-Rhlid R, Chabot S, Piché Y, Chênevert R (1993) Isolation and identification of flavonoids from Ri T-DNA transformed roots (*Daucus carota*) and their significance in vesicular-arbuscular mycorrhiza. Phytochem 33:1369-1371

Bianciotto V, Bonfante P (1993) Evidence of DNA replication in an arbuscular mycorrhizal fungus in the absence of the host plant. Protoplasma 176:100-105.

Bonfante P, Balestrini R, Mendgen K (1994) Storage and secretion processes in the spore of *Gigaspora margarita* Becker and Hall as revealed by high-pressure freezing and freeze substitution New Phytol 128:93-101

Bonfante-Fasolo P. 1984. Anatomy and morphology of VA mycorrhizae. In: Powell C L, Bagyaraj D J (eds) VA Mycorrhiza. CRC Press, Boca Raton, pp 5-33

Buée M (2000) Mise en évidence d'un facteur racinaire impliqué dans la croissance des champignons mycorhiziens à arbuscules. Ph. D. Thesis, Université Paul Sabatier, Toulouse, France

Buée M, Rossignol M, Jauneau A, Ranjeva R, Bécard G (2000) The pre-symbiotic growth of arbuscular mycorrhizal fungi is induced by a branching factor partially purified from plant root exudates. Mol Plant-Microbe Interact 13:693-698

Chabot S, Bel-Rhlid R, Chênevert R, Piché Y (1992). Hyphal growth promotion *in vitro* of the VA mycorrhizal fungus, *Gigaspora margarita* Becker & Hall, by the activity of structurally specific flavonoids compounds under CO_2-enriched conditions. New Phytol 122:461-467

Cooper K M (1984) Physiology of VA mycorrhizal associations. In: Powell C L, Bagyaraj D J (eds). VA Mycorrhiza. CRC Press, Boca Raton, pp 155-186

Cooke J C, Gemma J N, Koske R E (1987) Observations of nuclei in vesicular-arbuscular mycorrhizal fungi. Mycologia 79:331-333

Declerck S, Cranenbrouck S, Dalpé Y, Séguin S, Grandmougin-Ferjani A, Fontaine J, Sancholle M (2000) *Glomus proliferum* sp. nov.: a description based on morphological, biochemical, molecular and monoxenic cultivation data. Mycologia 92:1178-1187

Douds D D, Nagahashi G, Abney G D (1996) The differential effects of cell wall-associated phenolics, cell walls and cytosolic phenoloics of host and non-host roots on the growth of two species of AM fungi. New Phytol. 133:289-294

Elias K S, Safir G R (1987) Hyphal elongation of *Glomus fasciculatus* in response to root exudates. Appl Environ Microbiol 53:1928-1933

Friese C F, Allen M F (1991) The spread of VA mycorrhizal fungal hyphae in the soil: inoculum types and external hyphal architecture. Mycologia, 83:409-418.

Gaspar L, Pollero R, Cabello M (1997) Variation in lipid composition of alfalfa roots during colonization with the arbuscular mycorrhizal fungus *Glomus versiforme*. Mycologia 89:37-42

George E, Marschner H, Jakobsen I (1995) Role of arbuscular-mycorrhizal fungi in the uptake of phosphorus and nitrogen from soil. Crit Rev Biotechnol 15:257-270

Ghachtouli N E L, Paynot M, Martin-Tanguy J, Morandi D, Gianinazzi S (1996) Effect of polyamines and polyamines biosynthesis inhibitors on spore germination and hyphal growth of *Glomus mossea* Mycol Res 100:597-600

Gianinazzi-Pearson V, Branzanti B, Gianinazzi S (1989) *In vitro* enhancement of spore germination and early hyphal growth of a vesicular-arbuscular mycorrhizal fungus by host root exudates and plant flavonoids. Symbiosis 7:243-255

Giovannetti M, Sbrana C, Avio L, Citernesi A S, Logi C (1993) Differential hyphal morphogenesis in arbuscular mycorrhizal fungi during pre-infection stages. New Phytol. 125:587-593

Giovannetti M, Sbrana C, Citernesi A S, Avio L (1996). Analysis of factors involved in fungal recognition responses to host-derived signals by arbuscular mycorrhizal fungi. New Phytol. 133:65-71

Giovannetti M, Sbrana C (1998) Meeting a non-host: the behaviour of AM fungi. Mycorrhiza 8:123-130

Harrison M J, van Buuren M L (1995) A phosphate transporter from the mycorrhizal fungus *Glomus versiforme*. Nature 378:626-629

Hepper C M (1983) Limited independent growth of a vesicular-arbuscular mycorrhizal fungus *in vitro*. New Phytol 93:537-542

Hepper C M (1987) VAM spore germination and hyphal growth *in vitro*- prospects for axenic culture, p. 172-174 In DM Sylvia, LL Hung, JH Graham (ed), 7th NACOM. Mycorrhizae in the next decade. Practical application and research priorities, University of Florida, Gainesville

Jabaji-Hare S (1988) Lipid and fatty acid profiles of some vesicular-arbuscular mycorrhizal fungi: contribution to taxonomy. Mycologia 80:622-629

Jolicoeur M, Germette S, Gaudette M, Perrier M, Bécard G (1998) Intracellulaire pH in arbuscular mycorrhizal fungi: a symbiotic physiological marker. Plant Physiol 116:1279-88

Joner E J, Ravnskov S, Jakobsen I (2001) Arbuscular mycorrhizal phosphate transport under monoxenic conditions using radio-labelled inorganic and organic phosphate. Biotech Letters 22:1705-1708

Karandashov V, Kuzovkina I, Hawkins H J, George E (2000) Growth and sporulation of the arbuscular mycorrhizal fungus *Glomus caledonium* in dual culture with transformed carrot roots. Mycorrhiza 10:23-28

Koide R T, Kabir Z (2000) Extraradical hyphae of the mycorrhizal fungus *Glomus intraradices* can hydrolyse organic phosphate. New Phytol 148:511-517

Koske R E (1981) Multiple germination by spores of *Gigaspora gigantea* Trans Br Mycol Soc 76:320-330

Lammers PJ, Jun J, Abubaker J, Arreola R, Gopalan A, Bago, B., Hernández-Sebastiá C, Allen J, Douds DD, Pfeffer, P.E., Shachar-Hill, Y. (2001). The glyoxylate cycle in an arbuscular mycorrhizal fungus: gene expression and carbon flow . Plant Physiol (in press).

Lei J, Bécard G, Piché Y (1991) Plant factors stimulate ^{32}P uptake and plasmalemma ATPase activity in a vesicular-arbuscular mycorrhizal fungus, *Gigaspora margarita*. New Phytol 118:289-294

Logi C, Sbrana C, Giovannetti M (1998) Cellular events involved in survival of individual arbuscular mycorrhizal symbionts growing in the absence of the host App Environ Microbiol 64:3473-3479

Lösel D M, Cooper K M (1979) Incorporation of ^{14}C-labeled substrates by uninfected and VA mycorrhizal roots of onion. New Phytol 83:415-426
Martin F, Canet D, Marchal P (1985) ^{13}C NMR study of the mannitol cycle and trehalose synthesis during glucose utilization by the ectomycorrhizal ascomycete *Cenococcum graniforme*. Plant Physiol 77:449-502
Mosse B (1988) Some studies relating to 'independent' growth of vesicular arbuscular endophytes. Can J. Bot 66:2533-2540
Mosse B, Hepper C M (1975) Vesicular-arbuscular infections in root organ cultures. Physiol. Pl. Pathol. 5:215-223
Murphy D J (1991) Storage lipid bodies in plants and other organisms. Prog Lipid Res 29:299-324
Nagahashi G, Douds D D (1999) A rapid and sensitive bioassay with practical application for studies on interactions between root exudates and arbuscular mycorrhizal fungi. Biotechnol Tech 13:893-897
Nair M G, Safir G R, Siqueira JO (1991) Isolation and identification of vesicular-arbuscular mycorrhiza-stimulatory compounds from clover (*Trifolium ripens*) roots. Appl Environ Microbiol. 57:434-439
Pfeffer P E, Douds D D, Bécard G, Shachar-Hill Y (1999) Carbon uptake and the metabolism and transport of lipids in an arbuscular mycorrhiza. Plant Phys 120:587-598
Poulin M J, Bel-Rhlid R, Piché Y, Chênevert R (1993) Flavonoids released by carrot (*Daucus carota*) seedlings stimulate hyphal development of vesivular-arbuscular mycorrhizal fungi in the presence of optimal CO_2 enrichment. J Chem Ecol 19:2317-2327
Poulin M, Simard J, Catford J, Labrie F, Piché Y (1997) Response of symbiotic endomycorrhizal fungi to estrogens and antiestrogens. Mol. Plant-Microbe Interact 10:481-487
Powell C L (1976) Development of mycorrhizal infections from Endogone spores and infected root fragments. Trans Brit Mycol Soc 66:439-445
Sancholle M, Dalpé Y, Grandmougin-Ferjani A (2001) Lipids of mycorrhizae. In: Hock B (ed) The mycota IX: Fungal Associations. Berlin Heidelberg, Springer-Verlag, pp 63-93
Schreiner R P, Koide R T (1993) Stimulation of vesicular-arbuscular mycorrhizal fungi by mycotrophic and nonmycotrophic plant root systems. Appl Environ Microbiol 59:2750-2752
Shachar-Hill Y, Pfeffer P E, Douds D, Osman S F, Doner LW, Ratcliffe RG (1995) Partitioning of intermediate carbon metabolism in VAM colonized leek. Plant Physiol 108:7-15
Simon L, Bousquet J, Lévesque R C, Lalonde M (1993) Origin and diversification of endomycorrhizal fungi and coincidence with vascular land plants. Nature 363:67-69
Smith S E, Read D J (1997) Mycorrhizal Symbiosis. Academic Press, London
Smith F A, Smith S E (1997) Structural diversity in (vesicular)-arbuscular mycorrhizal symbioses. New Phytol 137:373-388
Solaiman M D, Saito M (1997) Use of sugars by intraradical hyphae of arbuscular mycorrhizal fungi revealed by radiorespirometry. New Phytol 136:533-538
St-Arnaud M, Hamel C, Vimard B, Caron M, Fortin J A (1996) Enhanced hyphal growth and spore production of the arbuscular mycorrhizal fungus *Glomus intraradices* in an *in vitro* system in the absence of host roots. Mycol Res 100:328-332
Strullu D G, Diop T A, Plenchette C (1997) Réalisation de collections in vitro de *Glomus intraradices* (Schenck et Smith) et de *Glomus versiforme* (Karsten et Berch) et proposition d'un cycle de développement. C. R. Acad. Sci. Paris 320: 41-47
Sward R J (1981) The structure of the spores of *Gigaspora margarita* I: the dormant spore. New Phytol 87:761-768
Tawaraya K, Watanabe S, Yoshida E, Wagatsuma T (1996) Effect of onion (*Allium cepa*) root exudates on the hyphal growth of *Gigaspora margarita*. Mycorrhiza. 6:57-59
Tsai S M, Phillips D A (1991) Flavonoids released naturally from alfalfa promote development of symbiotic *Glomus* spores *in vitro*. Appl Environ Microbiol. 57:1485-1488
Vierheilig H, Bago B, Albrecht C, Poulin M-J, Piché Y (1998) Flavonoids and arbuscular-mycorrhizal fungi. In: Buslig B, Manthey J (eds) Flavonoids in the living system. Plenum Press, New York, pp 9-33

Mycorrhizal Technology in Agriculture
ed. by S. Gianinazzi, H. Schüepp, J.M. Barea and K. Haselwandter

Arbuscular mycorrhizal fungal mycelium: from germlings to hyphal networks

M. Giovannetti[1], C. Sbrana[1] and L. Avio[2]

[1] *Dipartimento di Chimica e Biotecnologie Agrarie, Università di Pisa, Italy*
[2] *Centro di Studio per la Microbiologia del Suolo, C. N. R., I-56124 Pisa, Italy*

Keywords: Mycorrhizal mycelium, spore germination, anastomosis, hyphal networks, self-recognition

Introduction

Arbuscular mycorrhizal (AM) fungal spores are able to germinate in the absence of the host, but are unable to produce extensive mycelia and to complete their life cycle, without establishing a functional symbiosis with a host plant. The lack of host-regulated spore germination did not negatively influence the survival of these obligate biotrophs, which, according to fossil records and DNA sequence data, originated between 460 and 360 million years ago (Phipps and Taylor 1996; Pirozynski and Dalpé 1989; Redecker et al. 2000; Remy et al. 1994; Simon et al. 1993). Different efficient survival strategies have been inferred to be active in these ancestral organisms, to compensate for the lack of host-regulated spore germination, e. g. a wide host range (Smith and Read 1997), the regulation of infection structure differentiation (Giovannetti et al. 1994), the ability of multiple germination (Koske 1981b), an energy-saving mechanism operating when spores germinate in the absence of the host (Logi et al. 1998). In addition to these strategies, the ability to form wide hyphal networks by both pre-symbiotic and symbiotic mycelium may represent a fundamental mechanism for increasing the chances of AM symbionts to contact host roots (Giovannetti et al. 1999).

The aim of this chapter is to review the acquisitions and works which contributed to the understanding of cellular and molecular events in the life cycle of arbuscular mycorrhizal fungi (AMF) involved in the development of pre-symbiotic and symbiotic mycelium, from spore germination and hyphal growth in the absence of the host to infection structures differentiation and formation of extraradical mycorrhizal networks.

Spore germination

AM fungal spores are able to germinate and grow under different environmental conditions, and the main physical, chemical and microbiological factors influencing germination have been studied in different genera and species of AMF (Giovannetti 2000). Early studies on spore germination evidenced the phenomenon of spore dormancy, and in 1983 Tommerup tried to discriminate between dormancy and quiescence, defining 'dormant spores' as those failing to germinate although exposed to physical and chemical conditions supporting germination of apparently identical spores of the same species, which were identified as quiescent spores (Tommerup 1983). Many species of AMF show spore dormancy, such as *Glomus intraradices, Glomus clarum, Glomus caledonium, Glomus monosporum, Glomus coronatum, Acaulospora laevis, Acaulospora longula, Acaulospora* spp. (Gazey et al. 1993; Hepper 1979; Louis and Lim 1988; Tommerup 1983). On the other hand, some species do not show dormancy, such as *Gigaspora gigantea* and *Gigaspora margarita* (Koske 1981a; Siqueira et al. 1982; Sward 1981a). Since the experimental evidences have been obtained using different isolates in different parts of the world, it is difficult to correlate dormancy with species or genera, and the phenomenon remains to be completely understood.

The molecular signals able to relieve spore dormancy and activate the cell cycle are not known, though AMF spores are able to initiate the germination process in response to different environmental conditions, such as pH, temperature, moisture, mineral and organic nutrients, host plants, microorganisms.

Host-derived signals do not represent triggers of spore germination, since AMF spores are able to germinate in axenic culture in the absence of the host (Hepper and Smith 1976; Koske 1981a; Mosse 1959; Powell 1976; Tommerup 1983), although they have been shown to affect positively both spore germination and germling growth (Bécard and Piché 1989a; Gianinazzi-Pearson et al. 1989; Giovannetti et al. 1993a; Graham 1982; Nair et al. 1991; Tsai and Phillips 1991).

Early studies on spore germination reported the activation of the cell cycle in AMF growing in the absence of the host, by showing the development of dense regions containing normal cytoplasm and many dividing nuclei in germinating spores of *Acaulospora laevis* (Mosse 1970), and large numbers of nuclei with highly condensed chromatin and a prominent nucleolus in *Gigaspora margarita* spores after 24 h of incubation on water agar (Sward 1981a). Although some authors inferred the absence of nuclear division during and after germination (Burggraaf and Beringer 1989), other authors confirmed the early reports, showing both DNA replication and the occurrence of nuclear division in germlings of *Gigaspora margarita* (Bianciotto and Bonfante 1993). It is interesting to note that the number of nuclei in each spore was shown to decrease during the early days of germination (Bécard and Pfeffer 1993), confirming early reports on redistribution of spore cytoplasm and nuclei into germ tubes (Mosse 1959; Sward 1981b). This phenomenon was confirmed by the detection of nuclei migrating along germling hyphae of *Gigaspora rosea* and *Glomus caledonium* (Bago et al. 1998c; Logi et al. 1998), and of cytoskeletal components, both microtubules and microfilaments, in actively growing germlings of *Glomus mosseae* and *Glomus caledonium* (Åström et al. 1994; Logi et al. 1998).

Development of the pre-symbiotic mycelium

After spore germination germ tubes elongate and give rise to a coenocytic mycelial network the extension of which does not usually exceed 200 mm (Tab. 1). Mean growth rate of the mycelium during the early growth phase ranged from 1.97 µm/min in *Glomus caledonium* (Logi et al. 1998) to 2.64 µm/min in *Gigaspora rosea* (Giovannetti et al. 2000). The use of image analysis, associated to time-lapse and video-enhanced microscopy allowed to monitor the migration of protoplasm along two directions in hyphae during spore germination, and to show that nuclei, organelles and vacuoles moved at rates of 2.98–4.27 µm/s in *Glomus caledonium*. (Logi et al. 1998). These values, obtained in pre-symbiotic mycelium, are comparable to the values obtained by other authors in extraradical hyphae of *Glomus intraradices*, 1–5 µm/s (Nielsen and Jakobsen pers. comm.), and to those reported for nuclear migration in non-mycorrhizal fungi such as *Coprinus congregatus* and *Aspergillus nidulans* (Ross 1976; Suelmann et al. 1997).

Recent works reported that the fundamental mechanism allowing the development of large networks in pre-symbiotic mycelium is represented by anastomosis between compatible hyphae (Giovannetti et al. 1999). Time-lapse microscopy studies allowed to monitor the growth of hyphae of *Glomus mosseae* and *Glomus caledonium*, and showed that the complete fusion of hyphal walls was accomplished in 35 min.

The establishment of protoplasm continuity, the characteristic feature of true anastomoses, was evidenced by the visualization of succinate dehydrogenase activity and of many nuclei in the middle of fusion bridges (Giovannetti et al. 1999). The phenomenon of hyphal self-recognition was quantified by measuring the frequency of anastomosis between contacting hyphae belonging either to the same individual or to different individuals of the same isolate. Results showed that frequency ranged from 40% to 57% of total hyphal contacts in *Glomus mosseae*, from 34% to 55% in *Glomus caledonium* cultures, and from 59% to 90% in *Glomus intraradices* (Giovannetti et al. 1999).

The regular occurrence of hyphal fusions in mycelia originated from individually germinated spores suggests that the phenomenon may play an important role in maintaining the physiological continuity within each mycelium. Experimental pairings between *Glomus mosseae, Glomus caledonium* and *Gigaspora rosea* showed that hyphae of individuals belonging to one species never formed anastomoses with hyphae of the other, evidencing the ability of hyphae to discriminate self from nonself (Giovannetti et al. 1999; Giovannetti and Sbrana 2001). The occurrence of anastomosis in pre-symbiotic mycelium, as well as their functional significance should be further investigated, since hyphal fusions in AMF could represent a means for the exchange of genetic material, particularly if anastomoses occurred between hyphae derived from genetically different spores.

Table 1. Length of pre-symbiotic mycelium of AM fungal spores germinated in different experimental conditions

Fungal species	Experiment system, incubation times	Hyphal length per spore	Reference
Acaulospora laevis	in soil, 21d	98–134 mm	Tommerup 1984
	in vitro, 21d	0.33–17.6 mm	Tommerup 1983
Gigaspora calospora	*in vitro*, 21d	13.9–16 mm	Tommerup 1983
Gigaspora gigantea	*in vitro*, 18d	544 ± 43 mm	Douds et al. 1996
	in vitro, 10d	318 ± 12 mm	Buee et al. 2000
Gigaspora margarita	*in vitro*, 18d	189 ± 27 mm	Douds et al. 1996
	in vitro, 30d	55 mm	Siqueira et al. 1982
	in vitro, 9d	1.7–17.7 mm	Gianinazzi-Pearson et al. 1989
	in vitro, 20d	~75 mm	Poulin et al. 1993
	in vitro, 28d	~20–25 mm	Bécard and Piché 1989a, b
	in vitro, 5d	38.4 mm	Bécard et al. 1992
	in vitro, 7d	4.14–4.67 mm	Morandi et al. 1992
	in vitro, 25d	~75 mm	Balaji et al. 1995
Gigaspora ramisporophora	*in vitro*, 14d	16.3 ± 4.5–70.5 ± 7.6 mm	Ishii et a. 1997
Gigaspora rosea	*in vitro*, 8d	16.5–72.8 mm	Giovannetti et al. 2000
Glomus caledonium	in soil, 14d	90–104 mm	Tommerup 1984
	in vitro, 28d	66 ± 4 mm	Hepper 1979
	in vitro, 14d	13.4–16.2 mm	Tommerup 1983
	in vitro, 15d	10–40 mm	Logi et al. 1998
	in vitro, 6 m	70.83 ± 7.08 mm	Logi et al. 1998
Glomus epigaeum	*in vitro*, 7d	0.4–0.6 mm	Graham 1982
Glomus etunicatum	*in vitro*, 28d	0.14 mm	Schreiner and Koide 1993
	in vitro, 21d	~0.5 mm	Tsai and Phillips 1991
Glomus fasciculatus	*in vitro*, 14d	0.25–0.28 mm	Elias and Safir 1987
Glomus intraradices	*in vitro*, 49d	0.8–3.9 cm/cm^2	Bago et al. 1998b
Glomus monosporum	*in vitro*, 14d	11.4–13.1 mm	Tommerup 1983
Glomus mosseae	*in vitro*, 8d	>10 mm	Hepper and Smith 1976
	in grit, 20d	17.6 ± 2.8 mm	Giovannetti et al. 1993
	in soil, 43d	7.1–8.5 mm	Powell 1976
Glomus sp.	*in vitro*, 14d	~2 mm	Nair et al. 1991

Growth arrest in the absence of the host

When spores of AMF germinate in the absence of the host they arrest growth within 15–20 days of germination (Bécard and Piché 1989; Beilby and Kidby 1980; Hepper 1984; Koske 1981a; Logi et al. 1998; Mosse 1959). Such behaviour is neither correlated to any block in vital metabolic pathways nor caused by exhaustion of spore reserves (Beilby and Kidby 1980; Hepper 1979; Koske

1981b). Microscopic observations of germinating spores showed a progressive increase in the proportion of empty hyphae (Tommerup 1984) and time-lapse studies confirmed that during developmental arrest *Glomus caledonium* and *Gigaspora rosea* hyphae retracted protoplasm from the tips and produced cross walls separating peripheral empty hyphal tracts from viable ones. Hyphae devoid of cytoplasm, nuclei and organelles did not show any metabolic activity, whereas hyphal tracts proximal to the germinated spore were still viable, and infective, even after 6 months (Logi et al. 1998), confirming the long-term ability of AMF to retain infectivity in the absence of the host (Tommerup 1984). Video microscopy allowed to monitor protoplasm withdrawal from hyphal tips, which was accomplished at the average rate of 0.1–0.28 μm/min in *Glomus caledonium* and of 0.27–0.89 μm/min in *Gigaspora rosea* (Giovannetti et al. 2000). The migration of nuclei and cellular organelles from hyphal tips towards the mother spore has been envisaged as a process of resource reallocation, functional to maintaining the limited energy resources of germlings (Logi et al. 1998). This hypothesis is consistent with previous findings showing that spore reserves are never totally depleted when germlings cease growth (Beilby and Kidby 1980; Hepper 1979; Koske 1981a), and that fungal spores are able to germinate several times when grown in the absence of the host (Koske 1981b; Mosse 1959). It remains to be unravelled whether the consumption of a fixed pool of resources for germination acts as trigger for fungal growth arrest, thus representing an energy-saving mechanism operating when germinated spores of the obligately biotrophic AMF fail to contact a suitable host.

Host recognition and differentiation of infection structures

Molecules released by host roots are widely recognized to act as triggers of fungal developmental switches leading to the establishment of AM symbiosis, although their chemical nature remains to be unravelled. The first detectable event indicating fungal host recognition is represented by increased hyphal branching occurring in the presence of host roots or root exudates (Bécard and Fortin 1988; Elias and Safir 1987; Mosse 1988; Mosse and Hepper 1975; Powell 1976). Detailed investigations showed that the differential hyphal morphogenesis elicited in AM fungal mycelium by host-derived signals was always associated with host roots, and was accomplished in 24 h since the beginning of the plant/fungus interaction (Giovannetti et al. 1993b). The roots of non-mycorrhizal plants or of plant hosts of ecto-, arbutoid and ericoid mycorrhizas, did not show any effect on fungal morphogenesis, confirming that host-derived signals are the early cues to which AM symbionts respond, and through which they are able to discriminate unambiguously host from non-host (Giovannetti et al. 1994). Other experiments confirmed the wide occurrence of the phenomenon both *in vivo* and *in vitro*, in different genera and species of AMF (Buee et al. 2000; Giovannetti et al. 1996), and demonstrated that the signal molecules eliciting fungal recognition responses have a maximum molecular weight of 500 Da (Giovannetti et al. 1996).

The profuse hyphal branching and the arrest of apical growth occurring in AM hyphae in the presence of host-derived compounds have been interpreted as a developmental switch necessary for locating suitable infection sites on the root surface and may represent a sign of fungal commitment to the symbiotic status (Giovannetti 1997; Giovannetti and Sbrana 1998). Actually fungal infection structures, the appressoria, which are considered the most significant evidence indicating the successful recognition of a potential host plant (Staples and Macko 1980), were detected on the surface of host roots whereas they were not produced on the roots of all the non-host plants which did not elicit any differential morphogenesis (Giovannetti 1997; Giovannetti et al. 1994).

A successful AM infection is established when infection hyphae originating from appressoria colonize the root growing intracellularly and intercellularly and developing haustorial structures in the cortical cells, the arbuscules. Detailed information on intraradical fungal structures is reported by Smith and Smith (1997).

Extraradical hyphal networks

AMF living in symbiosis with host plants develop extensive extraradical hyphal networks which explore the soil, absorb mineral nutrients and translocate them to the roots, playing a major role in plant nutrient uptake (Smith and Read 1997). The extraradical hyphae have also been demonstrated to transfer carbon, nitrogen and phosphorus between plant species interconnected by a common mycorrhizal mycelium, thus representing important factors for the exploitation and redistribution of resources within plant communities (Chiariello et al. 1982; Francis and Read 1984; Read 1997). Moreover, the extraradical mycelium represents an essential inoculum source for rapid colonization of seedlings in natural environments (Addy et al. 1997), and contributes to the formation and stability of soil structure (Tisdall and Oades 1979; Wright and Upadhyaya 1998). Many studies, carried out after destructive extraction from the soil, have reported that the length of extraradical mycelium ranged from 1.1 to 54 m/g soil (Tab. 2).

Non-destructive investigations of AM extraradical mycelium, performed in root observation chambers and *in vitro* systems, allowed to detect its structure and development, before and after the establishment of the symbiosis (Bago et al. 1998b; Bécard and Fortin 1988). Nevertheless, virtually nothing is known of the mechanisms allowing the formation of extensive, three-dimensional hyphal networks through which nutrients are proposed to flow. Recent investigations found that the frequency of anastomoses per hyphal contact ranged from 64% to 78%, depending on the host plant (Giovannetti et al. 2001).

Further studies should monitor the developmental structure of the mycorrhizal network and the process of formation of hyphal interconnections, since

Table 2. Length of symbiotic extraradical mycelium of AMF under different experimental conditions

Fungal species	Experiment system, incubation times	Mycelial length	Reference
Acaulospora laevis	in soil, 7d and 14d	1.1–6.2 m/g dry soil 2.7–6.9 m/g dry soil	Jakobsen et al. 1992
Gigaspora margarita	*in vitro*, 24d	110 cm	Bécard and Piché 1989a
Gigaspora rosea	in soil, 135d	4.8 m/g soil	Schreiner et al. 1997
Glomus caledonium	in soil, 49d	3–5 m/g soil	Smith et al. 2000
Glomus etunicatum	in soil 135d	3.3 m/g soil	Schreiner et al. 1997
Glomus intraradices	monoxenic culture, 14 weeks	30 cm/cm^2	Bago et al. 1998a
	monoxenic culture, 120 d	25–100 cm/cm^2	Bago et al. 1998b
	monoxenic culture, 8–15 weeks	2.6 ± 0.3 mm/mm^2	St.-Arnaud et al. 1996
Glomus mosseae	in soil, 135d microcosm, 7d	8.1 m/g soil 5.3 m/cm^3 soil	Schreiner et al. 1997 Giovannetti et al. 1998
Glomus sp.	in soil, 7d and 14d	1.7–6.4 m/g dry soil 2.1–15.1 m/g dry soil	Jakobsen et al. 1992
Glomus sp.	field measurement	49 m/m root 7.8 m/g soil	Tisdall and Oades 1979
Scutellospora calospora	in soil, 49d in soil, 7d and 14d	9–10 m/g soil 1.3–4.6 m/g dry soil 2.5–9.2 m/g dry soil	Smith et al. 2000 Jakobsen et al. 1992

in natural situations the ability to form anastomosis by extraradical hyphae originating from different plants could represent, given the wide host range of AMF, the biological basis for the establishment of non-finite hyphal webs, linking together plants belonging to diverse families, genera and species (Chiariello et al. 1982; Molina et al. 1992; Perry et al. 1989).

References

Addy H D, Miller M H, Peterson R L (1997) Infectivity of the propagules associated with extraradical mycelia of two AM fungi following winter freezing. New Phytol. 135:745-753

Åström H, Giovannetti M, Raudaskoski M (1994) Cytoskeletal components in the arbuscular mycorrhizal fungus *Glomus mosseae*. Mol Plant Microbe Interaction 7:309-312

Bago B, Azcon Aguilar C, Goulet A, Piché Y (1998a) Branched absorbing structures (BAS): a feature of the extraradical mycelium of symbiotic arbuscular mycorrhizal fungi. New Phytol 139:375-388

Bago B, Azcon-Aguilar C, Piché Y (1998b) Architecture and developmental dynamics of the external mycelium of the arbuscular mycorrhizal fungus *Glomus intraradices* grown under monoxenic conditions. Mycologia 90:52-62

Bago B, Zipfel W, Williams R M, Chamberland H, Lafontaine J G, Webb W W, Piché Y (1998c) *In vivo* studies on the nuclear behavior of the arbuscular mycorrhizal fungus *Gigaspora rosea* grown under axenic conditions. Protoplasma 203:1-15

Balaji B, Poulin M J, Vierheilig H, Piché Y (1995) Responses of an arbuscular mycorrhizal fungus,

Gigaspora margarita, to exudates and volatiles from the Ri T-DNA- transformed roots of nonmycorrhizal and mycorrhizal mutants of *Pisum sativum*. Exp Mycol 19:275-283

Beilby J P, Kidby D K (1980) Biochemistry of ungerminated and germinated spores of the vesicular-arbuscular mycorrhizal fungus, *Glomus caledonium*: changes in neutral and polar lipids. J Lipid Res 21:739-750

Bécard G, Fortin J A (1988) Early events of vesicular-arbuscular mycorriza formation on Ri T- DNA transformed roots. New Phytol 108:211-218

Bécard G, Piché Y (1989a) Fungal growth stimulation by CO_2 and root exudates in vesicular-arbuscular mycorrhizal symbiosis. Appl Environ Microbiol 55:2320-2325

Bécard G, Piché Y (1989b) New aspects on the acquisition of biotrophic status by a vesicular-arbuscular mycorrhizal fungus, *Gigaspora margarita*. New Phytol 112:77-83

Bécard G, Douds D D, Pfeffer P E (1992) Extensive in vitro hyphal growth of vesicular-arbuscular mycorrhizal fungi in the presence of CO_2 and flavonols. Appl Environ Microbiol 58:821-825

Bécard G, Pfeffer P E (1993) Status of nuclear division in arbuscular mycorrhizal fungi during in vitro development. Protoplasma 174:62-68

Bianciotto V, Bonfante P (1993) Evidence of DNA replication in an arbuscular mycorrhizal fungus in the absence of the host plant. Protoplasma 176:100-105

Buee M, Rossignol M, Jauneau A, Ranjeva R, Bécard G (2000) The pre-symbiotic growth of arbuscular mycorrhizal fungi is induced by a branching factor partially purified from plant root exudates. Mol Plant Microbe Interact 13:693-698

Burggraaf J P, Beringer J E (1989) Absence of nuclear DNA synthesis in vesicular-arbuscular mycorrhizal fungi during *in vitro* development. New Phytol 111:25-33

Chiariello N, Hickman J C, Mooney H A (1982) Endomycorrhizal role for interspecific transfer of phosphorus in a community of annual plants. Science 217:941-943

Douds D D, Nagahashi G, Abney G D (1996) The differential effects of cell wall associated phenolics, cell walls, and cytosolic phenolics of host and non host roots on the growth of two species of am fungi. New Phytol 133:289-294

Elias K E, Safir G R (1987) Hyphal elongation of *Glomus fasciculatus* in response to root exudates. Appl Environ Microbiol 53:1928-1933

Francis R, Read D J (1984) Direct transfer of carbon between plants connected by vesicular-arbuscular mycorrhizal mycelium. Nature 307:53-56

Gazey C, Abbott L K, Robson A D (1993) VA Mycorrhizal Spores from 3 Species of *Acaulospora* - Germination, Longevity and Hyphal Growth. Mycol Res 97:785-790

Gianinazzi-Pearson V, Branzanti B, Gianinazzi S (1989) *In vitro* enhancement of spore germination and early hyphal growth of a vesicular-arbuscular mycorrhizal fungus by host root exudates and plant flavonoids. Symbiosis 7:243-255

Giovannetti M (1997) Host signals dictating growth direction, morphogenesis and differentiation in arbuscular mycorrhizal symbiont. In: Schenk H E A, Herrmann R G, Jeon K W, Muller N E (eds) Eukaryotism and symbiosis. Springer, Berlin, pp 405-411

Giovannetti M (2000) Spore germination and pre-symbiotic mycelial growth. In: Kapulnik Y, Douds D D (eds) Arbuscular mycorrhizas: physiology and function. Kluwer Academic Publishers, Dordrecht, pp 47-68

Giovannetti M, Avio L, Sbrana C, Citernesi A S (1993a) Factors affecting appressorium development in the vesicular- arbuscular mycorrhizal fungus *Glomus mosseae* (Nicol. & Gerd.) Gerd. & Trappe. New Phytol 123:114-122

Giovannetti M, Sbrana C, Avio L, Citernesi A S, Logi C (1993b) Differential hyphal morphogenesis in arbuscular mycorrhizal fungi during pre-infection stages. New Phytol 125:587-594

Giovannetti M, Sbrana C, Logi C (1994) Early processes involved in host recognition by arbuscular mycorrhizal fungi. New Phytol 127:703-709

Giovannetti M, Sbrana C, Citernesi A S, Avio L (1996) Analysis of factors involved in fungal recognition responses to host derived signals by arbuscular mycorrhizal fungi. New Phytol 133:65-71

Giovannetti M, Azzolini D, Citernesi A S, Logi C, Sbrana C (1998) Studies on mycorrhizal networks by using microcosms, microchambers and video-enhanced microscopy. Second International Conference on Mycorrhiza, ICOM 2, 5-10/VII, Uppsala, Sweden

Giovannetti M, Sbrana C (1998) Meeting a host: the behaviour of AM fungi. Mycorrhiza 8:123-130

Giovannetti M, Azzolini D, Citernesi A S (1999) Anastomosis formation and nuclear and protoplasmic exchange in arbuscular mycorrhizal fungi. Appl Environ Microbiol 65:5571-5575

Giovannetti M, Sbrana C, Logi C (2000) Microchambers and video-enhanced light microscopy for monitoring cellular events in living hyphae of arbuscular mycorrhizal fungi. Plant and Soil 226:153-159

Giovannetti M, Sbrana C (2001) Self and non-self responses in hyphal tips of arbuscular mycorrhizal fungi. In: Geitman A, Cresti M., Heath I B (eds) Cell biology of plant and fungal tip Growth. NATO Sciences Series, IOS Press, Amsterdam, pp. 221-231

Giovannetti M, Fortuna P, Citernesi A S, Morini S, Nuti M P (2001) The occurrence of anastomosis formation and nuclear exchange in intact arbuscular mycorrhizal networks. New Phytol 151:717-724

Graham J H (1982) Effect of citrus exudates on germination of chlamydospores of the vesicular-arbuscular mycorrhizal fungus, *Glomus epigaeum*. Mycologia 74:831-835

Hepper C M (1979) Germination and growth of *Glomus caledonium* spores: the effects of inhibitors and nutrients. Soil Biol Biochem 11:269-277

Hepper C M (1984) Inorganic sulphur nutrition of the vesicular-arbuscular mycorrhizal fungus *Glomus caledonium*. Soil Biol Biochem 16:669-671

Hepper C M, Smith G A (1976) Observation on the germination of *Endogone* spores. Trans Br Mycol Soc 66:189-194

Ishii T, Narutaki A, Sawada K, Aikawa J, Matsumoto I, Kadoya K (1997) Growth stimulatory substances for vesicular-arbuscular mycorrhizal fungi in Bahia grass (*Paspalum notatum* Flugge.) roots. Plant Soil 196:301-304

Jakobsen I, Abbott L K, Robson A D (1992) External hyphae of vesicular-arbuscular mycorrhizal fungi associated with *Trifolium subterraneum* L. New Phytol 120:371-380

Koske R E (1981a) *Gigaspora gigantea*: observations on spore germination of a VA- mycorrhizal fungus. Mycologia 73:288-300

Koske R E (1981b) Multiple germination by spores of *Gigaspora gigantea*. Trans Br Mycol Soc 76:328-330

Logi C, Sbrana C, Giovannetti M (1998) Cellular events involved in survival of individual arbuscular mycorrhizal symbionts growing in the absence of the host. Appl Environ Microbiol 64:3473-3479

Louis I, Lim G (1988) Effect of storage of inoculum on spore germination of a tropical isolate of *Glomus clarum*. Mycologia 80:157-161

Molina R, Massicotte H B, Trappe J M (1992) Ecological role of specificity phenomena in ectomycorrhizal plant communities: potentials for interplant linkages and guild development. In: Read D J, Lewis D H, Fitter A H, Alexander I J (eds) Mycorrhizas in ecosystems. CAB International, Wallingford, Oxon, pp 106-112

Morandi D, Branzanti B, Gianinazzi-Pearson V (1992) Effect of some plant flavonoids on in vitro behaviour of an arbuscular mycorrhizal fungus. Agronomie 12:811-816

Mosse B (1959) The regular germination of resting spores and some observations on the growth requirements of an *Endogone* sp. causing vesicular- arbuscular mycorrhiza. Trans Br Mycol Soc 42:273-286

Mosse B (1970) Honey-coloured, sessile *Endogone* spores. I. Life history. Arch Mikrobiol 70:167-175

Mosse B (1988) Some studies relating to 'independent' growth of vesicular- arbuscular endophytes. Can J Bot 66:2533-2540

Mosse B, Hepper C M (1975) Vesicular-arbuscular mycorrhizal infections in root organ cultures. Physiol Plant Pathol 5:215-223

Nair M G, Safir G R, Siqueira J O (1991) Isolation and identification of vesicular-arbuscular mycorrhiza- stimulatory compounds from clover (*Trifolium repens*) roots. Appl Environ Microbiol 57:434-439

Perry D A, Amaranthus M P, Borchers S L, Brainerd R E (1989) Bootstrapping in ecosystems. Biosci 39:230-237

Phipps C J, Taylor T N (1996) Mixed arbuscular mycorrhizae from the Triassic of Antarctica. Mycologia 88:707-714

Pirozynski K A, Dalpé Y (1989) Geological history of the Glomaceae with particular reference to mycorrhizal symbiosis. Symbiosis 7:1-36.

Poulin M J, Bel Rhlid R, Piché Y, Chenevert R (1993) Flavonoids released by carrot (*Daucus carota*) seedlings stimulate hyphal development of vesicular-arbuscular mycorrhizal fungi in the presence of optimal CO_2 enrichment. J Chem Ecol 19:2317-2327

Powell C L (1976) Development of mycorrhizal infections from *Endogone* spores and infected root fragments. Trans Br Mycol Soc 66:439-445

Read D (1997) Mycorrhizal fungi - The ties that bind. Nature 388:517-518

Redecker D, Morton J B, Bruns T D (2000) Ancestral lineages of arbuscular mycorrhizal fungi (Glomales). Mol Phylogenet Evolut 14:276-284

Remy W, Taylor T N, Hass H, Kerp H (1994) Four hundred-million-year-old vesicular arbuscular mycorrhizae. Proc Natl Acad Sci USA 91:11841-11843

Ross I A (1976) Nuclear migration rates in *Coprinus congregatus*: a new record? Mycologia 68:418-422

Schreiner R P, Koide R T (1993) Stimulation of vesicular-arbuscular mycorrhizal fungi by mycotrophic and nonmycotrophic plant root systems. Appl Environ Microbiol 59:2750-2752

Schreiner R P, Mihara K L, McDaniel H, Bethlenfalvay G J (1997) Mycorrhizal fungi influence plant and soil functions and interactions. Plant Soil 188:199-209

Simon L, Bousquet J, Levesque R C, Lalonde M (1993) Origin and diversification of endomycorrhizal fungi and coincidence with vascular land plants. Nature 363:67-69

Siqueira J O, Hubbell D H, Schenck N C (1982) Spore germination and germ tube growth of a vesicular-arbuscular mycorrhizal fungus in vitro. Mycologia 74:952-959

Smith F A, Smith S E (1997) Structural diversity in (vesicular)-arbuscular mycorrhizal symbioses. New Phytol 137:373-388

Smith S E, Read DJ (1997) Mycorrhizal simbiosis. London: Academic Press

Smith F A, Jakobsen I, Smith S E (2000) Spatial differences in acquisition of soil phosphate between two arbuscular mycorrhizal fungi in symbiosis with *Medicago truncatula*. New Phytol 147:357-366

St Arnaud M, Hamel C, Vimard B, Caron M, Fortin J A (1996) Enhanced hyphal growth and spore production of the arbuscular mycorrhizal fungus *Glomus intraradices* in an *in vitro* system in the absence of the host roots. Mycol Res 100:328-332

Staples R C, Macko V (1980) Formation of infection structures as a recognition response in fungi. Exp Mycol 4:2-16

Suelmann R, Sievers N, Fischer R (1997) Nuclear traffic in fungal hyphae: in vivo study of nuclear migration and positioning in *Aspergillus nidulans*. Mol Microbiol 25:757-769

Sward R J (1981)a The structure of the spores of *Gigaspora margarita*. III. Germ tube emergence and growth. New Phytol 88:667-673

Sward R J (1981)b The structure of the spores of *Gigaspora margarita*. II. Changes accompanying germination. New Phytol 88:661-666

Tisdall J M, Oades J M (1979) Stabilization of soil aggregates by the roots systems of ryegrass. Aust J Soil Res 17:429441-429441

Tommerup I C (1983) Spore dormancy in vesicular-arbuscular mycorrhizal fungi. Trans Br Mycol Soc 81:37-45

Tommerup I C (1984) Persistence of infectivity by germinated spores of vesicular- arbuscular mycorrhizal fungi in soil. Trans Br Mycol Soc 82:275-282

Tsai S M, Phillips D A (1991) Flavonoids released naturally from alfalfa promote development of symbiotic *Glomus* spores in vitro. Appl Environ Microbiol 57:1485-1488

Wright S F, Upadhyaya A (1998) A survey of soils for aggregate stability and glomalin, a glycoprotein produced by hyphae of arbuscular mycorrhizal fungi. Plant Soil 198:97-107

Mycorrhizal Technology in Agriculture
ed. by S. Gianinazzi, H. Schüepp, J.M. Barea and K. Haselwandter

Biolistic transformation of arbuscular mycorrhizal fungi: advances and applications

L.A. Harrier[1], S. Millam[2] and P. Franken[3]

[1] *Plant Science Division, Scottish Agricultural College, West Mains Road, Edinburgh, Scotland, UK*
[2] *Gene Expression Unit, Scottish Crop Research Institute, Invergowrie, Dundee, Tayside, Scotland, UK*
[3] *Max-Planck-Institut for Terrestrial Microbiology, Karl-von-Frisch-Straße, D-35043 Marburg, Germany*

Keywords: Transformation, biolistics, reporter genes

Introduction

The development of methods for both transient and stable gene delivery into any cell, tissue or organelle at will, is as fundamental to the genetic revolution as DNA sequencing and the polymerase chain reaction. Consequently, gene transfer methods have been developed for many important species in all kingdoms. Certain bacterial, fungal, plant and animal cells are routinely transformed in laboratories all over the world. However, despite the existence of diverse transformation techniques, gene delivery still remains a limiting step in several fields of research. The study of the obligate symbionts, arbuscular mycorrhizal fungi (AMF) in their symbiotic and asymbiotic stages of development is one such field of research that has been hindered by the inability to develop suitable methodology for the introduction of foreign DNA into them.

For the successful transformation of fungi there are two steps that are of critical importance: the delivery of the DNA into the fungal cells and the adaptation of the transformation vector containing a functional promoter and a translatable coding region to the cellular environment. A majority of the protocols for transforming fungi involve the preparation of fungal protoplasts (Fincham et al. 1989), however in the case of AMF this is not possible as the fungi have aseptate hyphae.

There are alternative methods to enable fungal competence to take up DNA which avoid the use of protoplasts (Hargreaves and Turner 1992). Among these whole cell techniques, the use of microprojectile bombardment (biolistics) has been used successfully on several fungal species (Tab. 1).

The biolistic process

In the standard biolistic protocol, purified DNA is precipitated onto microcarriers (usually gold or tungsten particles). The microcarriers are propelled at

Table 1. Species of fungi successfully transformed by biolistics

Fungal species	Reference
Neurospora crassa	Armaleo et al. 1990
Phytophthora spp.	Bailey et al. 1993
Gliocladium virens	Lorito et al. 1993
Trichoderma harzianum	Lorito et al. 1993
Botryotinia fuckeliana	Hilber et al. 1994
Aspergillus nidulans	Fungaro et al. 1995
Paxillus involutus	Bills et al. 1995
Metarhizium anisopliae	St Leger et al. 1995
Neocallimastix frontalis	Durand et al. 1997
Gigaspora rosea	Forbes et al. 1998b
Coccidioides immitis	Yu and Cole 1998
Laccaria bicolor	Bills et al. 1999
Erysiphe graminis	Chaure et al. 2000

high velocity toward the recipient cells by the biolistic device. Different biolistic mechanisms have been described by Sanford et al. (1987). These mechanisms include gas entrainment, transferred impulse, macroprojectile acceleration, centripetal acceleration and electrostatic acceleration. However, the widely applicable helium driven PDS-1000/He Device from Biorad was used in the transformation studies of AMF (Forbes et al. 1998a, b). Furthermore, this machine fulfils the criteria required for an efficient reproducible biolistic device. These criteria include: flexibility, repeatability, control over particle dispersal, adaptation to different organisms, minimal tissue damage, alternative particle velocities and safety.

The basic design of the PDS-1000/He Device was developed by Sanford et al. (1991) and a detailed schematic diagram of the machine is presented in Figure 1. It is a small high-pressure chamber, in which prior to firing, the air is removed by a vacuum pump. This reduces the frictional drag on the microcarriers. The chamber is temporarily sealed by a rupture disc, which is made out of Kaplon membrane that is laminated to burst at a designated pressure in this equipment from 450–2200 psi. The rupture disk is sealed against the base of the gas acceleration tube by the retaining cap. Helium gas is supplied by a high pressure tank and is the preferred gas as it expands much faster than other gases and thus imparts higher velocities on the microcarriers. When the critical gas pressure is reached the rupture disk bursts releasing a powerful shock wave.

In the path of the shock wave is the microcarrier launch assembly (Fig. 1B), whereby the shock wave hits the macrocarrier and it separates from its holder and flies until it impacts against the stopping screen. The microcarriers (DNA coated gold particles) are launched and continue downward until they impact and penetrate the target cells.

Many biological, microprojectile particles and microprojectile accelerator parameters require to be delimited for each particular species to be trans-

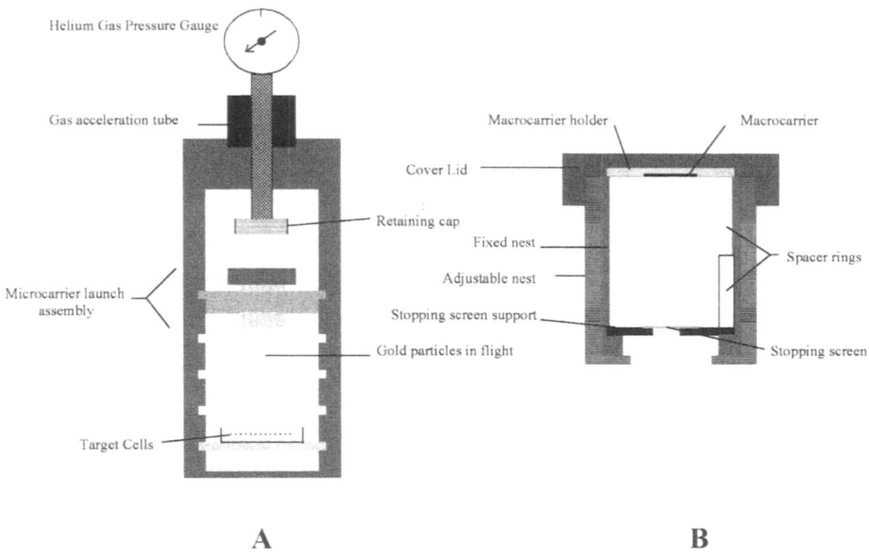

Figure 1. Schematic overview of the Biolistic PDS-100/Helium device. (A) Overview of chamber; (B) Enlarged view of the microcarrier launch assembly)

formed. Table 2 lists the transformation parameters successfully utilised on the AMF *G. rosea*. The study of Forbes et al. (1998b) demonstrated that a

Table 2. Particle bombardment parameters optimised for the AMF *Gigaspora rosea*

Parameter	Optimised parameter
Amount of plasmid utilised per shot	5 μg
Final calcium chloride concentration	1 M
Final spermidine concentration	16 μM
Amount of gold particles	1.2 mg
Size of gold particles	0.6 micron
DNA load per bombardment	625 μg
Macrocarrier gold load per bombardment	150 μg
Acceleration pressure	1350 psi
Chamber vacuum pressure	27 inches Hg
Gap distance[1]	7 cm
Macrocarrier travel distance[2]	2 cm
Target distance[3]	12 cm
Stopping plate aperture	0.8 cm

[1] distance between rupture disk and macrocarrier;
[2] distance between macrocarrier and stopping screen;
[3] distance between stopping screen and target tissue

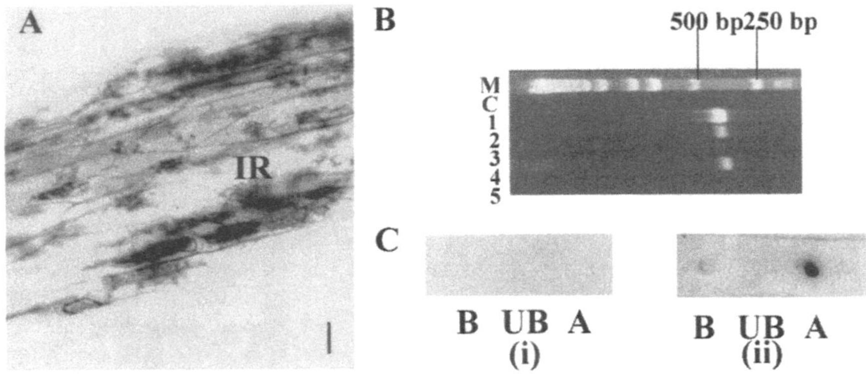

Figure 2. Study of the colonisation ability of transformed *Gigaspora rosea* and the molecular characteristics of second generation spores (A) A section of *A. porrum* root colonised by the transformed *G. rosea* and stained with methylene blue to show internal fungal structures (IR- intraradical hyphae; bar 50 μM). (B) Detection of the presence of the *GUS* gene in the second generation spores by PCR. (C- no DNA; 1 -purified plasmid pNOM102; 2- transformed spore DNA; 3- untransformed spore DNA; DNA extracted from plant root tissue colonised (4) by transformed *G. rosea* and uncolonised root system (5)). (C) Immuno-dot blot experiments with second generation spores. (Ci) Control immuno-dot blot experiment. The control was performed by omitting the anti-β-glucuronidase antibody. Detection of *GUS* gene expression in unbombarded spores (UB), bombarded spores (B) and in *Agrobacterium tumefaciens* (A) containing pBI101.3 using the anti-β-glucuronidase antibody (Cii). Ten μg of protein was blotted onto the membrane.

chimeric *gusA* gene based on the constitutive *gpd* promoter, expresses GUS transiently in the spores and hyphae of *G. rosea.* To establish whether this expression was transient and/or stable and that the transformation did not alter the ability of the fungus to colonise, pot cultures with *Allium porrum* as the host plant were established using the transformed spores of *G. rosea* as a source of inoculum. Colonisation of the host root system by the transformed fungi was obtained after 6 months (Fig. 2) and second generation spores were obtained for further molecular analysis.

It was found that the transformed *G. rosea* spores were able to colonise the roots of *A. porrum* (Fig. 2a). Furthermore the GUS gene was detectable via PCR using GUS specific oligonucleotide primers in the colonised roots and in the second generation spores (Fig. 2B). The presence of the GUS gene was not indicative of expression, therefore immunodetection of the GUS proteins in the spores was detected using the anti-β-glucuronidase antibody (Fig. 2Cii).

Future work

Future work will be required to embrace areas of biolistic technology of AMF to optimise the methods for subsequent applications. Possibly, the most significant area for development is the design and construction of improved reporter vectors. Furthermore, a more efficient selection/screening system for transformants is required.

AM fungal enhanced reporter vectors should include efficient homologous promoters, genome location targeting mechanism and a reliable reporter gene. In the study by Forbes et al. (1998b) the *Aspergillus nidulans* promoter glyceraldehyde 3-phosphate dehydrogenase was employed to drive expression of the *GUS* gene in the fungal tissue. However, future studies will require the utilisation of strong constitutive promoters. One possibility is the gene encoding the translation elongation factor EF1 alpha that has been recently cloned from a genomic library of *G. versiforme* (van Buuren et al. 1999). This promoter has been successfully used for transformation vectors in other fungi (Burmester 1995; van den Wymelenberg et al. 1997; Rosel and Kunze 1998; Kitamato et al. 1998).

Targeting of the reporter gene to a specific area of the genome for integration is an important factor to consider. Integration at a suitable genomic location will not only increase the efficiency of integration but also the expression of the reporter gene. Several different types of DNA sequences have been utilised to achieve this in other transformation systems. This includes the use of genetic elements (such as *seg1*), transposable elements and repetitive sequences to increase the efficiency of integration and expression (Burmester et al. 1990: Ruiz-Díez and Martínez-Suárez 1999; Mackenzie et al. 2000; Schilde et al. 2001). In addition, these types of different DNA sequences may enhance the stability of the transformants. A *Glomus mosseae* Ac-type transposase homologue mRNA was identified (AF072117 Accession number, Voegeli-Lange et al., unpublished) thus suggesting the presence of transposable elements in AMF, however, no detailed analysis has yet been completed. Genomes of AMF have been shown to contain repetitive elements (Zeze et al. 1996, 1999) and, the repetitive rDNA has been used effectively in the design of other fungal reporter vectors (Ruiz-Díez and Martínez-Suárez 1999; Mackenzie et al. 2000). The transformation efficiency of the human pathogenic fungus *Scedosporium prolificans* was significantly by incorporating a 600 bp region of rDNA complex (ITS1-5.8s-ITSII) into the vector used for transformation (Ruiz-Díez and Martínez-Suárez 1999).

The ideal genetic reporter system would not be endogenously expressed in the cell of interest and would have the characteristics of being sensitive, quantitative, rapid, easy, reproducible and safe. The most widely used reporter vectors include chloramphenicol acetyl-transferase (CAT), β-galactosidase, β-glucuronidase and bioluminescence based markers. In choosing a reporter gene for AMF, the reporter has not only to be detectable in fungal tissue, but also from within plant tissue, when the fungus is colonising the plant root. In using a fluorescent based reporter system there is often background fluorescence from the root system, for example GFP is often difficult to detect above background autofluorescence.

However, bioluminescence does not suffer from this and would allow for real time measurements to be taken. Bioluminescence is found in many phyla, particularly in the marine environment including bacteria, dinoflagellates etc. Each system has independently evolved biochemical systems for light production (Wood 1997). Luciferase genes cloned from bacteria, firefly and *Renilla* have found uses as indicators for gene expression. However, the firefly luciferase is the most commonly used of the bioluminescence reporters (Wood 1995). The popularity of firefly luciferase as a genetic reporter is due both to the sensitivity and ease of the enzyme assay, and to the tight coupling of protein synthesis with enzyme activity (Wood 1997). Furthermore, the gene encoding the luciferase is a cDNA and contains no introns that may reduce the efficiency of expression of the genetic reporter (Scherf and Wood 1995). This system would be conducive for usage in AMF for several reasons, real time measurements could be undertaken, there would be no background interference from plant root tissue, so measurements of the fungus in the symbiotic phase could be obtained and it would allow for rapid screening of transformants. However, a drawback to utilising this type of approach is specific and expensive equipment is needed, for measuring and visualising luciferase activity.

Fungal transformation systems based upon the expression of dominant selectable markers have been previously investigated. Several selectable markers have been developed in fungi based on resistance to the fungicide benomyl (Orbach et al. 1986; St Leger et al. 1995) and the antibiotics, oligomycin (Ward et al. 1986) and hygromycin (Volmer and Yanofsky 1986; Cullen et al. 1987; Mort-Bontemps and Fevre 1997). Hygromycin as a selectable marker for AMF has been tested for *G. rosea*, however, high concentrations of hygromycin were unable to inhibit the germination and subsequent growth and development of *G. rosea* when grown on water agar plates containing different concentrations of hygromycin (Fig. 3).

Suitable selectable markers for transformants in AMF are likely to be difficult to obtain due to the lack of knowledge of the physiology and reproductive strategies of these fungi, but there are several other antibiotics that are particularly effective in fungi, in particular phleomycin and zeocin. The later has many advantages over conventional antibiotics as it shows high toxicity and the gene is also small (355 bp) which enables smaller vectors to be constructed and provides more efficient transformations. However, the utilisation of the luciferase based reporter system allows for the rapid screening of transformed fungi using charged coupled device (CCD)-image enhanced microscopy. This technique provides very sensitive detection of luminescent organisms and increases the speed and sensitivity of screening transformed AMF spores. Furthermore, this technique can be used to image roots. Beauchamp et al. (1993) combined CCD imaging luminometry and dilution plate enumeration to demonstrate differences in colonisation of luminescence-marked rhizobacterial *Pseudomonads* on a range of plant roots.

The major drawback for screening transformants under non-selecting conditions is that transformation efficiency decreases. For instance, in plant systems it has been demonstrated that transformation of potato under non-selecting conditions is 1/400 of that under selective conditions (Kaya et al. 1990). Also, in

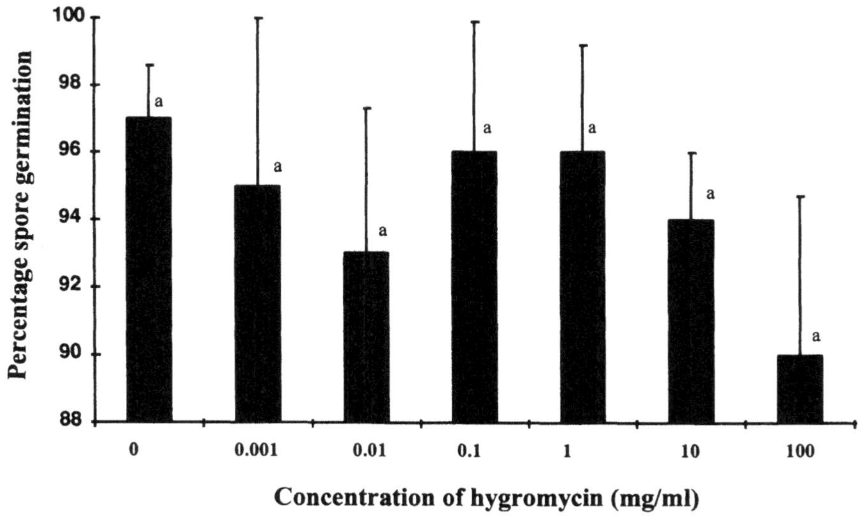

Figure 3. Percentage germination of *Gigaspora rosea* spores incubated on water agar plates containing varying concentrations of hygromycin. (Data are means ± standard error and letters denote significant differences ($P < 0.05$) between samples)

Arabidopsis without selection for transformation competence, the stable transformation frequency of shoots obtained after co-cultivation and regeneration on non-selective medium was shown to be below 0.5% (Buck et al. 2000). Moreover, when detection of the *GUS* gene presence and expression within third generation transformed *G. rosea* spores was evaluated, it was not possible to detect either the gene and/or the protein corresponding to the *GUS* gene.

Future applications of biolistic technology of AMF

The development and introduction of reporter genes into AMF has many applications in studying gene transcription, such as transcriptional control element testing and the identification of interacting proteins. Initially, it will be useful to study molecular events during presymbiotic stages, because transient expression after germination of bombarded spores seems to work relatively efficient as described above. Differential RNA display analysis has shown that RNA accumulation patterns of *G. rosea* do not change during germination and hyphal development in absence of the root (Franken et al. 2000). Addition of root exudates (Tamasloukht et al. 2000), however, results not only in an improved hyphal development, but also in the appearance and/or disappearance of novel cDNA fragments indicating up- and down-regulation of certain genes. The corresponding genes are currently being isolated, and constructs, where the respective promoters are combined with appropriate reporter gene coding regions, will be delivered into spores. After germination in the presence or absence of the inducing agent, monitoring reporter gene activity will tell, if gene regulation is due to induction or repression of the corresponding promoters. Further analysis can be targeted to the identification of regulative sequences, by constructing deletion derivatives of such inducible promoters. Furthermore, such transformed spores can be used as a system to screen for inducing components within root exudates, this provides a more sensitive approach than measuring hyphal length or calculating the number of branches. Another first target could be the *Glomus mosseae* gene encoding the phosphoglycerate kinase (Harrier et al. 1998). It has been shown that the protein is differentially expressed (Harrier and Sawczak 2000) and the promoter is available. Interestingly, in the promoter region there are two putative carbon regulated upstream activating sequences, which may have a key role to play in the appropriate responses of these fungi during the symbiotic and asymbiotic stages of development (Harrier 2001).

Another important aspect, which can be analysed, is gene function. Sequence homologies may indicate, in what cellular aspect a given gene is involved, but a final proof will always be the interruption of its functioning. Because of the coenocytic nature of the AMF, it is not possible to obtain mutants, since wild type alleles will readily complement such deficiencies quite easily. It is therefore necessary to use a strategy based on processes in the cytoplasm. Combining a strong promoter with the coding region of the gene under analysis in the opposite orientation will lead to the production antisense RNA. This molecule will bind to the original transcript in the cytoplasm and the newly formed double stranded RNA is recognised by the cellular machinery and digested. In this way, the expression of the corresponding protein can be interrupted and a newly appearing phenotype would point to the function of the gene. This approach has often been used in plants, where homologous recombination is not possible, and was recently reviewed (Kumria et al. 1998).

Moreover, vectors with reporter genes can be designed to monitor other cellular processes in addition to transcriptional gene regulation. For example, recombination events, gene targeting, RNA processing, protein secretion pathways and signal transduction pathways.

The study of recombination events is particularly pertinent to AMF as very little is known about their genetic variation and population structure. AMF are believed to reproduce clonally, but this lack of sexuality has not been confirmed by genetic studies and the absence of sexual structures may not be evidence for the lack of genetic recombination. Since the advent of the techniques of modern molecular biology, many methods have been utilised to try and determine genetic variation and population structure (Rosendahl et al. 1994; Rosendahl and Taylor 1997; Zeze et al. 1997; Hirjri et al. 1999; Hosny et al. 1999; Sanders 1999). However, to date these techniques have proved relatively uninformative due to factors such as the obligate nature of AMF and problems with deriving sufficient quantities of available DNA for detailed molecular analyses. Advances in gene transfer technologies such as biolistics provide increased opportunities to introduce recombination vectors (Herzing and Meyn 1993) into AMF which would allow for the *in situ* detection of recombination events and any products of unequal sister chromatid exchange. Moreover, the exchange of genetic information via the formation of anastomoses between species of the same and different AMF (Giovannetti et al. 1999) and/or zygospore differentiation (Tommerup and Sivasithamparam 1990) could be monitored by tracking transformed nuclei using, for instance, the luciferase or bioluminescent methods outlined above. These studies would provide data that may enhance our understanding of how these fungi maintain their genetic diversity within spores.

The ability to mark and/or tag AMF would greatly aid the study of the ecology of these fungi *in situ* and enable methodology for assessing the environmental safety of any introduced transgene. This technology may increase our understanding of the environmental factors controlling AM fungal growth, activity, survival and interactions in natural environments. Tagged fungi would also allow for the assessment of the growth, activity, and dispersal of introduced species within the environment. Moreover, these markers could be used to monitor the persistence of recombinant DNA and its potential for transfer to indigenous microflora after introduction into the environment. This technique may provide significant advantages for risk assessment studies in comparison to traditional methodologies. From a technical viewpoint, refinements to the physical parameters of biolistic transformation are constantly under development, and spin-offs from other studies could prove valuable to the uptake of such methods in AMF.

In the future, the possibility of transforming AMF could be used for applied aspects. Van der Heijden et al. (1998) demonstrated that using a combination of isolates as inoculum could optimally enhance the growth of plant populations. This may be due to functional complementarity of different AMF (Koide 2000). For example, using a compartmented soil system, *Glomus caledonium* was more efficient in taking up phosphate relatively far from the plant, while *Scutellospora calospora* preferentially obtained the nutrient close to the root (Smith et al. 2000). Numerous other publications have shown that different AM fungal isolates vary in their efficiency in aspects such as improving growth on poor soils, supplying tolerance against heavy metals or drought or inducing resistance against pathogens. When the molecular basis for these characteristics has been identified, transformation technology could be used to combine different features in key isolates. Moreover, features from non-Glomalean fungi could be introduced to enable exploitation of new substrates and/or to detoxify contaminated soils. This would provide a mechanism for the formulation of enhanced AM fungal inoculants.

Although methods for stable transformation of AMF are still under development, the potential application for both fundamental studies such as gene expression or the mode of action of reproductive systems, and for applied studies such as in environmental improvement, ensure that this area will be the focus of exciting developments in the future.

Acknowledgements
We thank SEERAD for financial support. pNOM102 was kindly provided by R.P. Oliver of the Carlsberg Research Centre, Denmark.

References

Armaleo D, Ye G N, Klein T M, Shark K B, Sanford J C, Johnston S A (1990) Biolistic nuclear transformation of *Saccharomyces cerevisiae* and other fungi. Curr. Genet. 19:97-103

Bailey A N, Mena G L, Herrera-Estrella L (1993) Transformation of four pathogenic *Phytophthora* sp. by microprojectile bombardment on intact mycelia. Curr. Genet. 23:42-46

Beauchamp J W, Kloepper J W, Lemke P A (1993) Luminometric analyses of plant root colonisation by bioluminescent pseudomonads Can. J. Microbiol. 1333:3099-3107

Bills S N, Richter D L, Podila G K (1995) Genetic transformation of the ectomycorrhizal fungus *Paxillus involutus* by particle bombardment. Mycol. Res. 99:557-561

Bills S N, Podila G K, Hiremath S (1998) Genetic engineering of the ectomycorrhizal fungus *Laccaria bicolour* for use as a biological control agent. Mycologia 91:237-242

Buck S, Wilde C, Montagu M, van Depicker A (2000) Determination of the T-DNA integration frequencies upon co-cultivation of *Arabidopsis thaliana* root explants Mol. Plant-Microbe Int. 13:658-665

Burmester A (1995) Analysis of the gene for the elongation factor-1-α from the Zygomycete *Absidia glauca* - use of the promoter region for constructions of transformation vectors Microbiol. Res. 150:63-70

Burmester A, Wöstemeyer A, Wöstemeyer J (1990) Integrative transformation of a zygomycete, *Absidia glauca* with vectors containing repetitive DNA. Curr. Genet. 235:166-172

Chaure P, Gurr S J, Spanu P (2000) Stable transformation of *Erysiphe graminis*, an obligate biotrophic pathogen of barley. Nature Biotech. 18:205-207

Cullen D, Leong S A, Wilson L J, Henner D J (1987) Transformation of *Aspergillus nidulans* with the hygromycin-resistance gene *hph*. Gene 57:21-26

Durand R, Rascle C, Fischer M, Fere M (1997) Transient expression of the bate-glucuronidase gene after biolistic transfomration of the anaerobic fungus *Neocallimastix frontalis*. Curr. Genet. 31: 158-161

Fincham J R S (1989) Transformation of fungi. Microbiol. Rev. 53:148-170

Fungaro M H P, Rech E, Muhlen G S, Vainstein M H, Pascon R C, Queiroz M V de, Pizzirani-Kleiner A A, Azevedo J L de (1995) Transformation of *Aspergillus nidulans* by microprojectile bombardment on intact conidia. FEMS Micro. Lett. 125:293-298

Forbes P, Millam S, Harrier L A, Gollotte A, Hooker J E (1998a) Transformation of the arbuscular mycorrhizal fungus *Gigaspora rosea* Nicolson et Schenck by Particle Bombardment. Proceedings of the Second International Conference on Mycorrhiza, July 5-10, SLU Service/Repro, Uppsala, pp 63-64

Forbes P J, Millam S, Hooker J E, Harrier L A (1998b) Transformation of the arbuscular mycorrhizal fungus *Gigaspora rosea* Nicolson et Schenck using particle bombardment. Mycol. Res. 102:497-501

Franken P, Requena N, Bütehorn B, Krajinski F, Kuhn G, Lapopin L, Mann P, Rhody D, Stommel M (2000) Molecular analysis of the arbuscular mycorrhiza symbiosis. Arch. Agron. Soil Sci. 45:271-286

Giovannetti M, Azzolini D, Citernesi A S (1999) Anastomosis formation and nuclear and protoplasmic exchange in arbuscular mycorrhizal fungi. Appl. Environ. Microbiol. 65:5571-5575

Giovannetti M, Citernesi A (1993) Time course of appressorium formation on host plants by arbuscular mycorrhizal fungi. Mycol. Res. 97:1140-1142

Hargreaves J A, Turner G (1992) Gene transformation in plant pathogenic fungi. In: Gurr S J, McPherson M J, Bowles D J (eds) Molecular plant pathology. Oxford University Press pp 79-97

Harrier L A (2000) Isolation and sequence analysis of the arbuscular mycorrhizal fungus *Glomus mosseae* (Nicol et Gerd) Gerdemann et Trappe 3-phosphoglycerate kinase (*PGK*) gene promoter region. DNA Seq. 11:463-473

Harrier L A, Sawczak J (2000) Detection of the 3-phosphoglycerate kinase protein of *Glomus mosseae* (Nicol et Gerd) Gerdemann et Trappe. Mycorrhiza, 10:81-86

Harrier L A, Wright F, Hooker J E (1998) Isolation of the 3-phosphoglycerate kinase (PGK) gene from the arbuscular mycorrhizal fungus *Glomus mosseae* (Nicol et Gerd). Gerdemann et Trappe. Curr. Genet. 34:386-392

Herzing L B K, Meyn M S (1993) Novel *lacZ*-based recombination vectors for mammalian cells. Gene 137:163-169

Hilber U W, Bodmer M, Smith F D, Köller W (1994) Biolistic transformation of conidia of *Botryotinia fuckeliana*. Curr. Genet. 25:124-127

Hijri M, Hosny M, van Tuinen D, Dulieu H (1999) Intraspecific ITS polymorphism in *Scutellospora castanea* (Glomales, Zygomycetes) is structured within multinucleate spores. Fun. Gen. Biol. 26:141-151

Hosny M, Hijri M, Passerieux E, Dulieu H (1999) rDNA units are highly polymorphic in *Scutellospora castanea* (Glomales, Zygomycetes). Gene 226:61-71

Kaya S, Imai T, Ishige M (1990) The efficiency of transformation by *Agrobacterium tumefaciens* without selection marker. Jpn. J. Breed. 40:82-83

Kitamoto N, Matsui J, Kawai Y, Kato A, Yoshino S, Ohmiya K, Tsukagoshi N (1998) Utilisation of the Tef1-α Gene (*Tef1*) promoter for expression of poly-galacturonase genes, *Pgaa* and *Pgab*, in *Aspergillus oryzae*. Appl. Microbiol. Biotech. 50:85-92

Koide R T (2000) Functional complementarity in the arbuscular mycorrhizal symbiosis. New Phytol. 147:233-235

Kumria R, Verma R, Rajam M V (1998) Potential applications of antisense RNA technology in plants. Curr. Sci. 74:35-41

Lorito M, Hayes C K, Di Pietro A, Harman G E (1993) Biolistic transformation of *Trichoderma harzianum* and *Gliocadium virens* using plasmid and genomic DNA. Curr. Genet. 24:349-356

Mackenzie D A, Wongwathanarat P, Carter A T, Archer D B (2000) Isolation and use of a homologous Histone H4 promoter and a ribosomal DNA region in a transformation vector for the oil producing fungus *Mortierella alpina*. Appl. Environ. Microbiol. 66:4655-4661

Mort-Bontemps M, Fevre M (1997) Transformation of the oomycete *Saprolegnia monoica* to hygromycin resistance. Curr. Genet. 31:272-275

Orbach M J, Porro E B, Yanofsky C (1986) Cloning and characterisation of the gene for beta-tubulin from a benomyl resistant mutant of *Neurospora crassa* and its use as a dominant selectable marker. Mol. Cell Biol. 6:2452-2461

Rosel H, Kunze G (1998) Integrative transformation of the dimorphic yeast *Arxula adeninivorans* Ls3 based on hygromycin B resistance. Curr. Genet. 33:157-16

Rosendahl S, Dodd J C, Walker C (1994) Taxonomy and phylogeny of the Glomales. In: Gianinazzi S, Schüepp H (eds) Impact of arbuscular mycorrhizas on sustainable agriculture and natural ecosystems Birkhäuser, Basel pp 1-12

Rosendahl S, Taylor J W. (1997) Development of multiple genetic markers for studies of genetic variation in arbuscular mycorrhizal fungi using AFLP™ Mol. Ecol. 6:821-829

Ruiz-Díez B, Martínez-Suárez J V (1999) Electrotransformation of the human pathogenic fungus *Scedosporium prolificans* mediated by repetitive rDNA sequences. FEMS Immuno. Med. Micro. 25:275-282

Sanders I R (1999) No sex please, we're fungi. Nature 399:737-738

Sanford J C, Klein T M, Wolf E D, Allen N (1987) Delivery of substances into cells and tissues using a particle bombardment process. J. Part. Sci. Tech. 5:27-37

Sanford J C, Devitt M L, Russel J A, Smith F D, Harpending P R, Roy M K, Johnston S A (1991) An improved helium driven biolistic device. Technique 3:3-16

Schenborn E, Groskreutz D (1999) Reporter gene vectors and assays. Mol. Biotech. 13:29-44.

Schilde C, Wöstemeyer J, Burmester A (2001) Green fluorescent protein as a reporter for gene expression in the mucoralean fungus *Absidia glauca*. Arch. Microbiol. 175:1-7

Sherf B A, Wood K V (1995) Firefly luciferase engineered for improved genetic reporting. Promega Notes 49:14-21

Smith F A, Jakobsen I, Smith S E (2000) Spatial differences in acquisition of soil phosphate between two arbuscular mycorrhizal fungi in symbiosis with *Medicago truncatula*. New Phytol. 147:357-366

St Leger R J, Shimizu S, Joshi L, Bidochka M J, Roberts D W (1995) Co-transformation of *Metarhizium anisopliae* by electroporation or using the gene gun to produce stable GUS transformants. FEMS Micro. Lett. 131:289-294

Tamasloukht B, Kuhn B, Becard G, Franken P (2000) RNA accumulation patterns of arbuscular mycorrhizal fungi during presymbiotic stages. In: Weber H C, Imhof S, Zeuske D (eds.) Programs, Abstracts and Papers of the Third International Congress on Symbiosis, Philipps, University of Marburg, Germany p 217

Tommerup I C, Sivasithamparam K (1990) Zygospores and asexual spores of *Gigaspora dicephens*, an arbuscular-mycorrhizal fungus. Mycol. Res. 94:897-900

van Buuren M L Lanfranco L, Longato S, Minerdi D, Harrison M J, Bonfante P (1999) Construction and characterization of genomic libraries of two endomycorrhizal fungi: *Glomus versiforme* and *Gigaspora margarita*. Mycol. Res. 103:955-960

van den Wymelenberg A J, Cullen D, Spear R N, Schoenike B, Andrews J H (1997) Expression of green fluorescent protein in *Aureobasidium pululans* and quantification of the fungus on leaf surfaces. BioTechniques 23:686-690

van der Heijden M G A, Klironomos J N, Ursic M, Moutoglis P, Streitwolfengel R, Boller T, Wiemken A, Sanders I R (1998) Mycorrhizal fungal diversity determines plant biodiversity, ecosystem variability and productivity. Nature 396:69-72

Volmer S J, Yanofsky C (1986) Efficient cloning of genes of *Neurospora crassa*. Proc. Natl. Acad. Sci. USA 83:4869-4873

Ward M, Wilkinson B, Turner G (1986) Transformation of *Aspergillus nidulans* with a cloned oligomycin resistant ATP synthase subunit 9 gene. Mol. Gen. Genet. 202:265-270

Wood K V (1995) Marker proteins for gene expression. Curr. Opin. Biotech. 6:50-58

Wood K V (1998) The chemistry of bioluminescent reporter assays. Promega Notes 65:14-18

Yu J, Cole G T (1998) Biolistic transformation of the human pathogenic fungus *Coccidioides immitis*. J. Micro. Meth. 33:129-141

Zeze A, Hosny M, Gianinazzi-Pearson V, Dulieu H (1996) Characterisation of a highly repeated DNA sequence (SC1) from the arbuscular mycorrhizal fungus *Scutellospora castanea* and its detection *in planta*. Appl. Environ. Microbiol. 62:2443-2448

Zeze A, Hosny M, Van Tuinen D, Gianinazzi-Pearson V, Dulieu H (1999) MYCDIRE, a dispersed repetitive DNA element in arbuscular mycorrhizal fungi. Mycol. Res.103:572-576

Zeze A, Sulistyowati E, Ophel-keller K, Barker S, Smith S (1997) Intrasporal genetic variation of *Gigaspora margarita*, a vesicular arbuscular mycorrhizal fungus, revealed by M13 minisatellite-primed PCR. Appl. Environ. Microbiol. 63:676-678

Mycorrhizal Technology in Agriculture
ed. by S. Gianinazzi, H. Schüepp, J.M. Barea and K. Haselwandter

Arbuscular mycorrhizal modifications to plant root systems: scale, mechanisms and consequences

G. Berta[1], A. Fusconi[2] and J.E. Hooker[3]

[1] *Dipartimento di Scienze e Tecnologie Avanzate, Università del Piemonte Orientale 'A. Avogadro', Corso Borsalino 54, I-15100, Alessandria, Italy*
[2] *Dipartimento di Biologia Vegetale, Università di Torino, Viale Mattioli 25, I-10125, Torino, Italy*
[3] *School of Applied Sciences, University of Glamorgan, Pontypridd, Mid-Glamorgan CF37 1UD United Kingdom*

Keywords: root system structure, root apices, woody plants, herbaceous plants, root mortality

Introduction

Plant root systems serve a variety of different functions at the plant, community and ecosystem scales. For the plant they provide stability for the shoot, an organ for uptake, transport and storage of water and nutrients and are frequently the site of nitrogen-fixing or mycorrhizal symbioses. At the community and ecosystem scale they provide a source of photo-autotrophic carbon, through symbiosis or exudation and root mortality, that is utilised by symbionts or rhizosphere organisms respectively.

In the foregoing discussion it is important to recognise two distinct types of plant: woody plants, that form woody tissues in above ground parts and herbaceous plants that do not. In both, the temporal and spatial distribution and form of plant root systems is to some extent determined genotypically, but is also subject to biotic and abiotic interactions that can have a marked influence. Unfortunately, information on how genetic, biotic or abiotic factors affect root systems is extremely limited. Nevertheless, it is known that one of the most important biotic interactions occurs with mycorrhizal fungi. These fungi form mutualistic associations with an estimated 240,000 plant species (Bonfante and Perotto 1995). Of these, the arbuscular mycorrhizal fungi are by far the most widespread, with an estimated 150 fungal species (Smith and Read 1997) forming mycorrhizas with the roots of over 80% of all plant species (Smith and Gianinazzi-Pearson 1988). Benefits to plants are known to include improved nutrition (Ravnskov and Jakobsen 1995; Stribley et al. 1980; see Clark and Zeto 2000 for review), protection from pathogens (Hooker et al. 1996; Trotta et al. 1996; Vigo et al. 2000) and enhanced tolerance of heavy metals (Heggo and Angle 1990) and drought (Allen and Allen 1986; Hernandez-Sebastia et al. 1999). Understanding the extent of these benefits, how they are achieved and the potential application of this knowledge to agriculture have been foci of research on arbuscular mycorrhizas over the past decade. The outcomes of this research have significantly increased understanding of the symbiosis, and the availability of molecular tools to complement physiological and ecological approaches now promises a rapid increase in knowledge. However, of the research to date amongst the most exciting findings is that arbuscular mycorrhizal fungi (AMF) can have a very significant impact on the development of root systems, with consequential effects on the anatomy and physiology of roots and the architecture and mortality of the root system. These changes, and their consequences for agricultural and natural plant communities, are the subject of this review.

Plant root systems

For many years it was considered that AMF did not affect the development of the root system (Harley and Smith 1983). This assumption is perhaps not surprising since although we now know that the modifications induced can be extensive, they are nevertheless difficult to identify and quantify without careful experimental design and analysis. This is in contrast to modifications induced by ectomycorrhizal fungi which can frequently be observed on visual inspection; the shorter mycorrhizal tips are clearly distinguishable. In the case of AMF the nature of the modifications make visual perception more difficult. Changes frequently result in a more branched root system, with a larger proportion of smaller diameter, higher order roots (some less than 0.1 mm). Analyses require a detailed comparative analysis of colonised and un-colonised root systems, with experimental protocols that ensure all roots of a plant are recovered, with their connections to neighbouring roots intact (Hooker et al. 1998 for protocols). Some researchers have utilized specialised techniques for extracting roots from the soil (Hooker et al. 1992) and image analysis techniques to aid quantification of root branching (Hooker and Atkinson 1992).

As a background to what follows, it is constructive to introduce a range of variables that can be utilised to describe variation in a root system. Atkinson (1992) categorised these as (i) quantitative (length or weight), (ii) structural (the way the roots that make up a root system are connected together), (iii) spatial (the arrangement of the roots within the soil) and (iv) temporal (when growth and development of the root system occurs). Root system architecture is the term commonly applied to describe structural changes, although the term morphology has also been applied (Schellenbaum et al. 1991; Hooker et al. 1992; Berta et al. 1995). To date, two main approaches have been applied to describe and quantify AMF-induced changes to root system architecture: topological or developmental. The former was applied originally to root systems by Fitter (1985). It uses a mathematical classification with two parameters, the magnitude and exterior path-length, to describe the form of the root system. The latter uses a traditional, and arguably more intuitive, approach to describe the root system based on chronology of development, where higher order laterals develop from a lateral of lower order. Both approaches have their own merits, and provide valuable ways of describing root systems. For a discussion of their respective virtues, see Atkinson et al. (1994) and Hooker and Atkinson (1996).

Applying these approaches, together with careful experimental protocols, it has been demonstrated widely that colonisation of roots by AMF can result in major changes to root system development and architecture. In the first report Berta et al. (1990) demonstrated increased branching of secondary roots of *Allium porrum* when colonised by the AMF *Glomus* sp. E3 (145% of non-colonised roots) but found no evidence for a change in topology. Shortly afterwards, Schellenbaum et al. (1991) provided data for the effects of the AMF *Glomus fasciculatum* on *Vitis vinifera*, reporting that branching of primary, secondary and tertiary roots colonised by the AMF was 140%, 200% and 266% respectively greater than non-colonised roots. Similarly, Tisserant et al. (1992) in a study using the same AMF species and *Platanus acerifolia* reported 145% increases in branching of secondary roots when colonised. Moreover, in contrast to the data of Berta et al. (1990), analyses of these data revealed that in both studies AMF colonised roots showed evidence of altered topology, with colonised *V. vinifera* root systems exhibiting a more random pattern and *P. acerifolia* a more dichotomous pattern.

Since these early reports there have been several others that also provide evidence for effects of AMF on root system development, with increased branching evident when roots are colonised (Berta et al. 1995; Citernesi et al. 1998; Hooker et al. 1992; Figure 1). However, there are also data suggesting that the response may not be universal.

Forbes (1996), for example, studied root development in *Plantago lanceolata* and reported no change in the branching of colonised root systems due to AMF when plants were the same size. And in *Lycopersicum esculentum* there are reports of a reduction (Trotta et al. 1996) or no change (Vigo et al. 2000) to the branching of colonised root systems. A report by Hetrick et al. 1988 also suggested a reduction in branching due to AMF. However, because analysis relied on topological analyses alone and did not directly measure branching we contend that this data is not conclusive.

Root system development is largely dependent on the collective behaviour of constituent root apices. An awareness of the impact of AMF on these apices is thus crucial to fully understand the extent of modifications. Unfortunately data is very scarce, but usefully benefits from some highly focused studies with *Allium porrum*.

Root apices

What is clear from these studies is that, when colonised by AMF (*Glomus* sp. E3) and grown in controlled conditions, adventitious roots have a determinate pattern of growth. The apices of all plants, colonised or not by AMF show, dur-

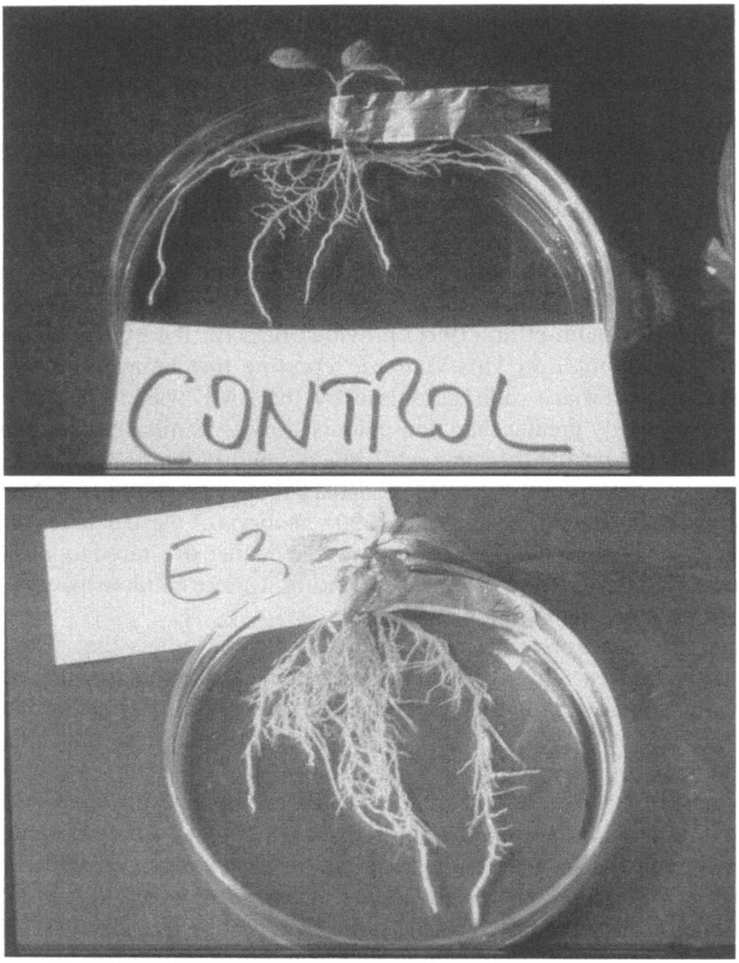

Figure 1. Micropropagated *Prunus cerasifera* plants, colonized and not colonized by the AM fungus *Glomus* sp. E3. In mycorrhizal plants (below), roots are more branched and with a larger diameter than in controls.

ing growth, a reduction of diameter and root cap size and a progressive reduction in the distance between the initials and the differentiated zone, along with a progressive chromatin condensation and a lowering of the mitotic activity. Subsequently these cells completely lose their meristematic activity and differentiate to resemble parenchyma cells; some apices then degenerate and become necrotic. In plants colonised by AMF the percentage of active apices decreases more rapidly than in non-colonised plants, and in highly-colonised plants most of the root apices are either inactive or necrotic (Berta et al. 1990;

Fusconi et al. 1986). The apices of adventitious roots of colonised plants have a similar structure to those of non-colonised plants but are considerably larger, have larger proximal meristems with more cell-files in the cortex and central cylinder, and larger root caps. This increase is due partly to a larger number of meristematic cells, and partly to an increase in cell size (Fusconi et al. 1994). Moreover, despite being larger, apices of colonised roots have a lower metabolic activity, as illustrated by the lower degree of labelling after ^{3}H-uridine incorporation (Fusconi et al. 1994; Figure.2), and the mitotic cycles of meristematic cells become longer with increasing colonisation (Berta et al. 1991). Specifically, interphase is longer in colonised roots, especially when highly infected, due to lengthening of G1 and S, and G2 is reduced (Berta et al. 1991). The fraction of the cells occupied by mitosis is similar in colonised and non-colonised roots, whereas the relative duration of the individual phases are altered markedly, with metaphase progressively longer with increasing colonisation (Berta et al. 1991; Fusconi et al. 1986). These data thus show that

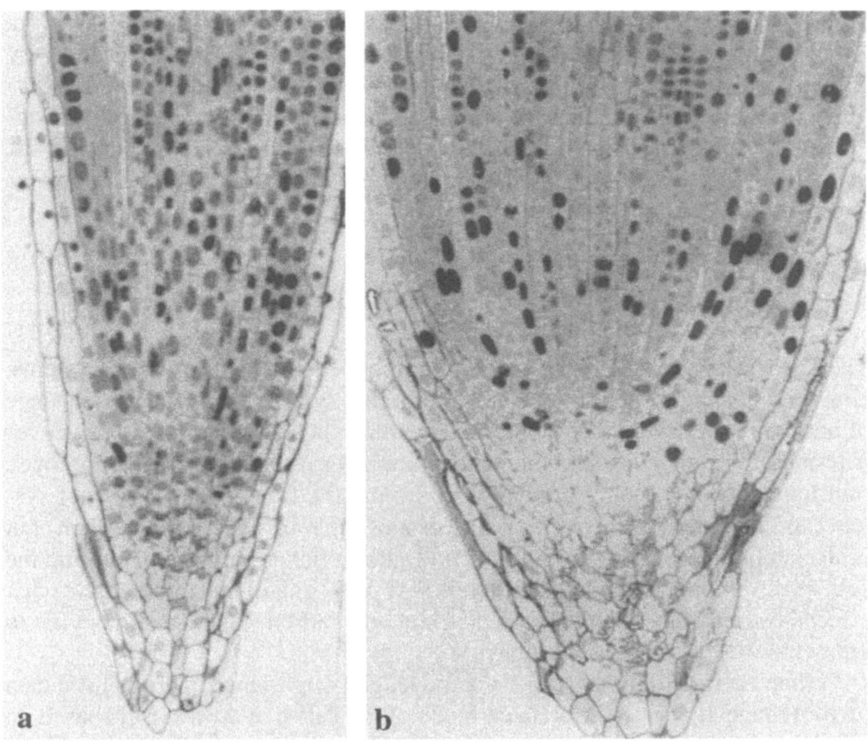

Figure 2. Median longitudinal sections of root apices of *Allium porrum* plants supplied with H-thymidine for 24 h. (a) control, (b) infected with *Glomus sp.* strain E3. Note the higher percentage of labelled nuclei in control than in mycorrhizal root apices, that also have larger diameter (a, b have the same magnification).

colonisation by AMF blocks meristematic activity in colonised plants and it is probable that this is the direct cause of modifications to the root system of *Allium porrum*. Interestingly, AMF can alter the structure and activity of root apices even in the absence of penetration (Berta et al. 1993). This block of meristematic activity observed in root apices of colonised *Allium porrum* has also been observed in another monocot, *Ornithogalum umbellatum* (Berta et al. 1993), but not in the dicotyledonous plants *Lycopersicon esculentum* (Fusconi et al. 1999) and *Prunus cerasifera* (Berta et al. 1995). However, in *P. cerasifera*, although the mitotic index is unaffected by AMF, the percentage of cells in metaphase increases, suggesting a blocking effect of the fungus on this phase of the cell cycle (Berta et al. 1995).

Another general effect of colonisation is an increase of the diameter of root apices, leading to thicker roots. Even if in some cases these increases are possibly a consequence of better conditions of growth leading to larger plants and organs, in others it seems to be a specific response to colonisation. This is the case, for example, with *L. esculentum*, where the diameter of root apices increases even in absence of a growth effect (Fusconi et al. 1999). Moreover, it appears to depend on the order of roots and the species of AMF involved. For example, in micropropagated plants of *P. cerasifera* colonisation by *Glomus mosseae* results in a greater growth enhancement than *Glomus intraradices*, but the diameter of primary, secondary and tertiary roots are significantly higher in plants colonised by *Glomus intraradices*; by contrast the diameter of quaternary roots are higher in plants colonised by *Glomus mosseae* (Berta et al. 1995).

Root system mortality

The discussion has so far focused on the development of root systems. However, at any point in time the form of a root system is the net consequence of not only root development but also root mortality. How long a root lives for is termed root longevity and this is the parameter usually reported. Moreover, root longevity is not only important in driving the final form of the root system (and determining the respiratory costs of the root system to the plant) but is also an important driver of nutrient and carbon fluxes between plants and the soil. In developing an understanding of how AMF and plants interact and what effects this may have on plant-soil nutrient and carbon fluxes it is thus vital to know how AMF affect root longevity.

Perhaps surprisingly, given its acknowledged importance, there is little data on root longevity *per se*. Moreover, much of the data that does exist is based on inaccurate methodology, only measuring differences in root biomass between sampling dates (Fogel 1985) and failing to take account of roots produced and lost between sampling dates. The only accurate method is to make direct observations of roots' life histories using, for example, minirhizotrons (see Hooker et al. 2000; Smit et al. 2000 for reviews of methodology). Some researchers have used this approach. Head (1966), for example, studied apple roots in a biotron and estimated mean root longevity as 3 to 4 weeks. Atkinson (1985) followed a similar strategy for the same species and, depending on the genotype, identified longevity as between 2 and 5 weeks. More recently, data obtained using observational methods shows plants with root longevity from 7 to 14 (Black et al. 1998; Watson et al. 2000) up to 540 days (Burton et al. 2000).

Data on the effects of AMF on root longevity is scant. Hooker et al. (1995) used minirhizotrons to make direct measures of poplar root life histories, and identified a significant decrease in mean longevity when roots were colonised by an AMF. Espeleta et al. (1999) also reported a difference in longevity of Volkamer lemon rootstock roots, if colonised by an AMF and droughted for 15 weeks but in this case longevity was increased. However, in contrast to Hooker et al. (1995), who inoculated the entire root system, Espelata et al. (1999) excavated a section of woody root, removed lateral roots and inoculated this section. The data is thus interesting in terms of understanding local events, induced by the localised presence of an AMF, but it doesn't provide data on root system behaviour in response to an AMF that has the opportunity to colonise the entire root system. A recent report by Gavito et al. (2001) also used minirhizotrons to study the effects of AMF on pea. In this case they identified few differences between colonised and non-colonised plants. Although contrary to the effects reported in poplar this data supports the hypothesis presented later in this paper. Clearly further studies are required before any firm conclusions can be drawn but the reports available to date provide useful data.

Mechanisms of modification

One of the mechanisms through which AMF modify root system development, architecture and longevity is undoubtedly through effects on nutrition. It is known, for example, that change in phosphorus supply will modify root system development and architecture (Amijee et al. 1989; Drew 1975; Fitter and Stickland 1991) and changes in nitrogen nutrition can also result in alterations (Drew 1975). Because increased supply of both, nitrogen and particularly phosphorus, are frequently a consequence of colonisation, it is reasonable to hypothesise that modifications are a consequence of changed nutrition. However, there is also evidence to suggest that other mechanisms are involved. For example, in a study with *Populus* and the AMF *Glomus sp.* E3, Hooker et al. (1992) carefully manipulated nutrition to ensure no nutritional benefits were provided by the AMF and plants of the same size and nutritional status were compared, but still measured marked increases in branching of roots when they were colonised by AMF. Thus, they argued that modifications were unlikely to be a consequence of enhanced nutrient supply alone and that other, AMF specific mechanisms must be involved.

Nevertheless, data from more recent studies questions this hypothesis. For example, Trotta et al. (1996) demonstrated that AMF did not induce modifications to the root system of *L. esculentum*. As already discussed, this is of interest in its own right because it represents an exception to the normal pattern of change exhibited by colonised roots. Additionally, because plants did not show the usual growth enhancement due to colonisation, and only very minor increases in leaf tissue concentrations of phosphorus we suggest this provides evidence for the role of phosphorus in mediating AMF-induced modifications to root systems. Further support is provided by Vigo et al. (2000) and Forbes et al. (1996) who reported effects of AMF on root systems of *L. esculentum* and *P. lanceolata* respectively, and present data that shows no effects of AMF on root system development. Crucially, in both cases plants were grown in continuous flow microcosms designed to ensure that plants were not deficient in nutrients and only direct effects of the AMF, not consequential effects due to changes in nutrient supply, would be measured.

However, given that the activity of root apices regulate the development of root growth, perhaps the most revealing evidence for the nature of mechanisms involved comes from studies of root apices. In *Allium porrum* + *Glomus sp.* E3 differences in total root length, adventitious root length and root branching of colonised plants decreased with increases in applied phosphorus concentration, and disappeared or were reversed at high phosphorus levels (Trotta et al. 1991a). These differences were related to the activity of their root apices, with phosphorus causing a lengthening of the mitotic cycle comparable to that induced by AMF at lower phosphorus concentrations, with some differences in the length of the phases. Specifically, S increased more with colonisation by AMF, whilst G1 became longer with the increase of phosphorus. However, the slowing down of the mitotic cycle did not block the mitotic activity of the root apices, as shown by the low percentage of active apices, and resulted in a slow and steady growth of the adventitious roots (Fusconi et al. 2000). Moreover, in *Allium porrum*, first and second order laterals do not respond to phosphorus nutrition and always are shorter in colonised plants (Berta et al. 1993).

The involvement of phytohormones in events at root apices is unclear, although they are probably involved in the morphogenesis of arbuscular mycorrhizas (Beyrle 1995). Data is scarce, but effects on some hormones are known. For example, AMF are known to alter proton extrusion by plant cells and ethylene release (Vierheilig et al. 1994), both involved in cell growth and senescence. Altered levels of the auxin IAA (Kaldorf and Ludwig-Müller 2000; Torelli et al. 2000), cytokinin (Torelli et al. 2000) and abscicic acid (Danneberg et al. 1992; Esch et al. 1994) have also been reported in plants colonised by AMF. The lengthening of the mitotic cycle following colonisation and supply of phosphorus could tentatively be related to an increase of abscicic acid (ABA) concentration, as ABA reduces the DNA synthesis rate (Jacqmard et al. 1995) and has an inhibitory effect on the progress in the cell cycle (Himmelbacch et al. 1998; Wang et al. 1998). Moreover, it has been found at higher levels in roots of *Zea mays* colonised by AMF (Danneberg et al. 1992) and is reported to be contained in the extraradical hyphae of two *Glomus sp.* isolates (Esch et al. 1994). However, ABA does not explain the block of the root apices and the increase of the metaphase index observed in colonised adventitious roots of *Allium porrum*, so this suggests that other factors, specific to the symbiosis, must be involved in the regulation of the cell cycle at the apex.

These data suggest that the morphological effects of AMF on the root system are not entirely phosphorus mediated, at least in *Allium porrum*. Conversely, they do not preclude (but do not conclude either) a major involvement of phosphorus in the modifications measured. They demonstrate that AMF bring about unique changes in root apices, not explained by phosphorus, that may or may not, be responsible for modifications to root system development. Further study is clearly needed and this will require the application of methodologies that permit discrimination between nutrient-driven and other mechanisms. Possible approaches include regulating nutritional factors around the root using a nutrient film (Vigo et al. 2000) and the use of axenic root cultures (see Kaska et al. 1999).

Consequences of modifications

Plant scale

AMF-induced modifications to root system development and architecture can have major impacts at the plant, community and ecosystem scales. At the plant scale, colonisation usually results in increased branching, resulting in a larger proportion of root systems being made up of smaller diameter, higher order roots. Importantly, this can also result in a changed topology, that may be more suited to exploration of scarce resources within a heterogeneous soil environment. The benefits of these modifications are clear, but such a strategy would suggest additional costs in terms of nutrient and carbon allocation, in addition to those arising as a direct consequence of the symbiosis. Clearly only a detailed cost-benefit analysis, unique to each situation, and taking account of temporal events, can provide an accurate assessment of net consequence. However, in appraising the usefulness of the strategy it is useful to consider some general assumptions. Firstly, in woody plants (although not all roots may develop secondary thickening) there is evidence to suggest that increased branching is accompanied by a reduction in mean longevity, since there is a general inverse relationship between diameter and longevity (Eissenstat et al. 2000; Wells and Eissenstat 2001) i.e. roots with a relatively small diameter tend to have shorter life-spans. Because colonisation by AMF will usually increase branching a further consequence should therefore, be a reduction in mean longevity of roots. Data by Hooker et al. (1995) confirms this expectation and is reviewed above. For herbaceous plants we can find no similarly comparative longevity data for roots of different diameters or age, using observational or other whole-root based approaches. Data on root longevity *per se* is also scarce, but is available for *Brassica olearacea* (70 days), *Allium porrum*

(130 days) (Smit and Zuin 1996) and sugar beet (between 60 and 130 days) (Van Noordwijk et al. 1994). However, a recent report by Gavito et al. (2001) shows that in pea, a herbaceous species, there is little impact of AMF on longevity. After reviewing these data we hypothesise that in woody plants, where roots are largely determinate, and the longevity of most roots is often relatively short, e.g. 7 to 14 days in poplar, the impact of AMF on root systems is to increase branching of roots, and (probably by altering the proportion of higher order roots) decrease mean root longevity. Conversely, in herbaceous plants, where roots are largely indeterminate and many live for virtually the life of the plant, colonisation by AMF often leads to an increase in root longevity. Evidence for the latter is provided by recent studies showing reductions in numbers of senescing nuclei in roots of *Allium porrum* (Lingua et al. 1999) and *L. esculentum* (unpublished data).

The value of these quite different strategies to the respective plant groups is uncertain. However, it is possible to hypothesise that optimal exploitation of a given volume of soil can be best achieved by either producing (i) a root system that is as long as possible and persists (consistent with an indeterminate strategy) or, (ii) a highly dynamic root system that is ephemeral (consistent with a determinate strategy) i.e. roots are produced and lost more often, resulting in a greater area of the soil being exploited by roots, and associated AMF, per unit time than would be exploited by a indeterminate root system of the same length. In both cases it is possible to see advantages conferred as a consequence of colonisation by AMF; the indeterminate type of root system would benefit from a larger root system that persists, and the determinate type of root would benefit from a more branched and more ephemeral root system. Although it would appear ostensibly that the determinate strategy is less energy efficient, requiring more carbon input, because of the root-building phases, there is evidence to suggest that this may not be the case. This apparent contradiction arises because both the respiratory costs of a root, and its capacity to uptake nutrients and water are not constant over time. Eissenstat et al. (2000), for example, defines root efficiency as the ratio of (nutrient) benefit to (respiratory) cost and has shown that the efficiency of apple roots for nutrient uptake can decrease dramatically with age; from 900 pmol g^{-1} s^{-1} in 2 week old roots to 300 pmol g^{-1} s^{-1} in 6 week old roots.

These data thus suggest that it may beneficial for a plant to have roots with a relatively short lifespan. Data from Burton et al. (2000) further supports this proposition. They found that longevity of roots was higher when nitrogen availability was greater and suggested the possibility that roots were kept alive as long as the benefit (nutrients) they provide outweighs the carbon cost of keeping them alive. Optimal root age needs to be established for the species of study, and the costs and benefits of AMF needs to be established, but these data suggest that AMF induced reductions in mean root longevity could well be advantageous or neutral. Finally, an increase in the proportion of higher order roots, a consequence of colonisation, would probably be advantageous to AMF in root systems where the lower order roots become woody, and thus less susceptible to colonisation by AMF. Modifications that increase branching, and numbers of higher order roots, would thus result in a host root system more susceptible to colonisation by AMF. This determinate *versus* indeterminate root hypothesis clearly needs testing, but we believe it forms a constructive focus for future research.

Other consequences relate to plant pathogens. It is recognized that AMF will generally reduce the severity of infection by root pathogens and we now know that some of these reductions are consequences of AMF-induced changes to plant roots. Most studied have been interactions between AMF and the root pathogen *Phytophthora* spp.; the basis of the hypotheses tested, that AMF induced changes to roots and root system architecture make roots less susceptible to infection by the pathogen or inhibit its spread. However, because *Phytophthora spp.* normally infect immediately behind the root tip, the increased branching normally associated with AMF suggests increased susceptibility. Norman et al. (1996) tested this hypothesis in strawberry, inoculated with *P. fragariae*, and showed that in plants not colonised by AMF this is what occurs i.e. infection increased with root branching. By contrast, in plants colonised by AMF, disease decreased as the number of root branching increased. A subsequent study showed that this was because root exudates, produced by AMF-colonised strawberry roots, stimulated the production of sporangia in the pathogen less than exudates from non-colonised strawberry roots (Norman and Hooker 2000). Studies with *L. esculentum* have suggested a similar involvement of root exudates in biocontrol of the pathogen *P. nicotianae* by AMF (Vigo et al. 2000), and also direct effects due to increased size of root apices (Fusconi et al. 1999).

Community and ecosystem

Recent experiments have demonstrated significant affects of AMF at the community scale, with consequences for relative performance of species and the species composition of plant communities (Van der Heijden et al. 1998). Despite this, most contemporary understanding of the impact of arbuscular mycorrhizal fungi at the community and ecosystem scale is largely based on assumptions, with specific data very limited. This is particularly true for root mediated consequences. Thus, although it is known that root distribution, exudation and mortality are important in plant:plant interactions, carbon and nutrient fluxes from the plant to soil phase and in driving heterotrophic activity in the rhizosphere direct Exp. evidence at the community or ecosystem scale is lacking. This is, at least partly, a consequence of difficulties encountered in scaling up experimental data obtained in pot or microcosm studies to the community and ecosystem level. More interactions take place as size increases and the biotic and abiotic make-up becomes increasingly complex and subject to random events. Scaling-up research findings from experiments carried out at the plant or small plot scale is thus complex, and requires a better understanding of scale-related processes in ecosystems. However, it is important to realize that this lack of understanding of consequences at these larger scales also exists for roots *per se* and is not only a problem for AMF-focused data. To progress this understanding will require significant scientific effort and approaches that integrate both modelling and experimental approaches and address questions that relate to scale.

In order to help focus research we propose that the arguments presented above, to explain the impact of AMF on plant root systems, suggest that one of the major consequence of AMF, at least in woody plants, is to facilitate co-existence between species and promote ecosystem stability. We hypothesise that they accomplish this by reducing competition between plants (individuals and species) by promoting spatial and temporal complimentarity between root systems. This is achieved through (i) increased branching, resulting in a root system more efficient in exploiting dispersed resources and less reliant on specific nutrient patches, (ii) reduced mean longevity, resulting in a larger proportion of younger roots with better uptake capacities, (iii) roots being more ephemeral, and thus less likely to be subject to competition, and (iii) increased carbon use efficiency, although this needs to be offset by carbon costs of the symbiosis (Peng et al. 1993). Co-existence will be further supported by benefits of AMF that are not limited to the host (see Vigo et al. 2000). Clearly, our AMF-led complimentarity hypothesis requires testing, but we propose it as a rational basis for further research that aims to better understand the impact of AMF at the community and ecosystem scales.

Acknowledgements
The authors are grateful for the comments of Dr G. Lingua on drafts of the manuscript.

References

Allen E B, Allen M F (1986) Water relations of xeric grasses in the field: interactions of mycorrhizas and competition. New Phytol. 104:559-571

Amijee F, Tinker P B, Stribley D (1989) The development of endomycorrhizal root systems VII: a detailed study of effects of soil phosphorus on colonization. New Phytol. 111:435-446

Atkinson D (1992) Tree root development: the role of models in understanding the consequences of arbuscular endomycorrhizal infection. Agronomie 12:817-820

Atkinson D (1985) Spatial and temporal aspects of root distribution as indicated by the use of a root observation laboratory. In: Fitter A H, Atkinson D, Read D J, Usher M B (eds) Ecological interactions in soil. Blackwell Scientific Publications, Oxford, UK pp 43-66

Atkinson D, Berta G, Hooker J E (1994) Impact of mycorrhizal colonisation on root architecture, root longevity and the formation of growth regulators. In: Gianinazzi S, Schüepp H (eds) Impact of arbuscular mycorrhizas on sustainable agriculture and natural ecosystems. Birkhäuser Verlag, Basel Switzerland pp 89-99

Berta G, Fusconi A, Trotta A, Scannerini S (1990) Morphogenetic modifications induced by the mycorrhizal fungus *Glomus* strain E3 in the root system of *Allium porrum* L. New Phytol. 114:207-215

Berta G, Tagliasacchi A M, Fusconi A, Gerlero D, Trotta A, Scannerini S (1991) The mitotic cycle in root apical meristems of *Allium porrum* L. is controlled by the endomycorrhizal fungus *Glomus* sp. strain E3. Protoplasma 161:12-16

Berta G, Fusconi A, Trotta A (1993) VA mycorrhizal infection and the morphology and function of root systems. Envir. Exp. Bot. 33:159-173

Berta G, Trotta A, Fusconi A, Hooker J E, Munro M, Atkinson D, Giovannetti M, Morini S, Fortuna P, Tisserant B, Gianinazzi-Pearson V, Gianinazzi S (1995) Arbuscular mycorrhizal induced changes to plant growth and root system morphology in *Prunus cerasifera*, Tree Physiol. 15:281-294

Beyrle H (1995) The role of phytohormones in the function and biology of mycorrhizas. In: Varma A, Hock B (eds) Mycorrhiza: structure, function, molecular biology and biotechnology. Springer-Verlag, Berlin Heidelberg New York pp 365-390

Black K E, Harbron C G, Franklin M, Atkinson D, Hooker J E (1998) Differences in root longevity of some tree species. Tree Physiol. 18:259-264

Bonfante P, Perotto S (1995) Strategies of arbuscular mycorrhizal fungi when infecting host plants. New Phytol. 130:3-21

Burton A J, Pregitzer K S, Hendrick R L (2000) Relationships between fine root dynamics and nitrogen availability in Michigan northern hardwood forests. Oecologia 125:389-399

Citernesi A S, Vitagliano C, Giovannetti M (1998) Plant growth and root system morphology of *Olea europaea* L. rooted cuttings as influenced by arbuscular mycorrhizas. J. Hort. Sci. Biotech. 73:647-654

Clark R B, Zeto S K (2000) Mineral acquisition by arbuscular mycorrhizal plants. J. Plant Nutr. 23:867-902

Danneberg G, Latus C, Zimmer W, Hundeshagen B, Schneider-Poetsch H, Bothe H (1992) Influence of vesicular-arbuscular mycorrhiza on phytohormone balance in maize (*Zea mays* L.) J. Plant Physiol. 141:33-39

Drew M C (1975) Comparison of the effects of a localised supply of phosphate, nitrate, ammonium and potassium. New Phytol. 75:479-490

Eissenstat D M, Wells C E, Yanai R D, Whitbeck J L (2000) Building roots in a changing environment: implications for root longevity. New Phytol. 147:33-42

Esch H, Hundeshagen B, Schneider-Poetsch H, Bothe H (1994) Demonstration of abscisic acid in spores and hyphae of the arbuscular-mycorrhizal fungus *Glomus* and the N_2-fixing cyanobacterium *Anabaena variabilis*. Plant Sci. 99:9-16

Espeleta J F, Eissenstat D M, Graham J H (1999) Citrus root responses to localized drying soil: a new approach to studying mycorrhizal effects on the roots of mature trees. Plant Soil 206, 1-10

Fitter A H (1985) Functional significance of root morphology and root system architecture. In: Fitter A H, Atkinson D, Read D J, Usher M B (eds) Ecological interactions in soil. Blackwell Scientific Publications, Oxford, UK pp 37-42

Fitter A H, Stickland T R (1991) Architectural analysis of plant root systems. 2. Influence of nutrient supply on architecture in contrasting plant species. New Phytol. 118:383-389

Fogel R (1985) Roots as primary producers in below-ground ecosystems. In: Fitter A H, Atkinson D, Read D J, Usher M B (eds) Ecological interactions in soil. Blackwell Scientific Publications, Oxford, UK pp 22-36

Forbes P J, Ellison C H, Hooker J E (1996) The impact of arbuscular mycorrhizal fungi and temperature on root system development. Agronomie 16:617-620

Fusconi A, Berta G, Scannerini S, Trotta A (1986) Meristematic activity in mycorrhizal and uninfected roots of *Allium porrum* L. In: Gianinazzi-Pearson V, Gianinazzi S.(eds) Physiological and genetical aspects of mycorrhizae: Proceedings of the 1st European Symposium on Mycorrhizae. INRA, Paris pp 667-671

Fusconi A, Tagliasacchi A M, Scannerini S, Trotta A, Gnavi E, de Padova S (1994) Root apical meristems of arbuscular mycorrhizae of *Allium porrum* L. Env. Exp. Bot. 43:181-193

Fusconi A, Gnavi E, Trotta A, Berta G (1999) Apical meristems of tomato roots and their modifications induced by arbuscular mycorrhizal and soilborne pathogenic fungi. New Phytol. 142:505-516

Fusconi A, Tagliasacchi A M, Berta G, Trotta A, Brazzaventre S, Ruberti F, Scannerini S (2000) Root apical meristems of *Allium porrum* L. as affected by arbuscular mycorrhizae and phosphorus. Protoplasma 214:219-226

Gavito M E, Curtis P S, Jakobsen I (2001) Neither mycorrhizal inoculation nor atmospheric CO_2 concentration has strong effects on pea root production and root loss. New Phytol. 149:283-290

Harley J L, Smith S E (1983) Mycorrhizal Symbiosis. Academic Press, London and New York pp 483

Head G C (1966) Estimating seasonal changes in the quantity of a white unsuberised root on fruit trees. J. Hort. Sci. 41:197-206

Heggo A, Angle J S (1990) Effects of vesicular-arbuscular fungi on heavy metal uptake by soybeans. Soil Biol. Biochem. 22:865-869

Hernandez-Sebastia C, Piche Y, Desjardins Y (1999) Water relations of whole strawberry plantlets *in vitro* inoculated with *Glomus intraradices* in a tripartite culture system. Plant Sci. 143:81-91

Hetrick B A D, Leslie J F, Wilson G T, Kitt D G (1988) Physical and topological assessment of effects of a vesicular arbuscular mycorrhizal fungus on root architecture of Big Bluestem. New Phytol. 110:85-96

Himmelbach A, Iten M, Grill E (1998) Signalling of abscisic acid to regulate plant growth. Phil. Trans. R. Soc. Lond. B 353:1439-1444

Hooker J E, Munro M, Atkinson D (1992a) Vesicular-arbuscular fungi induced alteration in poplar root system morphology. Plant and Soil 145:207-214

Hooker J E, Atkinson D (1992b) Application of computer-aided image analysis to studies of AM fungi effects on plant root system architecture and dynamics. Agronomie 12:821-824

Hooker J E, Black K E, Perry R L, Atkinson D (1995) Arbuscular mycorrhizal fungi induced alteration to root longevity of poplar. Plant and Soil 172:327-329

Hooker J E, Atkinson D (1996) Arbuscular mycorrhizal fungi-induced alteration to tree-root architecture and longevity. J. Pflanz. Bodenk. 159:229-234

Hooker J E, Berta G, Lingua G, Fusconi A, Sgorbati S (1998) Quantification of AMF-induced modifications to root system architecture and longevity. In: Varma A (ed) Mycorrhizal manual Springer Verlag pp 515-531

Hooker J E, Hendrick R, Atkinson D (2000) The measurement and analysis of fine root longevity. In: Smit A L, Bengough A G, Engels C, Noordwijk M van, Pellerin S, Geijn S C vande (eds) Root methods Springer-Verlag, Berlin, Heidelberg pp 274-304

Jacqmard A, Houssa C, Bernier G (1995) Abscisic acid antagonizes the effect of cytokinin on DNA-replication origins. J. Exp. Bot. 46:663-666

Kaldorf M, Ludwig-Muller J (2000) AM fungi affect the root morphology of maize by increasing indole-3-butyric acid biosynthesis. Physiol. Plant. 109:58-67

Kaska D D, Myllyla R, Cooper J B (1999) Auxin transport inhibitors act through ethylene to regulate dichotomous branching of lateral root meristems in pine. New Phytol. 142:49-58

Lingua G, Sgorbati S, Citterio A, Fusconi A, Trotta A, Gnavi E, Berta G (1999) Arbuscular mycorrhizal colonization delays nucleus senescence in leek root cortical cells. New Phytol. 141:161-169

Noordwick M, Brouwer G, Koning H, Meijboom F W, Grzebisz W (1994) Production and decay of structural root material of winter wheat and sugar-beet in conventional and integrated cropping systems; Agricul. Ecosyst. Environ.. 51:99-113.

Norman J R, Atkinson D, Hooker J E (1996) Arbuscular mycorrhizal fungal-induced alteration to root architecture in strawberry and induced resistance to the root pathogen *Phytophthora fragariae*. Plant and Soil 185:191-198

Norman J R, Hooker J E (2000) Sporulation of *Phytophthora fragariae* shows greater stimulation by exudates of non-mycorrhizal than by mycorrhizal strawberry roots Mycol. Res. 104:1069-1073

Peng S, Eissenstat D M, Graham J H, Williams K, Hodge N C (1993) Growth depression of mycorrhizal citrus at high phosphorus supply: analysis of carbon costs. Plant Physiol. 101:1063-1071

Ravnskov S, Jakobsen I (1995) Functional compatibility in arbuscular mycorrhizas measured as hyphal P transport to the plant. New Phytol. 129:611-618

Schellenbaum L, Berta G, Ravolanirina F, Tisserant B, Gianinazzi S, Fitter A H (1991) Influence of endo-mycorrhizal infection on root morphology in a micro-propagated woody plant species (*Vitis vinifera* L.) Ann. Bot. 68:135-141

Smit A L, Zuin A (1996) Root growth dynamics of Brussels sprouts (*Brassica oleracea* var. *gemmifera*) and leeks (*Allium porrum* L.) as reflected by root length, root colour and UV fluorescence. Plant and Soil 185:271-280

Smit A L, George E, Groenwold J (2000) Root observations and measurements at (transparent) interfaces with soil. In: Smit A L, Bengough A G, Engels C, Noordwijk M van, Pellerin S, Geijn S C van de (eds) Root methods. Springer-Verlag, Berlin, Heidelberg pp 235-271

Smith S E, Gianinazzi-Pearson V (1988) Physiological interactions between symbionts in vesicular-arbuscular mycorrhizal plants. Ann. Rev. Plant Physiol. 39:321-324

Smith S E, Read DJ (1997) Mycorrhizal Symbiosis. Academic Press pp. 605

Stribley D, Tinker P B, Snellgrove R C (1980) Effect of vesicular-arbuscular mycorrhizal fungi on the relations of plant growth, internal phosphorus concentration and soil phosphate analyses. J. Soil Sci. 31:655-672

Tisserant B, Schellenbaum L, Gianinazzi-Pearson V, Gianinazzi S, Berta G (1992) Influence of infection by an endomycorrhizal fungus on root development and root architecture in *Platanus acerifolia*. Allionia 30:171-181

Torelli A, Trotta A, Acerbi L, Arcidiacono G, Berta G, Branca C (2000) IAA and ZR content in leek (*Allium porrum* L.), as influenced by P nutrition and arbuscular mycorrhizae, in relation to plant development. Plant and Soil 226:29-35

Trotta A, Carminati C, Schellenbaum L, Scannerini S, Fusconi A, Berta G (1991) Correlation between root morphogenesis, VA mycorrhizal infection and phosphorus nutrition. In: Mc Michael B. L, Persson H (eds) Plant roots and their rnvironment. Elsevier, The Netherlands pp 333-339

Trotta A, Varese G C, Gnavi E, Fusconi A, Sampo S, Berta G (1996) Interactions between the soilborne root pathogen *Phytophthora nicotianae* var. *parasitica* and the arbuscular mycorrhizal fungus *Glomus mosseae* in tomato plants. Plant and Soil 185:199-209

Van der Heijden M G A, Klironomos J N, Ursic M, Moutoglis P, Streitwolf-Engel R, Boller T, Wiemken A, Sanders I R (1998) Mycorrhizal fungal diversity determines plant biodiversity, ecosystem variability and productivity. Nature 396:69-72

Vigo C, Norman J R, Hooker J E (2000) Biocontrol of the pathogen *Phytophthora parasitica* by arbuscular mycorrhizal fungi is a consequence of effects on infection loci. Plant Pathol. 49:509-514

Vierheilig H, Alt M, Mohr U, Boller T, Weimken A (1994) Ethylene biosynthesis and α-1, 3-glucanase in the roots of host and non-host plants of vesicular-arbuscular mycorrhizal fungi after inoculation with *Glomus mosseae*. J. Plant Physiol. 143:337-343

Wang H, Qi Q, Schorr P, Cutler A J, Crosby W L, Fowke L C (1998) ICK1, a cyclin-dependent protein kinase inhibitor from *Arabidopsis thaliana* interacts with both Cdc2a and CycD3,and its expression is induced by abscisic acid. Plant J. 15:501-510

Watson C A, Ross J M, Bagnaresi U, Minotta G F, Roffi F, Atkinson D, Black K E, Hooker JE (2000) Environment-induced modifications to root longevity in *Lolium perenne* and *Trifolium repens* Ann. Bot. 85:397-401

Wells E, Eissenstat D M (2001) Marked differences in survivorship among apple roots of different diameters. Ecology 82:882-892

Mycorrhizal Technology in Agriculture
ed. by S. Gianinazzi, H. Schüepp, J.M. Barea and K. Haselwandter

Plant genes involved in arbuscular mycorrhiza formation and functioning

A. Gollotte[1], L. Brechenmacher[2], S. Weidmann[2], P. Franken[3] and V. Gianinazzi-Pearson[2]

[1] *Crop Science Department, Scottish Agricultural College, West Mains Road, Edinburgh EH9 3JG, UK*
[2] *UMR INRA/Université de Bourgogne BBCE-IPM, CMSE-INRA, BP86510, 21065 Dijon Cedex, France*
[3] *Max-Planck Institut für terrestrische Mikrobiologie, Karl von Frisch Strasse, 35043 Marburg, Germany*

Keywords: plant mutants, signal transduction, recognition processes, functional genomics, expressed sequence tags (ESTs)

Introduction

Knowledge about that part of the plant genome involved in the establishment and functioning of the arbuscular mycorrhizal (AM) symbiosis is important for the basic understanding of this symbiosis. It is also essential for a 'genes to the field' approach based on the identification and exploitation of genes that could be central to developing sustainable plant production systems in the future.

Arbuscular mycorrhizas represent an ancient symbiosis as they were already formed by the first land plants at least 400 million years ago during the Ordovician- Devonian period (Simon et al. 1993; Remy et al. 1994; Redecker et al. 2000). The legume-*Rhizobium* symbiosis appeared much later, about 65 million years ago in the late Cretaceous era (Sprent 1994). This has led to the hypothesis that the cellular and molecular events occurring during the nodulation process may have evolved from those already established in the AM symbiosis (LaRue and Weeden 1994; Gianinazzi-Pearson 1997; Albrecht et al. 1999). Moreover, the similar pattern of plant cell colonisation by arbuscular mycorrhizal fungi (AMF) and biotrophic pathogenic fungi suggests that these compatible associations may have appeared through the evolution of a common ancestral plant-fungal interaction (Parniske 2000), and the possibility that some plant genes modulate the two types of interaction has been evoked (Ruiz-Lozano et al. 1999a). The AM symbiosis could therefore serve as a fundamental research model for better understanding plant-microbe interactions and their evolution in general.

In addition, the more rational use of AMF in sustainable systems of agriculture will require the selection of plants which can more efficiently exploit the AM symbiosis, together with the development of management practices which are compatible with the maximum expression of symbiosis efficiency. This will entail, in particular, manipulation of the symbiosis in the field in order to optimise root colonisation by efficient AMF and nutrient exchanges between the partners. At present, our knowledge of the cellular and molecular processes controlling the establishment and functioning of the AM symbiosis remains insufficient for this purpose. The identification of key plant genes involved in these mechanisms is therefore essential. Here, we review progress which has so far been achieved towards identifying such genes, through the use of mycorrhiza-defective plant mutants and analysis of gene expression in the AM symbiosis.

Polygenic variation in plant responsiveness to AMF

As in many biological processes, plant responsiveness to AMF must depend on many genes. This was first illustrated in wheat cultivars differing in their responsiveness to mycorrhization in terms of plant growth enhancement (Hetrick et al. 1995). Wheat intervarietal substitution lines, in which individual chromosomes of one less responsive cultivar were substituted with chromosomes from a more responsive cultivar, varied greatly in their growth response to AMF depending on the chromosome. Six chromosomes out of 21 had an effect on plant responsiveness to AMF, indicating that a polygenic component of AMF dependency exists in plants which could be exploited in plant breeding programmes.

Mycorrhiza-defective plant mutants

Evidence for the existence of individual key plant genes essential for symbiosis establishment and unctioning has come from the identification of mycorrhiza-defective plant mutants in different plant species. Such mutants have been obtained by chemical mutagenesis, gamma-irradiation, fast neutron bombardment or T-DNA/transposon transformation (Duc et al. 1989; Balaji et al. 1994; Sagan et al. 1995; Barker et al. 1998; Wegel et al. 1998) and their characteristics are summarised in Table 1. They have mainly been described in legumes amongst mutants altered in their interaction with *Rhizobium,* and 12 symbiosis-related genes (*sym*, *dmi*) have been identified as having pleiotropic effects on both nodulation and mycorrhization (Tab. 1).

Genetic analyses of several of these plant mutants have enabled the identification of complementation groups and the mapping of the corresponding genes on chromosomes (Kneen et al. 1994; Schneider et al. 1999; Huguet et al. unpublished results). The mutation is generally monogenic and recessive (Duc et al. 1989; Sagan et al. 1995, 1998; Barker et al. 1998; Wegel et al. 1998). Common genes are therefore involved in establishment of the AM symbiosis and the nodulation process, and studies carried out on *sym30* pea

Table 1. Characteristics of mycorrhiza-defective mutants identified in different plant species

Plant species	Mutant mycorrhizal phenotype	Other phenotype	Locus	References
Pisum sativum	app$^+$, inf$^-$ (Myc^{-1})	nod$^-$	*sym8, sym9, sym19, sym30*	Duc et al. 1989; Gianinazzi-Pearson et al. 1991a; Kolicheva et al. 1993; Balaji et al. 1994; Sagan et al. unpublished results
	app$^+$, inf$^+$, arb$^-$ (Myc^{-2})	nod$^{+/-}$, fix$^-$		Gianinazzi-Pearson et al. 1991a
Vicia faba	app$^+$, inf$^-$ (Myc^{-1})	nod$^-$		Duc et al. 1989
Medicago truncatula	app$^+$, inf$^-$ (Myc^{-1})	nod$^-$	*dmi1, dmi2, dmi3*	Sagan et al. 1995, 1998; Catoira et al. 2000
Phaseolus vulgaris	app$^+$, ep$^+$, inf$^-$(Myc$^-$)	nod$^{+/-}$, fix$^-$		Shirtliffe and Vessey 1996
Lotus japonicus	app$^+$, inf$^-$ (*mcbep*)	nod$^-$	*Ljsym72*	Senoo et al. 2000
	app$^+$, ep$^+$, hyp$^+$, inf$^-$ (*coi*$^-$)	nod$^-$	*Ljsym2, Ljsym3, Ljsym4*	Wegel et al. 1998
	app$^+$, inf$^+$, arb$^-$ (*mcbex*)	nod$^-$	*Ljsym71*	Senoo et al. 2000
Lycopersicon esculentum	app$^+$, inf$^{+/-}$, arb$^{+/-}$ (*rmc*)			Barker et al. 1998

coi$^-$: cortex invasion; *mcbep*: mycorrhizal colonisation blocked at epidermis; *mcbex*: mycorrhizal colonisation blocked at exodermis; *rmc*: reduced mycorrhizal colonisation mutants (P2, P53) (Sagan et al., unpublished results) have shown that this mutant is not affected in its interaction with pathogens including fungi, bacteria or nematodes (Gianinazzi-Pearson et al. 1994). This clearly indicates that the *sym30* gene must play a specific role in root symbioses. More recently, some mutants have also been identified in the non-legume plant, tomato (Barker et al. 1998). These mutants will be important for the characterisation of genes which are involved in the AM symbiosis independently of the legume-*Rhizobium* symbiosis. These different mycorrhiza-defective plant mutants provide not only evidence for the existence of key plant genes involved in the AM symbiosis but also important tools to determine if an effect of the symbiosis or expression of a gene requires a particular step in root colonisation. Phenotypic characterisation of the interactions between mycorrhiza-defective mutants and AMF can give some clues as to the possible function of the corresponding genes within the symbiosis.

Induction of plant defence reactions in Myc^{-1} mutants

The first described mycorrhiza-defective mutants were Myc^{-1} mutants which are characterised by the blocking of AMF development at the stage of appressorium formation on the root surface (Duc et al. 1989). Such mutants, which have since been identified in *P. sativum*, *V. faba*, *M. truncatula* and *P. vulgaris* (Tab. 1), can contribute to the understanding of very early recognition processes between plant roots and AMF.

Myc^{-1} mutants altered at the *sym30* locus in pea have been extensively studied. These mutants like Myc^{+} plants are able to induce differential hyphal morphogenesis, which occurs in an AMF before the symbionts come into contact with each other and leads to appressorium formation (Giovannetti 1997). It has been shown that root exudates from Myc^{+} and Myc^{-1} plants do not differ in their effect on protein synthesis in AMF during spore germination (Samra et al. 1996). Similarly, these mutants are able to induce expression of *nod* genes in *Rhizobium leguminosarum*, resulting in the synthesis of Nod factors. However, *Rhizobium* does not induce root hair curling in these mutants, contrary to this typical host response to the bacterium in symbiotic interactions (Sagan et al. 1994).

Molecular cytology studies have revealed that wall appositions are elicited in the epidermal cells of Myc^{-1} pea mutants in contact with AMF appressoria (Gianinazzi-Pearson et al. 1991a; Gollotte et al. 1993). These wall deposits contain molecules characteristic of plant defence reactions such as phenolic compounds, callose and PR1 proteins (Gollotte et al. 1993, 1996). In addition, expression of some defence-related genes and accumulation of salicylic acid is enhanced in roots of these mutants inoculated with either AMF or *R. leguminosarum* (Blilou et al. 1999; Ruiz-Lozano et al. 1999b). This suggests that the *sym30* gene, in Myc^{+} plants, is somehow involved in the control of plant defence reactions during symbiosis establishment. However, it cannot be concluded that this type of reaction occurs in all early mutants. In the case of a Myc^{-1} *dmi2* mutant of *M. truncatula*, wall appositions containing phenolic compounds are formed in plant cells in contact with appressoria, but these are less extensive than in pea and they do not contain callose, suggesting that the *dmi2* gene may have other functions essential to symbiosis development (Calantzis et al. 2001).

Myc^{-2} mutant

In contrast to Myc^{-1} mutants, AMF penetrate into the roots of a Myc^{-2} pea mutant but arbuscules are not fully developed compared to those formed in the Myc^{+} plants (Gianinazzi-Pearson et al. 1991a). *R. leguminosarum* can induce root hair curling in this mutant and infection threads are formed but are aborted before they reach the base of root hairs (Sagan et al. 1994). This mutant opens the possibility of identifying genes which are involved in later stages of the AM symbiosis involving arbuscule formation and functioning of the symbiotic interface. For example, H^{+}-ATPase is active in the periarbuscular membrane in a fully established symbiosis (Gianinazzi-Pearson et al. 1991b), whereas no ATPase activity is detected in the plant membrane surrounding aborted arbuscules in the Myc^{-2} pea mutant (Gianinazzi-Pearson et al. 1995). This indicates that the mutation has altered cell functions at the intracellular plant-fungus interface.

Recently, Lapopin et al. (1999) have shown that a gene, *Psam4*, coding for a repetitive proline-rich protein is down-regulated in Myc^{+} plants in response to AMF whilst it is upregulated by the pathogenic fungus *Aphanomyces euteiches*. Interestingly, the basal level of expression of *Psam4* is higher in the Myc^{-2} mutant than in a Myc^{+} plant, and wall material deposited around hyphae in root cortical cells persists in a Myc^{-2} mutant compared to what happens in Myc^{+} plants (Lherminier 1993). Since proline-rich proteins are considered to be components of plant cell walls (Showalter 1993), it is possible that *Psam4* may play a role in the control of defence reactions in Myc^{+} plants which does not occur in the Myc^{-2} mutant. The actual localisation of the PSAM4 proline-rich protein however remains to be determined.

Other mutants

A variety of other mycorrhiza-defective phenotypes has been obtained in mutants of the plant species *L. japonicus, P. vulgaris* and tomato. The mutations affect different steps of root colonisation, and the corresponding mutants will be useful in the identification of additional genes involved in the AM symbiosis.

One *L. japonicus* mutant, *mcbep* (<u>m</u>ycorrhizal <u>c</u>olonisation <u>b</u>locked at <u>ep</u>idermis), is characterised by the formation of abnormal appressoria and of runner hyphae on the root surface which do not penetrate the root tissues (Senoo et al. 2000). In the Myc⁻ mutant of *P. vulgaris*, the AMF forms appressoria on roots but its development is restricted to the outer root layers (Shirtliffe and Vessey 1996). In the case of three cor⁻ mutants in *L. japonicus* (*Ljsym2*, *Ljsym3*, *Ljsym4-1*), root colonisation is also blocked at the level of the rhizodermis and occasional normal development of arbuscules occurs in the root cortex, whilst the mutant *Lsym4-2* was always characterised by an arrest of the AMF at the level of root epidermis (Wegel et al. 1998; Bonfante et al. 2000). Detailed analysis of the *Ljsym4-2* mutant has shown that hyphal penetration of epidermal cells is associated with the death of the colonised plant cell (Bonfante et al. 2000). In another *L. japonicus* mutant, *mcbex* (mycorrhizal colonisation blocked at exodermis), the phenotype is unstable and AMF penetration of the root occurs with occasional formation of abnormal arbuscules (Senoo et al. 2000). This is similar to that in a mycorrhiza-deficient *M. sativa* genotype reported by Bradbury et al. (1993). Finally, in the case of tomato *rmc* (reduced mycorrhizal colonisation) mutants, AMF can form appressoria but generally fail to penetrate the root system although occasional normal development of intraradicular hyphae can occur (Barker et al. 1998).

Another group of interesting mutants are the supernodulation mutants of *P. sativum* and *M. truncatula,* corresponding to the genes *sym28*, *sym29* and *Mtsym12,* respectively. These mutants are characterised by an increase in root colonisation by AMF as well as increased nodulation and nitrate tolerance (Morandi et al. 2000). They may be important for understanding the mechanisms by which plants limit AMF development within root systems.

Gene cloning

The existence of different mycorrhiza-defective plant mutants indicates that numerous genes in different root tissues control root colonisation by AMF. Cloning of the symbiosis genes which have been altered in the mutants will be essential towards understanding their role in symbiosis establishment and functioning. This could be achieved through a map-based strategy, like that successfully adopted for other genes such as plant disease resistance genes (Hammond-Kosack and Jones 1997). However, this method is difficult to use on plants which have a large genome such as *P. sativum* and *V. faba*. In contrast, *M. truncatula*, *L. japonicus* and *L. esculentum* have smaller genomes, which are more amenable for genetic studies (Barker et al. 1990; Jiang and Gresshoff 1997; Barker et al. 1998). Once the mutated genes have been characterised, microsynteny genome analysis or PCR-based techniques could be used to clone the corresponding genes in agricultural crops such as pea or alfalfa. Recent transposon tagging and promoter trapping techniques could also represent approaches for cloning symbiosis genes (Schauser et al. 1999; Webb et al. 2000).

Whilst obtaining plant mutants enables identification of genes that regulate symbiosis establishment and functioning, investigations of the symbiotic transcriptome will provide information about gene expression involved in the symbiosis.

Mycorrhiza-related transcription profiling

Targeted genes

Probes corresponding to plant genes that regulate metabolic processes or that are differentially modulated during other plant-microbe interactions have been employed to identify molecular events in the AM symbiosis. Such probes were already available or were obtained by screening of cDNA libraries or by PCR using degenerated primers. Gene expression has been evaluated by Northern analyses, RT-PCR, *in situ* hybridisation and the use of transgenic plants carrying promoter/reporter gene constructs.

Arbuscule differentiation in root cells causes important modifications in the organisation of host cell contents and transcriptional activation of a gene encoding α-tubulin has been demonstrated during cytoskeleton rearrangement in tobacco cells containing arbuscules (Bonfante et al. 1996). Osmotic fluctuations in the host cytoplasm caused by arbuscule formation may be buffered by activation of genes encoding aquaporins which facilitate water transport across the plant tonoplast. Such a gene (*Mtaqp1*) is induced by mycorrhiza development in roots of *M. truncatula* (Krajinski et al. 2000).

In parsley, a similar gene *Pcrb7*, is induced in roots by AMF and also in cell suspension cultures by an elicitor from the pathogenic fungus *Phytophthora sojae* (Roussel et al. 1997). The products of both *Mtaqp1* and *Pcrb7* genes belong to the tonoplast intrinsic proteins, but PcRB7 groups into the delta-family while MtAQP1 is located in the gamma-family (Fig. 1). Whether this is mirrored by different functions is not clear up to now. Modifications in the host cell also lead to the formation of a symbiotic interface across which nutrient exchange can occur between fungus and plant. Arbuscule-containing cells are considered to be the principal site of phosphate transfer from fungal to plant cells and this process necessitates the existence of transport systems within plant cell membranes of the symbiotic tissues. Two out of eight proton pump ATPase plant genes studied in tobacco are specifically activated in these cells (Gianinazzi-Pearson et al. 2000). The corresponding enzyme, which is active in the plant membrane associated with the symbiotic interface (Gianinazzi-Pearson et al. 1991b), could provide a necessary driving force for transmembrane transport of phosphate into the plant cell after release from the AMF. Two genes *MtPT1* and *MtPT2*, which code for phosphate transporters, have been isolated from *M. truncatula*. However, transcript levels of *MtPT1* and *MtPT2* increase with phosphate starvation and decrease during mycorrhiza establishment (Liu et al. 1998), which indicates that they are probably not involved in the pathway of phosphate uptake by *M. truncatula* via the symbiotic fungus. In addition, cDNA presenting sequence similarity to a plant mem-

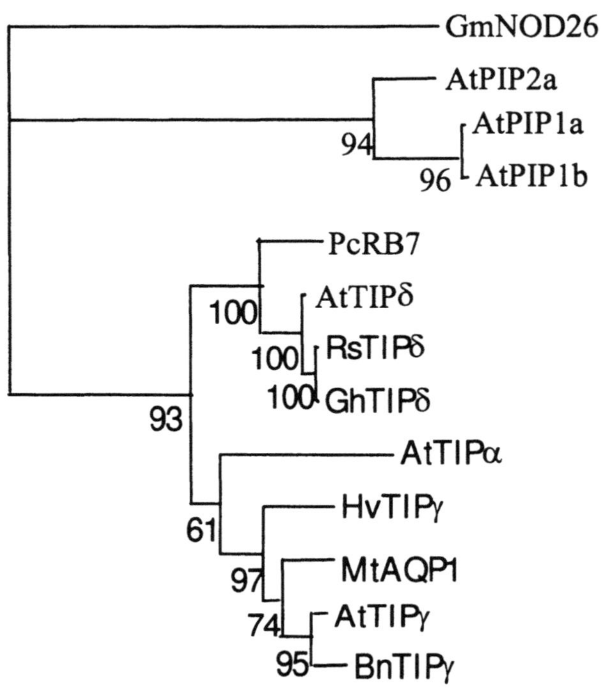

Figure 1. Alignment of the deduced amino acid sequences of the *Medicago truncatula* gene *Mtaqp*-1 and the *Petroselinum crispum* gene *Pcrb7* by the program ClustalW to different tonoplast intrinsic proteins from *Arabidopsis thaliana* (AtTIPα, γ and δ), *Raphanus sativa* (RsTIPδ), *Gossypium hirsutum* (GhTIPδ), *Hordeum vulgare* (HvTIPγ) and *Brassica napus* (BnTIPγ), as well as to three plasma membrane intrinsic proteins of *A. thaliana* (AtPIP2a, 1a and 1b) and NOD26 from *Glycine max* (GmNOD26). The dendrogram of the result was constructed using the programs PUZZLE and TREE-VIEW. Quartet puzzling support values of 1000 replicates are indicated.

brane hexose transporter, *Mtst1,* has been isolated from a *M. truncatula/G. versiforme* cDNA library (Harrison 1996). *Mtst1* transcripts accumulate in arbuscule-containing tissues suggesting that the corresponding gene may be active in the process of facilitating hexose transport towards colonised plant cells for uptake by fungal structures.

Most of the genes from other plant-microbe interactions which have been investigated in arbuscular mycorrhizas are those associated with defence responses to pathogens. Transcripts coding hydroxyproline rich glycoprotein (HRGP) and anionic peroxidase (POX), which are involved in cell wall reinforcement, increase in roots during colonisation by an AMF (Franken and Gnädinger 1994). Varying observations have been reported for genes coding enzymes of the phenyl propanoid pathway, depending on the host plant-AMF combinations. For example, genes encoding the enzymes phenylalanine ammonia-lyase (*PAL*), chalcone synthase (*CHS*), chalcone isomerase (*CHI*) and isoflavone reductase (*IFR*) either show little modifications in expression or are activated in later stages of mycorrhiza development. Likewise, the expression of pathogenesis-related (PR) genes varies in different arbuscular mycorrhizas from no significant modification to transient activation (Gianinazzi-Pearson et al. 1996; Dumas-Gaudot et al. 2000). Recent detailed analyses of the expression profile of eight different chitinase genes in *M. truncata* showed mycorrhiza-specific activation of two class III chitinase genes. It was suggested that these could be involved in the regulation of plant defence reactions in the later stages of AM development (Salzer et al. 2000). The transcripts of many of these different genes, as well as those of a glutathione-S-transferase gene, have been localised almost exclusively within arbuscule containing cells of mycorrhizal roots (Harrison and Dixon 1994; Blee and Anderson 1996; Gianinazzi-Pearson et al. 1996; Strittmatter et al. 1996; Franken et al. 2000). Furthermore, transcripts corresponding to the pea defence-related genes *PI 206, PI 49, PI 176* accumulate in response to both mycorrhizal and pathogenic fungi but their function remains unknown (Ruiz-Lozano et al. 1999b).

The fact that similarities exist in the genetic control of nodulation and arbuscular mycorrhizas has prompted research for genes which may be commonly expressed during root/microbe interactions in the two symbioses. Transcripts corresponding to lectinlike glycoprotein (*PsNlec1*) are accumulated during nodulation and in response to arbuscule development in pea root cells (Balestrini et al. 1999). The late nodulin gene *LbVf29* and the early nodulin genes *PsENOD5*, *PsENOD12*, *MtENOD12 MtENOD11* and *ENOD40,* which are induced by *Rhizobium*, have also been shown to be activated in arbuscular mycorrhizas formed by roots of $Nod^{+}Myc^{+}$ plants, and *MtENOD11* and *MtENOD12* expression has been localised in arbuscule-containing cells (Gianinazzi-Pearson 1996; van Rhijn et al. 1997; Albrecht et al. 1998; Schneider et al. 1999; Journet et al. 2001). In contrast, early nodulin genes are not induced by either *Rhizobium* or AMF in the *sym8* and *sym19* pea mutants or a *dmi2 M. truncatula* mutant, reinforcing the hypothesis that some signal transduction pathways are common to the two symbioses (Albrecht et al. 1998; Schneider et al. 1999; Vernoud et al. 2000). A pharmacological approach aimed at characterising signal transduction pathways leading to their expression during nodulation events has shown the involvement of G proteins (Pingret et al. 1998; Vernoud et al. 2000). However, application of mastoparan, an agonist of G proteins, to roots restored *MtENOD11* gene expression in the *dmi2 M. truncatula* mutant (Vernoud et al. 2000). This indicates that the *dmi2* mutation has occurred in a gene normally acting upstream of G proteins. These results are a first step towards a better understanding of the function of symbiosis genes during early recognition processes between plant roots and their microsymbionts.

Expression profiling and EST sequencing

Although targeted approaches are important and have enabled the identification of several plant genes involved in the AM symbiosis, their drawback is that they do not lead to identification of genes which are specific to the AM symbiosis, or which may not be specific but still undiscovered in other biological systems. Analyses of the functional genomics of AM interactions have therefore been undertaken using non-targeted differential screening techniques and, more recently, large-scale sequencing of cDNA expressed sequence tags (ESTs).

Five plant genes (psam 1–5) which are up- or down-regulated in AM interactions have been isolated from *Glomus mosseae*-inoculated roots of *P. sativum* using the differential RNA display technique (DDRT-PCR) (Martin-Laurent et al. 1997). The *psam1* gene is induced during mycorrhiza development, and the corresponding protein of unknown function is localised in the cytoplasm of arbuscule-containing cells (Martin-Laurent et al. 1998). The gene corresponding to *psam3*, which shows similarity to a human *Alu* sequence, is also induced in roots by AMF (Martin-Laurent 1998). In contrast, expression of *psam2* is repressed in arbuscular mycorrhizas but the gene is upregulated in plant roots infected by the pathogenic fungi *Chalara elegans* and *Aphanomyces euteiches*; the putative protein has similarity to animal secretory membrane proteins (Krajinski et al. 1998). Likewise, transcript accumulation of *psam4* which codes for a proline-rich protein is reduced in the AM symbiosis and during nodulation, but increased in response to the pathogenic fungus *A. euteiches* (Lapopin et al. 1999). The *psam5* gene, which shows similarity to a Clp serine protease gene, differs in that it is transiently induced very early during mycorrhiza and nodule interactions and its expression decreases in the later phases (Roussel 1999; Roussel et al. 2001).

Differential screening of cDNA libraries is another non-targeted approach which has been adopted to identify genes regulated during the AM symbiosis. Tahiri-Alaoui and Antoniw (1996) isolated cDNA fragments from *Glomus mosseae*-colonised tomato roots, corresponding to five plant genes (*MI-4, MI-13, MI-3, MI-1* and *MR-1*). The first four genes, which are induced in AM tissues, correspond to genes encoding a ribosomal protein, a cullin protein family, mannitol dehydrogenase and phosphoenolpyruvate carboxylase, respectively. *MR-1* represents a repressed gene in arbuscular mycorrhiza and has no similarity to known sequences. Isolated cDNA fragments of two other transcripts, *BMR6* and *BMR78*, accumulate to a higher level in mycorrhizal barley roots (Murphy et al. 1997). *BMR6* shows no similarity to known sequences, whereas *BMR78* probably codes for a proton pump ATPase. Differential screening of a *M. truncatula*/*G. versiforme* cDNA library has yielded a fragment (*Mt4*) similar to a phosphate starvation-inducible gene, which is repressed in AM tissues and by phosphate fertilisation (Burleigh and Harrison 1997). Four other cDNA clones have been isolated by subtractive hybridisation of the same cDNA library with RNA or cRNA from non-mycorrhizal roots. The corresponding plant genes, which are induced during symbiosis development, probably encode a xyloglucan endotransglycosylase-related protein, a putative arabinogalactan (AGP) in arbuscule-containing cells, a putative homologue of the mammalian p110 subunit of initiation factor 3 (eIF3) and an endoplasmic reticulum protein translocation complex (van Buuren et al. 1999, 2000).

In conclusion, those gene fragments isolated using differential expression techniques can be predicted to be genes involved in important aspects of plant cell function. These include protein synthesis/degradation, cell cycle, membrane-associated protein translocation/secretion, nutrient transport, wall synthesis/structure and primary metabolism. However in all, only approximately fifty plant genes regulated during AM interactions have been identified using the above-mentioned targeted approaches or non-targeted differential cDNA screening, and therefore other techniques are being employed in order to obtain a more global view of symbiosis-related genes.

Suppression subtractive hybridisation (SSH) and cloning (Diatchenko et al. 1996), a technique which combines advantages of targeted and non-targeted approaches, is currently being used to identify genes which are up-regulated in a fully established AM symbiosis in *M. truncatula*. The subtraction has been performed between cDNAs isolated from *Glomus mosseae*-colonised roots and non-mycorrhizal roots in order to obtain ESTs corresponding to genes with enhanced expression in the former (Brechenmacher et al. 2000). Fifty percent of the cloned ESTs represent genes which are significantly up-regulated in the AM symbiosis, and about one third of these are only expressed in mycorrhizal tissues. The large majority of these ESTs are of plant origin. Out of 46 more closely analysed ESTs, seven correspond to genes also activated during nodulation (Weidmann 2000), underlying again the existence of common events between the root symbioses with *Rhizobium* and AMF. Most of the sequenced ESTs have given no similarities with known genes at the nucleic acid or protein level, and remaining ones show homologies to genes encoding 18S ribosomal RNA, enzymes involved in primary metabolism, secondary and hormone metabolites, cysteine protease, nodulin 26 and PR1 protein.

Large-scale EST sequencing, which is a powerful tool for functional genomics, has recently been applied to the molecular study of root symbioses in the model legume *M. truncatula* by research groups in France and in the USA (Harrison et al. 2000; van Tuinen et al. 2000). Three cDNA libraries corresponding to non-mycorrhizal, mycorrhizal and nodulated roots respectively have been constructed by French researchers and 15,000 ESTs randomly sequenced. After classification of the sequences, more than 4000 ESTs appear as singletons in the libraries whilst there are about 1800 clusters of sequences present in more than one copy. In total, the EST libraries represent about 6000 distinct *M. truncatula* genes. Sequences are being annotated after systematic nucleic acid and protein similarity searches in order to provide the scientific community with a biologically informative database accessible on a web site (http://sequence.toulouse.inra.fr/Mtruncatula.html). The presence of ESTs with a significant differential representation between the non-mycorrhizal, mycorrhizal and nodulation cDNA libraries indicates a number of candidate genes which are up-regulated in the nodule and/or AM symbiotic tissues.

Conclusions

The cloning and identification of plant genes which control crucial steps in AM symbiosis establishment and functioning have become feasible objectives for the near future with the isolation of mycorrhiza-defective mutants in several model plant species. These genes, which could provide useful markers in plant breeding for highly symbiotic genotypes, must control expression of a multitude of other genes which are responsible for the development and functioning of the symbiosis. Few such genes have so far been fully characterised. However, the development of functional genomics through large scale EST sequencing and expression profiling is beginning to provide a more global view of plant genes regulated in arbuscular mycorrhizas. The isolation of novel genes and the identification of their function will be a major challenge to research in the next few years. The use of mycorrhiza-defective plant mutants and the application of high density array technology should provide rapid access to the subsets of genes which are essential to cell programmes governing signal transduction pathways in the different steps of plant-fungal interactions in arbuscular mycorrhizas.

Acknowledgements
Work cited in this chapter was partly supported by the Conseil Régional de Bourgogne (grant INRA/CRB), by the Deutsche Forschungsgemeinschaft (SFB 395), by the PROCOPE Germany – France research program (Contract n° 93134), by Genoscope (Evry, F) and an INRA/CNRS Genome project.

References

Albrecht C, Geurts R, Lapeyrie F, Bisseling T (1998) Endomycorrhizae and rhizobial Nod factors both require SYM8 to induce the expression of the early nodulin genes *PsENOD5* and *PsENOD12A*. Plant J 15:605-614

Albrecht C, Geurts R, Bisseling T (1999) Legume nodulation and mycorrhizae formation; two extremes in host specificity meet. EMBO J 18:281-288

Balaji B, Ba A M, LaRue T, Tepfer D, Piche Y (1994) *Pisum sativum* mutants insensitive to nodulation are also insensitive to invasion in vitro by the mycorrhizal fungus *Gigaspora margarita*. Plant Sci 102:195-203

Balestrini R, Perotto S, Gasverde E, Dahiya P, Guldmann L L, Brewin N J, Bonfante P (1999) Transcription of a gene encoding a lectinlike glycoprotein is induced in roots cells harboring arbuscular mycorrhizal fungi in *Pisum sativum*. Mol Plant-Microbe Interact 12:785-791

Barker D G, Bianchi S, Blondon S, Dattee Y, Duc G, Flament P, Gallusci P, Genier P, Guy P, Muel X, Tourneur J, Denarie J, Huguet T (1990) *Medicago truncatula*, a model plant for studying the molecular genetics of the *Rhizobium*-legume symbiosis. Plant Mol Biol Rep 8:40-49

Barker S J, Stummer B, Gao L, Dispain I, Oconnor P J, Smith S E (1998) A mutant in *Lycopersicon esculentum* Mill. with highly reduced VA mycorrhizal colonization: isolation and preliminary characterisation. Plant J 15:791-797

Blee K A, Anderson A J (1996) Defense-related transcript accumulation in *Phaseolus vulgaris* L. colonized by the arbuscular mycorrhizal fungus *Glomus intraradices* Schenck & Smith. Plant Physiol 110:675-688

Blilou I, Ocampo J A, Garcia-Garrido M (1999) Resistance of pea roots to endomycorrhizal fungus or *Rhizobium* correlates with enhanced levels of endogenous salicylic acid. J Exp Bot 50:1663-1668

Bonfante P, Bergero R, Uribe X, Romera C, Rigau J, Puigdomenech P (1996) Transcriptional activation of a maize α-tubulin gene in mycorrhizal maize and transgenic tobacco plants. Plant J 9:737-743

Bonfante P, Genre A, Faccio A, Martini I, Stougaard J, Schauser L, Webb K J, Parniske M (2000) The *Lotus japonicus Ljsym4* gene is required for the passage of root epidermis cells by rhizobia and arbuscular mycorrhizal fungi. Mol Plant-Microbe Interact 13:1109-1120

Bradbury S M, Peterson R L, Bowley S R (1993) Colonization or three alfalfa (*Medicago sativa* L.) nodulation genotypes by indigenous vesicular-arbuscular mycorrhizal fungi from soil. Symbiosis 15:207-215

Brechenmacher L, Weidmann S, van Tuinen D, Chatagnier O, Franken P, Gianinazzi-Pearson V (2000) Analysis of *Medicago truncatula* mycorrhiza expressed sequence tags by suppression subtractive hybridization. In: Weber H C, Imhof S, Zeuske D (eds) Proceedings of the Third International Congress on Symbiosis, Marburg, Germany, August 2000, p 43

Burleigh S H, Harrison M J (1997) A novel gene whose expression in *Medicago truncatula* roots is suppressed in response to colonization by vesicular-arbuscular mycorrhizal (VAM) fungi and to phosphate nutrition. Plant Mol Biol 34:199-208

Calantzis C, Morandi D, Arnould C, Gianinazzi-Pearson V (2001) Cellular interactions between *G. mosseae* and a Myc⁻ mutant in *Medicago truncatula*. Symbiosis 30:97-108

Catoira R, Galera C, de Billy F, Penmetsa R V, Journet E P, Maillet F, Rosenberg C, Cook D, Gough C, Dénarié J (2000) Four genes of *Medicago truncatula* controlling components of a Nod factor transduction pathway. Plant Cell 12:1647-1665

Diatchenko L, Chris L Y F, Campbell A P, Chenchik A, Moqadam F, Huang B, Lukyanov K, Gurskaya N D S E, Siebert P D (1996) Suppression subtractive hybridization: a method for generating differentially regulated or tissue-specific cDNA probes and libraries. Proc Natl Acad Sci USA 93:6025-6030

Duc G, Trouvelot A, Gianinazzi-Pearson V, Gianinazzi S (1989) First report of non-mycorrhizal plant mutants (Myc⁻) obtained in pea (*Pisum sativum* L.) and fababean (*Vicia faba* L.). Plant Sci 60:215-222

Dumas-Gaudot E, Gollotte A, Cordier C, Gianinazzi S, Gianinazzi-Pearson V (2000) Modulation of host defense systems. In: Kapulnik Y, Douds D D (eds) Arbuscular mycorrhizae: Molecular biology and physiology. Kluwer Academic Publishers, Dordrecht pp 173-200

Franken P, Gnädinger F (1994) Analysis of parsley arbuscular endomycorrhiza: Infection development and mRNA levels of defense-related genes. Mol Plant-Microbe Interact 7:612-620

Franken P, Requena N, Bütehorn B, Krajinski F, Kuhn G, Lapopin L, Mann P, Rhody D, Stommel M (2000) Molecular analysis of the arbuscular mycorrhiza symbiosis. Arch Agronom Soil Sci 45:271-286

Gianinazzi-Pearson V (1996) Plant cell responses to arbuscular mycorrhizal fungi: getting to the roots of the symbiosis. Plant Cell, 8:1871-1883

Gianinazzi-Pearson V (1997) Have common plant systems co-evolved in fungal and bacterial root symbioses? In: Legocki A, Bothe H, Puhler A (eds) Biological fixation of nitrogen for ecology and sustainable agriculture. Springer-Verlag, Berlin Heidelberg, pp 321-324

Gianinazzi-Pearson V, Gianinazzi S, Guillemin J P, Trouvelot A, Duc G (1991a) Genetic and cellular analysis of the resistance to vesicular-arbuscular (VA) mycorrhizal fungi in pea mutants. In: Hennecke H, Verma D P S (eds) Advances in molecular genetics of plant-microbe interactions. Kluwer Academic Publishers, Boston, pp 336-342

Gianinazzi-Pearson V, Smith S E, Gianinazzi S, Smith F A (1991b) Enzymatic studies on the metabolism of vesicular arbuscular mycorrhizas. V. Is H^+-ATPase a component of ATP-hydrolysing enzyme activities in plant-fungus interfaces? New Phytol 117:61-74

Gianinazzi-Pearson V, Gollotte A, Dumas-Gaudot E, Franken P, Gianinazzi S (1994) Gene expression and molecular modifications associated with plant responses to infection by arbuscular mycorrhizal fungi. In: Daniels M J, Downie J A, Osbourn A E (eds) Advances in molecular genetics of plant-microbe interactions. Kluwer Academic Publishers, Boston London, pp 179-186

Gianinazzi-Pearson V, Gollotte A, Lherminier J, Tisserant B, Franken P, Dumas-Gaudot E, Lemoine M C, van Tuinen D, Gianinazzi S (1995) Cellular and molecular approaches in the characterisation of symbiotic events in functional arbuscular mycorrhizal associations. Can J Bot 73:S526-S532

Gianinazzi-Pearson V, Dumas-Gaudot E, Gollotte A, Tahiri-Alaoui A, Gianinazzi S (1996) Cellular and molecular defence-related root responses to invasion by arbuscular mycorrhizal fungi. New Phytol 133:45-57

Gianinazzi-Pearson V, Arnould C, Oufattole M, Arango M, Gianinazzi S (2000) Differential activation of H^+-ATPase genes by an arbuscular mycorrhizal fungus in root cells of transgenic tobacco. Planta 211:609-613

Giovannetti M (1997) Host signals dictating growth direction, morphogenesis and differentiation in arbuscular mycorrhizal symbiosis. In: Schenk H, Herrmann R, Jeon K, Muller N and Schwemmler W (eds) Eukaryotism and symbiosis. intertaxonic combination versus symbiotic adaptation. Springer-Verlag, Berlin Heidelberg, pp 405-411

Gollotte A, Gianinazzi-Pearson V, Giovannetti M, Sbrana C, Avio L, Gianinazzi S (1993) Cellular localization and cytochemical probing of resistance reactions to arbuscular mycorrhizal fungi in a locus a myc⁻ mutant of *Pisum sativum* L. Planta 191:112-122

Gollotte A, Lemoine M C, Gianinazzi-Pearson V (1996) Morphofunctional integration and cellular compatibility between endomycorrhizal symbionts. In: Mukerji K G (ed) Concepts in mycorrhizal research. Kluwer Acad. Publishers, Dordrecht, pp 91-111

Hammond-Kosack K E, Jones J P (1997) Plant disease resistance genes. Ann Rev Plant Physiol Plant Mol Biol 48:575-607

Harrison M J, Dixon R A (1994) Spatial patterns of expression of flavonoid/ isoflavonoid pathway genes during interactions between roots of *Medicago truncatula* and the mycorrhizal fungus *Glomus versiforme*. Plant J 6:9-20

Harrison M J (1996) A sugar transporter from *Medicago truncatula*: altered expression pattern in roots during vesicular-arbuscular (VA) mycorrhizal associations. Plant J 9:491-503

Harrison M J, Liu J, Dewbre G R, Blaylock L A, Zhao L (2000) Toward an understanding of the development and functioning of an arbuscular mycorrhiza: molecular and genetic approaches. In: Weber H C, Imhof S, Zeuske D (eds) Proceedings of the Third International Congress on Symbiosis, Marburg, Germany, August 2000, p 84

Hetrick B A D, Wilson G W T, Gill B S, Cox T S (1995) Chromosome location of mycorrhizal responsive genes in wheat. Can J Bot 73:891-897

Jiang Q, Gresshoff P M (1997) Classical and molecular genetics of the model legume *Lotus japonicus*. Mol Plant-Microbe Interact 10:59-68

Journet E P, El-Gachtouli N, Vernoud V, de Billy F, Pichon M, Dedieu A, Arnould C, Morandi D, Barker D, Gianinazzi-Pearson V (2001) *MtENOD11*: a novel RPRP-encoding early nodulin gene is expressed in mycorrhizal arbuscular cells. Mol Plant-Microbe Interact. 14:737-748

Kneen B E, Weeden N F, LaRue T A (1994) Non-nodulating mutants of *Pisum sativum* (L.) cv. Sparkle. J Heredity 85:129-133

Kolicheva A N, Jacobi L M, Borisov A Y, Filatov A A, Tikhonovich I A, Muromtsev G S (1993) Pea gene *sym8* affects symbiosis both with *Rhizobium* and with endomycorrhizal fungi. Pisum Genetics 25:22

Krajinski F, Martin-Laurent F, Gianinazzi S, Gianinazzi-Pearson V, Franken P (1998) Cloning and analysis of *psam2*, a gene from *Pisum sativum L.* regulated in symbiotic arbuscular mycorrhiza and pathogenic root-fungus interactions. Physiol Mol Plant Pathol 52:297-307

Krajinski F, Biela A, Schubert D, Gianinazzi-Pearson V, Kaldenhoff R, Franken P (2000) Arbuscular mycorrhiza development regulates the mRNA of *Mtaqp*1 encoding a mercury-insensitive aquaporin of *Medicago truncatula*. Planta 211:85-90

Lapopin L, Gianinazzi-Pearson V, Franken P (1999) Comparative differential RNA display analysis of arbuscular mycorrhiza in *Pisum sativum* wild type and a mutant defective in late stage development. Plant Mol Biol 41:669-677

LaRue T A, Weeden N F (1994) The symbiosis genes of the host. In: Kiss G B, Endre G (eds) Proceedings of the 1st European Nitrogen Fixation Conference, Officina Press, Szeged, Hungary, pp 147-151

Lherminier J (1993) Apport de la cytologie moléculaire à l'étude des relations hôte-microorganismes non cultivables: mycoplasma-like-organisms et champignons endomycorhizogènes. PhD Thesis, Université de Bourgogne, Dijon, France

Liu H, Trieu A T, Blaylock L A, Harrison M J (1998) Cloning and characterization of two phosphate transporters from *Medicago truncatula* roots: Regulation in response to phosphate and to colonization by arbuscular mycorrhizal (AM) fungi. Mol Plant-Microbe Interact 11:14-22

Martin-Laurent F (1998). Recherche de gènes de la plante impliqués dans le processus d'endomycorhization entre le pois (*Pisum sativum L.*) et le champignon endomycorhizogène à arbuscules *Glomus mosseae*. PhD Thesis, Université de Bourgogne, Dijon, France.

Martin-Laurent F, Van Tuinen D, Dumas-Gaudot E, Gianinazzi-Pearson V, Gianinazzi S, Franken P (1997) Differential display analysis of RNA accumulation in arbuscular mycorrhiza of pea and isolation of a novel symbiosis-regulated plant gene. Mol Gen Genet 256:37-44

Martin-Laurent F, Arnould C, Chatagnier O, van Tuinen D, Franken P, Gianinazzi S, Gianinazzi-Pearson V (1998) Cellular localization of a plant protein PSAM1 in arbuscular mycorrhizas of *Pisum sativum*. Planta 207:153-157

Morandi D, Sagan M, Prado-Vivant E, Duc G (2000) Influence of genes determining supernodulation on root colonization by the mycorrhizal fungus *Glomus mosseae* in *Pisum sativum* and *Medicago truncatula* mutants. Mycorrhiza 10:37-42

Murphy P J, Langridge P, Smith S E (1997) Cloning plant genes differentially expressed during colonization of roots of *Hordeum vulgare* by the vesicular-arbuscular mycorrhizal fungus *Glomus intraradices*. New Phytol 135:291-301

Parniske M 2000 Intracellular accomodation of microbes by plants: a common developmental program for symbiosis and disease. Curr Opin Plant Biol 3:320-328

Pingret J L, Journet E P, Barker D G (1998) *Rhizobium* Nod factor signaling: evidence for a G protein-mediated transduction mechanism. Plant Cell 10:659-671

Redecker D, Kodner R, Graham L E (2000) Glomalean fungi from the Ordovician. Science 289:1920-1921

Remy W, Taylor TN, Hass H, Kerp H (1994) Four hundred-million-year-old vesicular arbuscular mycorrhizae. Proc Acad Sci USA 91:11841-11843

Roussel H (1999). Expression de gènes de la plante associés à l'endomycorhization. PhD Thesis, Université de Bourgogne, Dijon, France.

Roussel H, Bruns S, Gianinazzi-Pearson V, Hahlbrock K, Franken P (1997) Induction of a membrane intrinsic protein-encoding mRNA in arbuscular mycorrhiza and elicitor-stimulated cell suspension cultures of parsley. Plant Sci 126:203-210

Roussel H, van Tuinen D, Franken P, Gianinazzi S, Gianinazzi-Pearson V (2001) Signalling between arbuscular mycorrhizal fungi and plants: identification of a gene expressed during early interactions by differential RNA display analysis. Plant and Soil 232:13-19

Ruiz-Lozano J M, Gianinazzi S, Gianinazzi-Pearson V (1999a) Genes involved in resistance to powdery mildew in barley differentially modulate root colonization by the mycorrhizal fungus *Glomus mosseae*. Mycorrhiza 9:237-240

Ruiz-Lozano J M, Roussel H, Gianinazzi S, Gianinazzi-Pearson V (1999b) Defense genes are differentially induced by a mycorrhizal fungus and *Rhizobium* sp in wild-type and symbiosis-defective pea genotypes. Mol Plant-Microbe Interact 12:976-984

Sagan M, Messager A, Duc G (1994) Phenotypic characterization and classification of nodulation mutants of pea. Plant Sci 100:59-70

Sagan M, Morandi D, Tarenghi E, Duc G (1995) Selection of nodulation and mycorrhizal mutants in the model plant *Medicago truncatula* (Gaertn.) after γ-ray mutagenesis. Plant Sci 111:63-71

Sagan M, Delarambergue H, Morandi D (1998) Genetic analysis of symbiosis mutants in *Medicago truncatula*. In: Elmerich C (ed) Biological nitrogen fixation for the 21st Century, Kluwer Academic Publishers, Dordrecht pp 317-318

Salzer P, Bonanomi A, Beyer K, Vögeli-Lange R, Aeschbacher R A, Lange J A W, Kim D, Cook D R Boller T (2000) Differential expression of eight chitinase genes in *Medicago truncatula* roots during mycorrhiza formation, nodulation, and pathogen infection. Mol Plant-Microbe Interact 13:763-777

Samra A, Dumas-Gaudot E, Gianinazzi-Pearson V, Gianinazzi S (1996) Soluble proteins and polypeptide profiles of spores of arbuscular mycorrhizal fungi. Interspecific variability and effects of host (myc^+) and non-host (myc^-) *Pisum sativum* root exudates. Agronomie 16:709-719

Schauser L, Roussis A, Stiller J, Stougaard J (1999) A plant regulator controlling development of symbiotic root nodules. Nature 402:191-195

Schneider A, Walker S A, Poyser S, Sagan M, Ellis T H N, Downie J A (1999) Genetic mapping and functional analysis of a nodulation-defective mutant (*sym19*) of pea (*Pisum sativum* L.). Mol Gen Genet 262:1-11

Senoo K, Solaiman Z, Kawaguchi M, Imaizumi-Anraku H, Akao S, Tanaka A, Obata H (2000) Isolation of two different phenotypes of mycorrhizal mutants in the model legume plant *Lotus japonicus* after EMS-treatment. Plant Cell Physiol 41:726-732

Shirtliffe S J, Vessey J K (1996) A nodulation (Nod$^+$/Fix$^-$) mutant of *Phaseolus vulgaris* L has nodule-like structures lacking peripheral vascular bundles (Pvb$^-$) and is and is resistant to mycorrhizal infection (Myc$^-$) Plant Sci 118:209-220

Showalter A M (1993) Structure and function of plant cell wall proteins. Plant Cell 5:9-23

Simon L, Bousquet J, Levesque R C, Lalonde M (1993) Origin and diversification of endomycorrhizal fungi and coincidence with vascular land plants. Nature 363:67-69

Sprent J I (1994) Evolution and diversity in the legume-rhizobium symbiosis: chaos theory? Plant Soil 161:1-10

Strittmatter G, Gheysen G, Gianinazzi-Pearson V, Hahn K, Niebel A, Rohde W, Tacke E (1996) Infections with various types of organisms stimulate transcription from a short promoter fragment of the potato *gst1* gene. Mol Plant-Microbe Interact 9:68-73

Tahiri-Alaoui A, Antoniw J F (1996) Cloning of genes associated with the colonization of tomato roots by the arbuscular mycorrhizal fungus *Glomus mosseae*. Agronomie 16: 699-707

van Buuren M L, Maldonado-Mendoza I E, Trieu A T, Blaylock L A, Harrison M J (1999) Novel genes induced during an arbuscular mycorrhizal (AM) symbiosis formed between *Medicago truncatula* and *Glomus versiforme*. Mol Plant-Microbe Interact 12:171-181

van Buuren M L, Trieu A T, Blaylock L A, Harrison M J (2000) Isolation of genes induced during arbuscular mycorrhizal associations using a combination of subtractive hybridization and differential screening. In: Podila G K, Douds D (eds) Current advances in mycorhizae research. APS Press, St Paul pp 91-99

van Rhijn P, Fang Y, Glili S, Shaul O, Atzmon N, Wininger S, Eshed Y, Lum M, Li Y, Fujishige N, Kapulnik Y, Hirsch M (1997) Expression of early nodulin genes in alfalfa mycorrhizae indicates that signal transduction pathways used in forming arbuscular mycorrhizae and *Rhizobium*-induced nodules may be conserved. Proc Natl Acad Sci USA 94:5467-5472

van Tuinen D, Journet E P, Crespeau H, Gouzy J, Jaillon O, Niebel A, Carreau V, Chatagnier O, Kahn D, Gianinazzi-Pearson V, Gamas P (2000) *Medicago truncatula* root symbiosis explored by large-scale EST sequencing. In: Weber H C, Imhof S, Zeuske D (eds) Proceedings of the Third International Congress on Symbiosis, Marburg, Germany, August 2000, p 220

Vernoud V, Pingret J L, Chabaud M, Dedieu A, de Carvalho Niebel F, Journet E P, Barker DG (2000) Nod factor signal transduction in *Medicago truncatula* Nod$^-$/Myc$^-$ mutants TR25/26. In: de Wit P J G M, Bisseling T, Stiekema W J (eds) Biology of plant-microbe interactions. International Society for Molecular Plant-Microbe Interactions, St Paul, pp 114-119

Webb K J, Skot L, Nicholson M N, Jorgensen B, Mizen S (2000) *Mesorhizobium loti* increases root-specific expression of a calcium-binding protein homologue identified by promoter tagging in *Lotus japonicus*. Mol Plant-Microbe Interact 13:606-616

Wegel E, Schauser L, Sandal N, Stougaard J, Parniske M (1998) Mycorrhiza mutants of *Lotus japonicus* define genetically independent steps during symbiotic infection. Mol Plant-Microbe Interact 11:933-936

Weidmann S (2000) Caractérisation d'ESTs représentant des gènes différentiellement exprimés lors de la mycorhization chez *Medicago truncatula*. DEA de Biochimie, Biologie Cellulaire et Moléculaire, Université de Bourgogne, France

Mycorrhizal Technology in Agriculture
ed. by S. Gianinazzi, H. Schüepp, J.M. Barea and K. Haselwandter

Plant defense responses induced by arbuscular mycorrhizal fungi

M.J. Pozo[1], S. Slezack-Deschaumes[2], E. Dumas-Gaudot [3], S. Gianinazzi[3] and C. Azcón-Aguilar[1]

[1] *Departamento de Microbiología del Suelo y Sistemas Simbióticos, Estación Experimental del Zaidín (CSIC), Profesor Albareda 1, E-18008 Granada, Spain*
[2] *Laboratoire de Phytobiologie Cellulaire, UFR Sciences Vie, Université de Bourgogne, BP 47870, 21078, France*
[3] *UMR INRA/Université de Bourgogne BBCE-IPM, CMSE-INRA, BP86510, 21065 Dijon Cedex, France*

Keywords: Resistance to pathogens, defense mechanisms

Introduction

Plants in their environment daily face many organisms such as fungi, bacteria, mycoplasms, viruses, nematodes, etc. Many of them are potential pathogens; in fact thousands of microorganisms are known to cause plant diseases. Despite this large number of deleterious microorganisms, most of the plants are resistant to their attack since they have developed effective mechanisms to protect themselves.

Defense responses in plants

The defense mechanisms developed by plants can take many different forms, ranging from the elaboration of preformed chemical barriers, which provide nonspecific protection against a wide range of organisms, to more active host-specific responses that provide resistance (Hammond-Kosack and Jones 1996). The active mechanisms imply the recognition of the pathogen. Plants can recognize pathogens by different types of molecules, such as substances secreted by the foreign organism, or compounds released from fungal or plant cell walls during the early phases of the pathogen attack. These molecules are called elicitors. Whatever the source of the elicitor, the mechanism of signal perception appears to rely on the presence of specific receptors on the plant cell surface. Its union activates ion channels leading to changes in membrane polarity, and in the intracellular levels of different ions. This initiates a cascade of signaling processes that activates plant defenses, including production of reactive oxygen species (ROS) known as the "oxidative burst", and the activation of defense related genes. Among them, we can find: (i) those coding for the enzymes of the phenylpropanoid pathway, which results in a wide range of signal or antifungal molecules and also in the reinforcement of the cell wall, and (ii) those coding for the so-called pathogenesis-related (PR) proteins (reviewed by Somssich and Hahlbrock 1998). PR proteins are low molecular weight polypeptides that have been shown to play a role in plant defense. Some of them such as proteases, chitinases, chitosanases and β-1,3-glucanases can act directly on the pathogen (Van Loon et al. 1997).

Plant defense responses during arbuscular mycorrhizal symbiosis

Diseases are not the only outcome of plant-microbe interactions. In fact, mutually beneficial relationships are frequent in nature, arbuscular mycorrhizas being among the most widespread symbiotic associations all over the world. Arbuscular mycorrhizal fungi (AMF) penetrate the roots of the majority of the plants, leading to an extensive colonization of the root cortex. Penetration into the root and intercellular growth of the AMF appears to be permitted by the plant. It seems that AMF do not elicit defense mechanisms, as otherwise their growth would be limited (Gianinazzi-Pearson 1996). However, the host plant must exert a control over the fungal colonisation within the root since fungal proliferation is restricted to the cortex (Bonfante-Fasolo 1984). The establishment of this mutualistic association involves a series of complex interactions based on a continuous cellular and molecular dialogue between both symbionts, resulting in their mutual recognition and in the regulation of the fungal growth inside the root (Gianinazzi 1991). AM functioning should imply a high level of co-ordination and morphological and physiological integration. There are evidences supporting that plant defense related proteins may play an essential role in the control of intraradical fungal colonization (Lambais 2000).

AMF do not lack elicitor factors. They are able to initiate resistance events in non-mycotrophic plants (Allen et al. 1989), and in plant mutants unable to establish mycorrhizal symbiosis (Gollote et al. 1993). Therefore, it has been proposed that AMF should suppress, avoid or overcome the defense response of host plants to become established within the root tissues. Different models have been proposed (Gianinazzi-Pearson et al. 1996; Salzer and Boller 2000; Dumas-Gaudot et al. 2000a), but, obviously, AMF can not block the plant defense system, since this would render the plant more susceptible to pathogens (Barker et al. 1998). This is not the case, and, furthermore, many studies have shown that mycorrhizal plants are more resistant to soil-borne pathogens than the non-mycorrhizal ones (reviewed by Linderman 1994; St Arnaud et al. 1995; Azcón-Aguilar and Barea 1996; Azcón-Aguilar et al. in this book). Thus, the plant should identify the AM fungus as a mutualistic symbiont, or it should block the defense reaction at any new cellular contact, leading to localized repression or modulation of the defense mechanisms (Barker et al. 1998; Dumas-Gaudot et al. 2000a).

Therefore, the equilibrium between partial inhibition of plant defense mechanisms and plant control over fungal development seems to be a key point for a functional compatibility in the association. This is why plant defense reactions during mycorrhizal symbiosis establishment have deserved a lot of interest during the last ten years. Using different plant-AMF systems several studies have been carried out trying to investigate compounds and genes known to be induced during plant-pathogen interactions.

In relation to enzymes involved in ROS metabolism, Palma et al. (1993) described the induction of new superoxide dismutase isoforms in mycorrhizal clover roots and peroxidase activities have also been shown to increase during AM formation (Spanu and Bonfante-Fasolo 1988; Mathur and Vyas 1995; Blee and Anderson 2000; Schwob et al. 2000).

Concerning cell wall components, no changes have been found at the cellular level during the first stages of mycorrhizal colonization, i. e. during appressoria formation and colonization of external tissues, in contrast to the common reactions to pathogen attack. No accumulation of callose phenolics or lignin has been detected during mycorrhization (Gianinazzi-Pearson et al. 1996). Enhanced expression levels of genes coding for hydroxyproline-rich-glycoproteins (HRGP) have been shown (Franken and Gnädinger 1994), and such proteins have been detected around arbuscular hyphae (Bonfante-Fasolo et al. 1991; Balestrini et al. 1997). Transient increases in peroxidase activity (as mentioned above) are usually considered to be involved in cell wall reinforcement during plant defense reactions against pathogens (Iiyama et al. 1994).

In relation to the phenylpropanoid pathway, transient activation of phenylalanine-ammonia-lyase and chalcone synthase has been found, with accumulation in cells containing arbuscules (Harrison and Dixon 1994; Blee and Anderson 1996). Recently, higher levels of transcripts coding for chalcone-isomerase and transcinnamic acid 4-hydrolase have been shown in AM inoculated roots (Ruíz-Lozano et al. 1999). In fact, accumulation of phenolic compounds such as phytoalexins and flavonoids does occur (Morandi 1996), but it is weaker and appears later than in pathogenic interactions (Harrison 1997).

Among PR proteins a local accumulation of PR-1 proteins has been shown around cells containing arbuscules (Gianinazzi-Pearson et al. 1992; Franken and Gnädinger 1994; Cordier et al. 1998). A progressive increase in endoproteolytic activities has been observed during root colonization (Slezack et al. 1999). Concerning chitinases and β-1,3-glucanases, transient increases of total activity were reported during the early stages of the interaction (Spanu et al. 1989; Lambais and Mehdy 1993; Volpin et al. 1994; Kapulnik et al. 1996) and a local accumulation of mRNAs coding for these enzymes around arbuscules have been shown by *in situ* hybridization (Blee and Anderson 1996; Lambais and Mehdy 1995, 1998). By using polyacrylamide gel electrophoresis it has also been shown the induction of specific isoforms of chitinases, β-1,3-glucanase and chitosanases in fully established symbiosis (Dumas-Gaudot et al. 1992, 1994, 2000b; Pozo et al. 1996, 1998, 1999, Slezack et al. 2000).

As a general conclusion of all these works it is accepted that AMF induce defense responses in the host plant, but these reactions are transient, localized and weaker than those induced by pathogenic fungi (Gianinazzi-Pearson 1996). Some of these reactions could be the result of non-specific defense responses, but some others, such as the induction of specific isoforms, appear to be specialized, different to those induced by fungal pathogens (Dassi et al. 1996; Dumas-Gaudot et al. 1999; Pozo et al. 1996, 1998, 1999). Later it was shown that chitinase gene expression in mycorrhizal roots was different from that during interactions with pathogens and *Rhizobium* and involved the expression of two mycorrhiza-specific chitinase genes (Salzer et al. 2000).

In any case, the mechanisms by which the plant controls fungal development within the roots remain unclear. The induction of symbiosis-specific genes has been proposed to be responsible of the repression of a typical defense response during AM formation. The product of these genes would avoid the induction of the defense response or block it once initiated (Gianinazzi-Pearson et al. 1996). These genes could code for receptors or molecules able to activate receptors for signal molecules from the symbiont. Upon recognition, local control of the defense reaction would be possible. Thus, resistance phenotype in mutants may reflect alterations in the symbiosis-specific gene functions. For this reason mutant plants unable to establish mycorrhizas have became a useful tool for studying the mechanisms acting during the symbiosis. As many of these mutants are unaffected by the interactions with other root-infecting microorganisms, they provide a proof that specific host genes modulate root symbioses. These plants represent model plant material for identifying genes involved in the early events of AM establishment. As an example of these studies, mutation of the sym 8, sym 9 and sym 30 genes in pea has been shown to result in resistance to AMF and rhizobia, indicating a multigenic control of early interactions (Gianinazzi-Pearson 1996). In fact, common plant responses seem to be associated with host-microbe interactions in mycorrhiza and nodulation symbioses (Gianinazzi-Pearson and Dénarié 1997). Further evidences for common signaling pathways between nodulation and mycorrhization have been recently shown in *Medicago truncatula* mutants (Catoira et al. 2000). In relation with defense related genes, the early induction of similar plant defense genes in response to AMF and rhizobia has been shown (Ruíz-Lozano et al. 1999). The possible role of plant specific chitinases and chitosanases as perception systems has already been discussed (Pozo et al. 1998). It has been proposed that during evolution plant symbionts could have developed mechanisms to use host defense-recognition systems for symbiotic signal perception (Koide and Schreiner 1992).

Defense responses during AM and plant protection against pathogens

From a practical point of view, the induction of a partial or controlled defense response is a promising research area with regard to application in crop protection. Certain non-pathogenic microorganisms, such as some plant growth promoting rhizobacteria (PGPR) have been shown to evoke in plants a resistance response phenotypically similar to the Systemic Acquired Resistance (SAR) induced by pathogens. This phenomenon has been called Induced Systemic Resistance (ISR) (Van Loon et al. 1998). Beneficial organisms able to induce SAR or ISR are powerful biological control agents (Van Driesche and Bellows 1996).

Partial induction of plant defense reactions by AMF during root colonisation is also one of the mechanisms considered to be involved in plant protection by these fungi (St Arnaud et al. 1995; Azcón-Aguilar and Barea 1996; Azcón-Aguilar et al. in this book). Recent studies indicate that the bioprotection exerted by AMF appears to be the result of a combination of local and systemic mechanisms. Cordier et al. (1998) showed by immunocytochemical studies that arbuscule-containing cortical cells of *G. mosseae*-mycorrhizal tomato plants were immune to the pathogen *Phytophthora parasitica* and exhibited a localised resistance. They also evidenced systemic resistance in non-mycorrhizal root parts, characterised by the elicitation of host wall thickenings in reaction to the intercellular hyphae of the pathogen, and also by the formation of callose-rich encasement material around *P. parasitica* hyphae that are penetrating root cells. Systemic effects have been also shown with regard to lytic activity against *Phytophthora* cell walls. Extracts from non-colonised roots of *G. mosseae*-mycorrhizal plants exhibited lytic activity when infected by the pathogen whereas this activity was not found in the non-mycorrhizal plants inoculated with the pathogen (Pozo et al. 2002).

Differences in local and systemic patterns of gene expression for defense related enzymes during AM interactions have been evidenced and the possibility of multiple ways of signal exchanges has been proposed (Lambais and Mehdy 1995, 1998). In a recent work Shaul et al. (1999) suggest the existence of regulatory processes, initiated in the roots of mycorrhizal plants, that modify disease-symptom development and gene expression in their leaves. Long distance effects were also described by Fieschi et al. (1992), since they observed hyperpolarization of transmembrane potential in non-colonised cells of mycorrhizal root systems when compared with root cells from non-mycorrhizal plants. Long distance signals such as the plant hormones auxins, cytokinins, abscisic acid and ethylene have been shown to present altered levels in mycorrhizal plants (Allen et al. 1980, 1982; Danneberg et al. 1992; Dugassa et al. 1996; Hirsch et al. 1997). It is already known that changes in the hormonal balance in plants can modulate the expression of defense-related genes (Petruzzelli et al. 1999).

The overall conclusion from the available information, indicates that AMF are able to induce systemic protection against pathogens. From a biochemical point of view, ISR and SAR present many differences, although they are phenotypically similar. SAR require salicylic acid accumulation and it is modulated by ethylene and jasmonic acid, while ISR seems not to depend on salicylic acid, but it does depend on ethylene perception as a key step in signal transduction for starting ISR. Moreover, in contrast to SAR, which is associated with PR accumulation, ISR does not induce activation of PR genes (Pieterse et al. 1996). Certain metabolic changes (increase in peroxidase activity, phytoalexin accumulation, PR accumulation, etc.) have been related to ISR but none of them were consistently associated with the induced resistance status in the different biological systems studied. In spite of that, structural modifications (cell wall reinforcements, phenolic compounds accumulation, and papilla formation) have been described extensively in ISR expressing plants. Parallel aspects to these described for rhizobacteria-mediated ISR have been found for mycorrhiza-induced defense response in plants (Pozo et al. 2002). It is known that salicylic acid accumulation during AM colonisation is weak and transient, occurring only in early stages of the symbiosis establishment (Blilou et al. 1999); thus, it does not seem to be involved in a systemic resistance. In the present chapter we have summarized that the main changes on defense-related gene expression have been shown to occur locally, although systemic alterations have been described. Moreover, the structural modifications described at cytological level using the split root system (Cordier et al. 1998) are similar to those described in plants colonised by ISR-inducing rhizobacteria.

In view of all the remarks made above it is deduced that colonisation by some AMF lead to ISR, similar to that widely described in rhizobacteria-induced plants. Understanding the underlying mechanisms and the signals involved in the transmission of the resistance would be a key point in mycorrhizal research and will open new insights in the application of biocontrol techniques against a wide range of plant pathogens in the context of sustainable practices aimed at preserving environmental quality.

Acknowledgements
Some of the experimental work referred to in this chapter was sponsored by the project 1FD97-0763-C03-02. MJ Pozo was supported by a post-doctoral fellowship from the Spanish Council of Scientific Research (CSIC).

References

Allen M F, Moore T S, Christensen M (1980) Phytohormone changes in *Bouteloua gracilic* infected by vesicular-arbuscular mycorrhizae. I. Cytokenin increases in the host plant. Can. J. Bot. 58:371-374

Allen M F, Moore T S Christensen M (1982) Phytohormone changes in *Bouteloua gracilic* infected by vesicular-arbuscular mycorrhizae. II. Altered levels of gibberlin-like substances and abscisic in the hos plant. Can. J. Bot. 60:468-471

Allen M F, Allen E B, Frise C F (1989) Responses of the non-mycotrophic plant *Salsola kali* to invasion by vesicular-arbuscular mycorrhizal fungi. New Phytol. 111:45-49

Azcón-Aguilar C, Barea J M (1996) Arbuscular mycorrhizas and biological control of soil-borne plant pathogens - An overview of the mechanisms involved. Mycorrhiza 6:457-464

Balestrini R, Josä-Estanyol M, Puigdomänech P, Bonfante P (1997) Hydroxyproline-rich glycoprotein mRNA accumulation in maize root cells colonized by an arbuscular mycorrhizal fungus as revealed by in situ hybridization. Protoplasma 198:36-42

Barker S J, Tagu D, Delp G (1998) Regulation of root and fungal morphogenesis in mycorrhizal symbioses. Plant Physiol. 116:1201-1207

Blee K A, Anderson A J (1996) Defense-related transcript accumulation in *Phaseolus vulgaris* L colonized by the arbuscular mycorrhizal fungus *Glomus intraradices* Schenck and Smith. Plant Physiol. 110:675-688

Blee K A, Anderson A J (2000) Defense responses in plants to arbuscular mycorrhizal fungi. In: Podila G K, Douds D D (eds) Current advances in mycorrhizae research. APS Press, Minnesota pp 27-44

Blilou I, Ocampo J A, Garcia-Garrido J M (1999) Resistance of pea roots to endomycorrhizal fungus or *Rhizobium* correlates with enhanced levels of endogenous salicylic acid. J. Exp Botany. 1663-1668

Bonfante-Fasolo P (1984) Anatomy and morphology of VA mycorrhizae In: Powell C L, Bagyaraj D J (eds) VA Mycorrhiza. CRC Press, Boca Raton, Florida. pp 35-46

Bonfante-Fasolo P, Tamagnone L, Peretto R, Esquerré-Tugayé M T, Mazau D, Mosiniak M, Vian B (1991) Immunocytochemical locationof hydroxyproline rich glycoproteins at the interface between a mycorrhizalfungus and its host plants. Protoplasma 165:127-138

Catoira R, Galera C, de Billy F, Penmetsa V, Journet E P, Maillet F, Rosengerg C, Cook D, Gough C, Denarie J (2000) Four genes of *Medicago truncatula* controlling components of a NOd factor transduction pathway. The Plant Cell 12:1647-1665

Cordier C, Pozo M J, Barea J M, Gianinazzi S, Gianinazzi-Pearson V (1998) Cell defense responses associated with localized and systemic resistance to *Phytophthora* induced in tomato by an arbuscular mycorrhizal fungus. Mol. Plant-Microbe Interact. 11:1017-1028

Danneberg G, Latus C, Zimmer W, Hundeshagen B, Schneider-Poetsch H J, Bothe H (1992) Influence of vesicular arbuscular mycorrhiza on phytohormone balances in maize (*Zea mays* L.). J. Plant Physiol. 141:33-39

Dassi B, Dumas-Gaudot E, Asselin A, Richard C and Gianinazzi S (1996) Chitinase and β-1,3-glucanase isoforms expressed in pea roots inoculated with arbuscular mycorrhizal or pathogenic fungi. Eur. J Plant Pathol. 102:105-108

Dugassa G D, von Alten H, Schonbeck F (1996) Effects of arbuscular mycorrhiza (AM) on health of *Linum usitatissimum* L infected by fungal pathogens. Plant Soil. 185:173-182

Dumas-Gaudot E, Grenier J, Furlan V, Asselin A (1992) Chitinase, chitosanase and β-1,3-glucanase activities in *Allium* and *Pisum* roots colonized by *Glomus* species. Plant Sci. 84:17- 24

Dumas-Gaudot E, Asselin A, Gianinazzi-Pearson V, Gollotte A, Gianinazzi S (1994) Chitinase isoforms in roots of various pea genotypes infected with arbuscular mycorrhizal fungi. Plant Sci. 99:27-37

Dumas-Gaudot E, Bestel-Corre G, Gianinazzi S, Gianinazzi-Pearson V (1999). Differentially displayed proteins during mycorrhization and nodulation in *Medicago truncatula* roots. In : H Trippenhuis (Eds.) *Second Medicago truncatula workshop*. Amsterdam. p. 27

Dumas-Gaudot E, Gollotte A, Cordier C, Gianinazzi S, Gianinazzi-Pearson V (2000a) Modulation of host defense systems. In: Douds D, Kapulnik Y (eds) Arbuscular mycorrhizas: physiology and function. Kluwer Academic Press, Dortrecht Netherlands pp 173-200

Dumas-Gaudot E, Slezack S, Negrel J, Bestel-Corre G, Anstett A, van Tuinen D, Gianinazzi-Pearson V, Gianinazzi S (2000b) Chitinases and chitosanases in arbuscular and rhizobial symbioses : common or different events ? In : HC Weber, S Imhof et D Zeuske (Eds.) *Third International Congress on Symbiosis*. vol. Philips University of Marburg, Marburg. p. 68.

Fieschi M, Alloati G, Sacco S, Berta G (1992) Membrane potential hyperpolarisation in vesicular arbuscular mycorrhizae of *Allium porrum* L.: a non-nutritional long-distance effect of the fungus. Protoplasma 168: 136-140

Franken P, Gnädinger F (1994) Analysis of parsley arbuscular endomycorrhiza: Infection development and mRNA levels of defense-related genes. Mol. Plant-Microbe Interact. 7:612-620

Gianinazzi S (1991) Vesicular arbuscular (endo-) mycorrhiza:. Cellular, biochemical and genetic aspects. Agric. Ecos. Env. 35:105-119

Gianinazzi-Pearson V (1996) Plant cell responses to arbuscular mycorrhizal fungi: Getting to the roots of the symbiosis. The Plant Cell. 8:1871-1883

Gianinazzi-Pearson V, Tahiri-Alaoui A, Antoniw J F, Gianinazzi S, Dumas E (1992) Weak expression of the pathogenesis related PR-b1 gene and localization of related protein during symbiotic endomycorrhizal interactions in tobacco roots. Endocytobiosis & Cell Res 8:177-185

Gianinazzi Pearson V, Denarie J (1997) Red carpet programs for root endosymbioses. Trends Plant Sci 2:371-372

Gianinazzi-Pearson V, Dumas-Gaudot E, Gollotte A, Tahiri-Alaoui A, Gianinazzi S (1996) Cellular and molecular defense-related root responses to invasion by arbuscular mycorrhizal fungi. New Phytol. 133:45-57

Gollote A, Gianinazzi-Pearson V, Giovanetti M, Sbrana C, Avio L, Gianinazzi S (1993). Cellular localization and cytochemical probing of resistance reactions to arbuscular mycorrhizal fungi in a "locus a" myc mutant of *Pisum sativum* L. Planta 191:112-122

Hammond-Kosack K E, Jones J D G (1996) Resistance gene-dependent plant defense responses. The Plant Cell. 8:1773-1791

Harrison M J (1997) The arbuscular mycorrhizal symbiosis: An underground association. Trends Plant Sci 2:54-60

Harrison M J, Dixon R A (1994) Spatial patterns of expression of flavonoid/isoflavonoid pathway genes during interactions between roots of *Medicago truncatula* and the mycorrhizal fungus *Glomus versiforme*. Plant J. 6:9-20

Hirsch A M, Fang Y, Asad S, Kapulnik Y (1997) The role of phytohormones in plant-microbe symbioses. Plant Soil 194:171-184

Iiyama K, Lam T B T, Stone B A (1994) Covalent cross-links in the cellwall. Plant Physiol. 104:315-320

Kapulnik Y, Volpin H, Itzhaki H, Ganon D, Galili S, David R, Shaul O, Elad Y. Chet I, Okon Y. (1996). Suppression of defense responses in mycorrhizal alfalfa and tobacco roots. New Phytol. 133:59-64

Koide R T, Schreiner R P (1992) Regulation of the vesicular-arbuscular mycorrhizal symbiosis. Annu. Rev. Plant Physiol. Plant Mol. Biol. 43:557-581

Lambais M R (2000) Regulation of plant defense-related genes in arbuscular mycorrhizae. In: Podila G K, Douds D D (eds) Current advances in mycorrhizae research. APS Press, Minnesota pp 45-59

Lambais M R, Mehdy M C (1993) Suppression of endochitinase, β-1,3 endoglucanase, and chalcone isomerase expression in bean vesicular-arbuscular mycorrhizal roots under different soil phosphate conditions. Mol. Plant-Microbe Interact. 6:75-83

Lambais M R Mehdy M C (1995) Differential expression of defense-related genes in arbuscular mycorrhiza. Can. J. Bot 73:533-540

Lambais M R, Mehdy M C (1998) Spatial distribution of chitinases and β-1,3-glucanase transcripts in bean arbuscular mycorrhizal roots under low and high soil phosphate conditions. New Phytol. 140:33-42

Linderman R G (1994) Role of VAM fungi in biocontrol. In: Pfleger F L, Linderman R G (eds) Mycorrhizae and plant health. The American Phytopathological Society Press, St Paul pp 1-27

Mathur N, Vyas A (1995) Changes in isozyme patterns of peroxidase and polyphenol oxydase by VAM fungi in roots of ziziphus species. J. Plant Physiol. 145:498-500

Morandi D (1996) Occurrence of phytoalexins and phenolic compounds in endomycorrhizal interactions, and their potential role in biological control. Plant Soil 185:241-251

Palma J M, Longa M A, del Rio L A, Arines J (1993) Superoxide dismutase in vesicular arbuscular mycorrhizal red clover plants. Physiol. Plant. 87:77-83

Petruzelli L, Kunz C, Waldvogel R, Meins F, Leubner-Mertzger G (1999) Distinct ethylene and tissue-specific regulation of β-1,3-glucanases and chitinases during pea seed germination. Planta, 209:195-201

Pieterse C M J, van Wees S C M, Hoffland E, van Pelt J A, van Loon L C (1996) Systemic resistance in *Arabidopsis* induced by biocontrol bacteria is independent of salicylic acid accumulation and pathogenesis-related gene expresion. The Plant Cell. 8:1225-1237

Pozo M J, Dumas-Gaudot E, Slezack S, Cordier C, Asselin A, Gianinazzi S, Gianinazzi-Pearson V, Azcón-Aguilar C, Barea J M (1996) Detection of new chitinase isoforms in arbuscular mycorrhizal tomato roots: possible implications in protection against *Phytophthora nicotianae* var. *parasitica*. Agronomie 16:689-697

Pozo M J, Dumas-Gaudot E, Azcón-Aguilar C, Barea J M (1998) Chitosanase and chitinase activities in tomato roots during interactions with arbuscular mycorrhizal fungi or *Phytophthora parasitica*. J. Exp. Botany 49:1729-1739

Pozo M J, Azcón-Aguilar C, Dumas-Gaudot E, Barea J M (1999) β-1,3-glucanase activities in tomato roots inoculated with arbuscular mycorrhizal fungi and/or *Phytophthora parasitica* and their possible involvement in bioprotection. Plant Sci. 141:149-157

Pozo M J, Cordier C, Dumas-Gaudot E, Gianinazzi S, Barea J M, Azcón-Aguilar C (2002) Localized *vs* systemic effect of arbuscular mycorrhizal fungi on defense responses to *Phytophthora* infection in tomato plants. J. Exp. Botany. In press

Ruiz Lozano J M, Roussel H, Gianinazzi S, Gianinazzi-Pearson V (1999) Defense genes are differentially induced by a mycorrhizal fungus and *Rhizobium* sp. in wild type and Symbiosis-defective pea genotypes. Mol. Plant-Microbe Interact. 12:976-984

Salzer P, Boller T (2000) Elicitor-induced reactions in mycorrhizae and their suppression. In: Podila G K, Douds D D (eds) Current advances in mycorrhizae research. APS Press, Minnesota pp 1-10

Salzer P, Bonanomi A, Beyer K.; VogeliLange R, Aeschbacher R A, Lange J, Wiemken A, Kim D, Cook D R, Boller T (2000) Differential expression of eight chitinase genes in Medicago truncatula roots during mycorrhiza formation, nodulation, and pathogen infection. Mol. Plant-Microbe Interact. 3:763-777

Schwob I M, Ducher A, Coudret (2000) An arbuscular mycorrhizal fungus (*Glomus mosseae*) induces a defense-like response in rubber tree (*Hevea brasiliensis*) roots. J. Plant Physiol. 156:284-287

Shaul O, Galili S, Volpin H, Ginzberg I, Elad Y, Chet I, Kapulnik Y (1999) Mycorrhiza-induced changes in disease severity and PR protein expression in tobacco leaves Mol. Plant-Microbe Interact. 12:1000-1007

Slezack S, Dumas-Gaudot E, Rosendahl S, Kjoller R, Paynot M, Negrel J, Gianinazzi S (1999) Endoproteolytic activities in pea roots inoculated with the arbuscular mycorrhizal fungus *Glomus mosseae* and/or *Aphanomyces euteiches* in relation to bioprotection. New Phytol. 141:517-529.

Slezack S, Dumas-Gaudot E, Paynot M, Gianinazzi S (2000) Is a fully established arbuscular mycorrhizal symbiosis required for bioprotection of *Pisum sativum* roots against *Aphanomyces euteiches*?. Mol. Plant-Microbe Interact. 238-241

Spanu P, Bonfanfe-Fasolo P, (1988) Cell-wall bound peroxidase activity in roots of mycorrhizal *Allium porrum.* New Phytol. 109:119-124

Spanu P, Boller T, Ludwig A, Wiemken A, Faccio A, Bonfante-Fasolo P (1989) Chitinase in roots of mycorrhizal *Allium porrum*: regulation and localization. Planta. 177:447-455

Somssich I E, Hahlbrock K (1998) Pathogen defense in plants - a paradigm of biological complexity. Trends Plant Sci. 3:86-90

St-Arnaud M, Hamel C, Caron M, Fortin A (1995) Endomycorhizes VA et sensibilité des plantes aux maladies: Synthèse de la littérature et mécanismes d'interaction potentiels. La Symbiose Mycorrhizienne, État des Connaissances. Frelighsburg, Orbis Publishing. pp. 51-87

Van Driesche R G, Bellows T S (1996) *Biological control.* Chapman & Hall, New York, USA

Van Loon L C (1997) Induced resistance in plants and the role of pathogenesis-related proteins. Eur J Plant Pathol 103:753-765

Van Loon L C, Bakker P A H M, Pieterse C M J (1998) Systemic resistance induced by rhizosphere bacteria. Annu. Rev. Phytopathol. 26:453-483

Volpin H, Elkind Y, Okon Y, Kapulnik Y (1994) A vesicular arbuscular mycorrhizal fungus (*Glomus intraradix*) induces a defense response in alfalfa roots. Plant Physiol. 104:683-689

Mycorrhizal Technology in Agriculture
ed. by S. Gianinazzi, H. Schüepp, J.M. Barea and K. Haselwandter

Arbuscular mycorrhiza induced ATPases and membrane nutrient transport mechanisms

N. Ferrol[1], S. Gianinazzi[2] and V. Gianinazzi-Pearson[2]

[1] *Departamento de Microbiología del Suelo y Sistemas Simbióticos, Estación Experimental del Zaidín (CSIC), Profesor Albareda 1, E-18008, Granada, Spain*
[2] *Unité Mixte de Recherche INRA/Université de Bourgogne Biochimie, Biologie Cellulaire et Ecologie des Interactions Plantes/Micro-organismes, CMSE-INRA, BP 86510, F-21065 Dijon cedex, France*

Keywords: ATPases, nutrient transport, symbiotic interfaces, bidirectional nutrient exchange, gene expression

Introduction

The evolutionary success of arbuscular mycorrhizal (AM) symbiosis reflects the unique combination of a superior biotrophic mode of fungal carbon acquisition and the ability of the living plant to absorb nutrients, especially phosphorus, from the fungal partner (Jakobsen 1999). This mutualistic way of life must require controlled expression of a large set of membrane transport systems active in phosphate uptake from the soil by the extraradical hyphae, its transfer to the host plant across a symbiotic interface, and coupled to transport of photosynthates in the opposite direction. The implied membrane transporters are therefore integral systems in the functioning of the symbiosis. Very little is as yet known of the mechanisms which determine the controlled nutrient exchange between partners at the symbiotic interfaces. Membrane transport proteins responsible for movement of phosphate and carbon compounds across the individual plant and fungal membranes at the symbiotic interfaces have not so far been identified. However, increasing information about plasma membrane H^+-ATPases, a family of enzymes that drive active secondary transport systems, is being obtained from the use of molecular biology and cytology techniques. In this chapter, we summarize current knowledge on plant and fungal plasma membrane ATPases and discuss their possible implication in transport mechanisms operating at the membranes of both symbionts in arbuscular mycorrhiza.

Symbiotic interfaces

One of the prerequisites for nutrient movement between AM symbionts is the development of a distinctive interface which is bordered by fungal and plant plasma membranes and which is adapted, both structurally and functionally, to promote transfer. The structure and function of plant-microbe interfaces in biotrophic associations have been reviewed by Smith and Smith (1990), Bonfante and Perotto (1995), Gianinazzi-Pearson (1996), Smith and Read (1997) and Harrison (1999). Different types of interface are created during AM interactions depending on whether the fungus penetrates or not the host cell. Intercellular interfaces are formed as hyphae grow between cells of the root cortex, and intracellular interfaces are created when hyphae penetrate the wall of the host cell and grow within it, forming structures such as hyphal coils or arbuscules. As a fungal hypha penetrates a root cell, a newly synthesised plant cell membrane extends from the plasma membrane to surround it so that the fungus is topologically in the cellular apoplast. When a hypha differentiates to form an arbuscule in an inner cortical cell, this extended host membrane, which is referred to as the periarbuscular membrane, is separated from the arbuscule wall by a narrow interfacial matrix. Intracellular fungal development is frequently characterized by a reduction in wall-material in the interface, compared with that present in intercellular interfaces. Since arbuscules increase the surface area of contact between symbionts, and the plant plasma membrane is in close contact with the fungal cell wall at arbuscular interfaces, it has generally been assumed that nutrient exchange between the partners preferentially takes place here.

Smith and Read (1997) have suggested the presence of different interfaces which may offer the possibility of specialization of function and spatial separation of different transport processes in AM symbiosis. However, intra- and intercellular interfaces have the same basic structure at the cellular level. The fungal symbiont colonizes a compartment outside the plant protoplast, so that interfaces are always composed of the membranes of both partners separated by an apoplastic region. Nutrient transfer across a symbiotic interface involves efflux from the donor partner into, and uptake by the receiver partner from the interfacial apoplast. Although composition and structure of the apoplast can affect nutrient transfer between symbionts, nutrient movement between partners must be controlled by the bordering plant and fungal membranes, and more specifically by the plasma membrane proteins that mediate controlled exchange of solutes into and out of the symbiont cells.

Basic principles of membrane transport processes

Nutrient transport across plant and fungal membranes generally involves both active and passive processes (Marger and Saier 1993; Garril 1995; Tanner and Caspari 1996; Maathuis and Sanders 1999; Saier 1999). Active nutrient transport across plasma membranes is driven by an electrochemical gradient of protons generated by plasma membrane H^+-ATPases (Serrano 1989; Michelet and Boutry 1995; Rao and Slayman 1996; Palmgren and Harper 1999). These primary transporters pump protons out of the cell, thereby creating pH and electrical potential differences across the plasma membrane. The H^+ electrochemical gradient provides an energy gradient that can drive secondary active transport of solutes coupled to re-entry of H^+. This secondary active transport can be inwards (symport) or outwards (antiport) and is itself electrogenic (charge carrying) when the flux of a neutral compound such as glucose or sucrose is linked to that of H^+. Passive movement of solutes is driven by their concentration gradients or electrochemical potential gradients if the solute is charged, and it is mediated by a diffusion facilitator, such as a channel protein or uniport.

Knowledge about the distribution of plasma membrane H^+-ATPases is essential in determining mechanisms underlying the transport of a specific nutrient. Considerable progress has been made in understanding the physiological role of plasma membrane H^+-ATPases through the application of molecular biology methods. Molecular studies have shown that plasma membrane H^+-ATPases are encoded by multigene families, which include at least 7 genes in tomato (Ewing and Bennett 1994), 10 in *Arabidopsis* (Harper et al. 1994), 9 in *Nicotiana plumbaginifolia* (Michelet et al. 1994; Moriau et al. 1999; Ouffatole et al. 2000), 2 in *Saccharomyces cerevisiae* (Schlesser et al. 1988; Serrano et al. 1986), 2 in *Schizosaccharomyces pombe* (Ghislain and Goffeau 1991) and 5 in *Glomus mosseae* (Ferrol et al. 2000a). It is still not known why there are so many H^+-ATPase isoforms, but it has been proposed that the different isozymes may differ in their biochemical or regulatory characteristics, providing pools of functionally distinct H^+ pumps. For example, two *S. cerevisiae* ATPases are under different transcriptional control (Ghislain and Goffeau 1991) and present different biochemical properties (Supply et al. 1993). In plants, enzyme isoforms seem to have tissue-specific expression, showing highest expression levels in cells where intense active transport is expected to occur (Michelet and Boutry 1995; Palmgren and Harper 1999), and some appear to have distinct biochemical properties (Palmgren and Christensen 1994).

Plasma membrane ATPases in the AM symbiosis

Woolhouse hypothesised already in 1975 that the controlled exchange of nutrients between AM symbionts should involve membrane transport processes, based on the activity of the plasma membrane H^+-ATPase. He suggested that phosphate is passively released across the fungal plasma membrane into the interfacial apoplast and then absorbed into the host by a carrier-mediated active transport mechanism at the plant plasma membrane. Woolhouse (1975) further proposed that carbohydrates are passively released into the interfacial apoplast across the host plasma membrane and then actively taken up into the fungus by a sugar-H^+ transport system at the fungal plasma membrane. In this section, this hypothesis will be discussed in the light of data that has since been obtained concerning the regulation of plasma membrane ATPases in the AM symbiosis.

Host plasma membrane ATPase

The first analyses of plasma membrane ATPase distribution in AM roots consisted of cytochemical staining for ATP hydrolysis (Marx et al. 1982; Gianinazzi-Pearson et al. 1991). ATPase activity was detected along the plant membrane surrounding arbuscule branches in cortical root cells, but no ATP-dependent enzyme activity was detected in the peripheral membrane of colonized cortical cells, around senescent hyphae, in non-colonized cells or in cortical parenchyma cells of non-mycorrhizal roots. Furthermore, such ATPase activity is absent from the host membrane surrounding intracellular hyphal coils developing in epidermal or hypodermal root cells (Gianinazzi-Pearson and Gianinazzi 1988). Based on these observations, it was proposed that the periarbuscular membrane could be specifically involved in host absorption of solutes that have been transported from the soil by the fungus and released into the symbiotic interface of parenchyma root cells, through an H^+ symporter activated by the electrochemical gradient generated by a periarbuscular H^+-ATPase. In this way, colonised parenchyma cells of host roots acquire a new function of nutrient absorption which is generally restricted to the outer root cell layers. Moreover, the finding that no ATPase activity is present in root cortical cells of a *Glomus intraradices*-inoculated myc^{-2} pea mutant, which is defective in complete arbuscule formation, suggests that development of intracellular arbuscules is a signal event for specialization of the surrounding plant membrane (Gianinazzi-Pearson et al. 1995).

Although these cytochemical studies were based on the localization of H^+-ATPase activities in the absence and presence of different enzyme inhibitors, some authors have suggested that plasma membrane H^+-ATPase can be inactivated by both lead and fixatives used in the cytochemical procedure (Katz et al. 1988). However, Gianinazzi-Pearson et al. (2000) have recently corroborated the cytochemical data by immunolocalisation of H^+-ATPase protein using specific polyclonal antibodies. These observations have confirmed that H^+-ATPase protein is present in the plant membrane at the arbuscule interface and absent from the periphery of colonized cells and from adjacent non-colonized cortical root cells (Fig. 1), suggesting a specific synthesis of the enzyme during the symbiotic interaction.

Biochemical studies using membrane vesicles have provided further evidence that plasma membrane H^+-ATPase activity is regulated in mycorrhizal roots (McArthur and Knowles 1993; Bago et al. 1997; Benabdellah et al. 1999). These studies showed that whilst mycorrhizal colonization did not affect the specific activity (per mg of protein) of plasma membrane H^+-ATPase, there was an increase in enzyme activity on a fresh weight basis and in the rate of H^+ transport driven by the H^+-ATPase with no change in ATP hydrolysis. Based on these data, it was suggested that AM formation might regulate plasma membrane H^+-ATPase by increasing the coupling efficiency between H^+ transport and ATP hydrolysis.

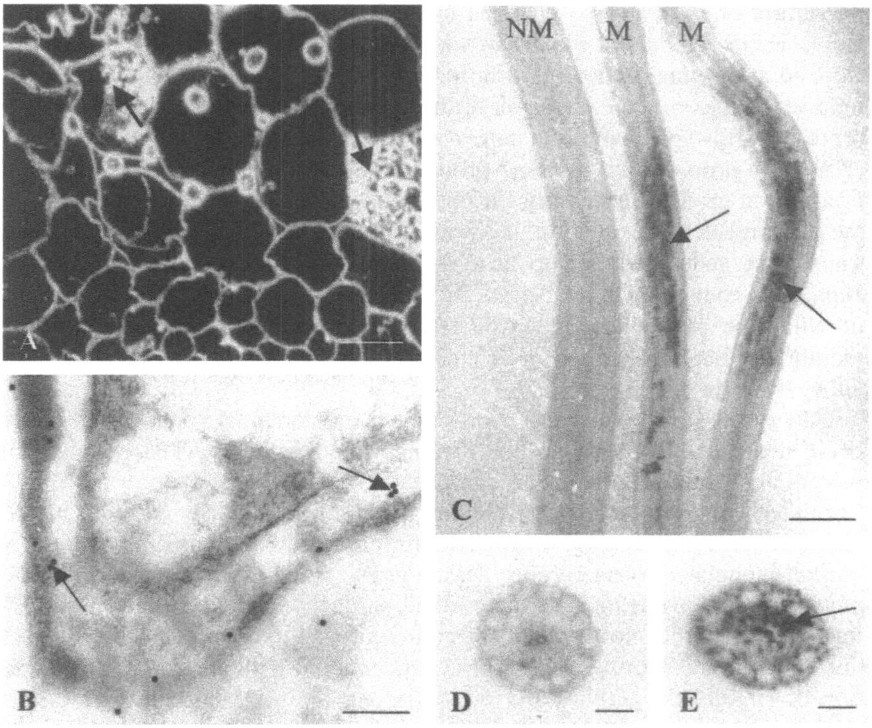

Figure 1. Localisation (arrows) of H^+-ATPase protein by immunocytochemistry (A, B) and of H^+-ATPase gene activation (C, D, E) in arbuscule-containing cells of mycorrhizal roots of tobacco. Indirect immunogold labelling is observed by light microscopy after silver enhancement (A) or directly by electron microscopy (B). Promoter activation of the H^+-ATPase gene is revealed in mycorrhizal roots by histochemical staining for GUS-reporter gene activity (C, E); no promoter-driven GUS activity is observed in non mycorrhizal roots (C, E). Bar = 15 μm (A), 0.2 μm (B), 0.5 mm (C), 1 mm (D), 1 mm (E).

As previously stated, plant plasma membrane H^+-ATPases are encoded by multigene families and there is evidence that some enzyme isoforms are expressed in specific cell, tissue or developmental stages. Since neither polyclonal antibodies used for ATPase localization nor biochemical ATPase assays distinguish between different isoforms, molecular approaches have been adopted to assess H^+-ATPase regulation in the AM symbiosis (Ferrol et al. 2000b). Murphy et al. (1997) first reported up-regulation of an H^+-ATPase-like gene during AM development in barley roots. Ferrol et al. (2000c) found that the expression of two out of three H^+-ATPase genes that can be active in tomato was differentially regulated in roots of wild-type and mycorrhiza-defective tomato mutants after inoculation with *Glomus mosseae*. In wild-type plants, mycorrhizal development induced down-regulation of *lha1*, up-regulation of *lha2* and no changes in the accumulation of *lha4* transcripts. However, in the mutant plants, in which development of the symbiosis was stopped at the stage of appressoria formation, AM fungal inoculation did not induce changes in the expression of these genes. Down-regulation of *lha1* in roots of wild-type plants but not in roots of the mutants led Ferrol et al. (2000c) to propose that the *lha1* protein could be involved in the generation of a proton gradient necessary for uptake of phosphate by epidermal cells in the non-mycorrhizal situation, in a way similar to two high affinity phosphate transporter genes reported to be down-regulated in mycorrhizal *Medicago truncatula* roots (Liu et al. 1998). The *lha2* gene selectively induced in roots of wild-type plants may be involved in energization of the periarbuscular membrane, but localization of H^+-ATPase gene expression is necessary for either hypothesis.

Cell type H^+-ATPase gene expression in mycorrhizal roots has recently been investigated in tobacco plants transformed with H^+-ATPase gene promoters and GUS reporter gene constructs (Gianinazzi-Pearson et al. 2000). Promoters of only two out of eight different H^+ATPase genes from *N. plumbaginifolia* (*pma* genes) showed differential activation in *Glomus fasciculatum*-inoculated tobacco roots. Strong *gusA* expression driven by *pma2* and *pma4* promoters was localised in cortical parenchyma cells colonized by living arbuscules, but not in adjacent non-colonized cells nor in the cortex of non-mycorrhizal roots (Fig. 1). These data concord with the specific localization of H^+-ATPase activity and protein along the periarbuscular membrane (Gianinazzi-Pearson et al. 1991; Gianinazzi-Pearson et al. 2000). It has been concluded from all these data that the H^+-ATPase activity on the plant membrane surrounding arbuscules in cortical parenchyma root cells is due to *de-novo* synthesis of the enzyme, as a result of plant gene activation by the symbiotic fungus. Induction of two different isoforms in arbuscule-containing cells suggests that H^+-ATPases could participate in processes other than membrane energization for active transport processes in the symbiotic interface. Gianinazzi-Pearson et al. (2000) propose that their activation could also be a component of a signal transduction pathway that triggers plant defence responses, as invoked for plant-pathogen interactions (Xing et al. 1996; Schaller and Oecking 1999), and so contribute to the localized induction of plant defence genes in arbuscule-containing cells (Harrison and Dixon 1994; Blee and Anderson 1996; Gianinazzi-Pearson et al. 1996).

It can be concluded from these different studies on barley, tomato and tobacco roots that AM fungal colonization modulates the expression of H^+-ATPase genes that can also be active in non-symbiotic tissues. However, Krajinski et al. (2000) have recently reported mycorrhiza-specific activation of an H^+-ATPase gene in roots of *M. truncatula*, but further studies are necessary to ascertain whether induction of a mycorrhiza-specific H^+-ATPase gene is a general phenomenon of the symbiotic interaction in different plant species.

Fungal plasma membrane ATPase

Current knowledge about regulation of the fungal plasma membrane H^+-ATPase in the different developmental stages of arbuscular mycorrhizal fungi (AMF) is scarce, because studies have been hampered by the obligately biotrophic nature of these organisms. As in the case of the host plasma membrane H^+-ATPase, the first evidence of ATPase activity in the presymbiotic or symbiotic stages of AMF came from cytochemical investigations. An ATPase activity was detected along the plasma membrane in apices of hyphae growing from spores of *Gigaspora margarita* germinating in the presence of root exudates or CO_2, and a positive correlation between this activity and ^{32}P uptake by germ tubes was observed (Lei et al. 1991). In the symbiotic stage, hyphae outside the root frequently have ATPase activity associated with their membranes, suggesting that uptake of mineral nutrients by external hyphae involve active membrane transport processes (Gianinazzi-Pearson et al. 1991). In accordance with this hypothesis, Harrison and van Buuren (1995) found expression of a high-affinity phosphate transporter gene in the extraradical hyphae of *Glomus versiforme*. Within roots, the intercellular hyphae, intracellular hyphal coils and arbuscular trunk hyphae frequently have plasma membrane-associated ATPase activity while this can be lacking from fine arbuscule branches (Marx et al. 1982; Gianinazzi-Pearson and Gianinazzi 1988; Gianinazzi-Pearson et al. 1991, 1995). These observations have led to the suggestion that carbon uptake from host cells might occur preferentially via the intercellular hyphae rather than across the arbuscular interface, as generally assumed. However as Gianinazzi-Pearson et al. (1991) pointed out, this hypothesis must remain tentative until more clear evidence on fungal H^+-ATPase distribution is obtained by other methods.

Genes encoding five putative plasma membrane H^+-ATPase isoforms have been partially cloned from *Glomus mosseae* (Ferrol et al. 2000a) and molecular characterization of these genes is underway. Results from these studies will provide insight into the physiological role of the different plasma membrane H^+-ATPase isoforms and into the fungal membranes involved in active transport processes. However, knowledge of the precise symbiotic interface over which carbon is transferred from the plant to the fungus requires localization of the sugar transport protein in the AM fungal membrane. Fungi have active and passive sugar transport systems (Lagunas 1993) and, as pointed out by Harrison (1999), it is unclear whether uptake of carbon by AMF requires an active transport mechanism or whether concentrations of carbon at the interfaces are sufficient for uptake by facilitated diffusion. By comparison, a monosaccharide/H^+ symporter is probably the transporter responsible for hexose uptake by both free-living and symbiotic stages of the ectomycorrhizal fungus *Amanita muscaria* (Nehls et al. 1998; Wiese et al. 2000). Molecular characterization of the membrane proteins involved is an essential step towards identifying the mechanisms by which AMF take up carbon compounds from the apoplastic interface, the molecular nature of the transferred carbon and the interface across which transfer occurs.

Conclusion

Although progress has been made over the last few years towards understanding the role and regulation of plasma membrane H^+-ATPases within the AM symbiosis, there still remains a vast unknown. On the plant side, future research needs to elucidate the regulatory mechanisms affecting gene expression and activity of each H^+-ATPase isoform within the symbiosis, and determine whether the H^+-ATPase acts as a primary transporter or as an intermediate in certain signal transduction pathways. On the fungal side, expression and functional analyses of the plasma membrane H^+-ATPase genes will provide insights into the physiological role of H^+-ATPase isoforms in the different developmental stages of AMF. However, as in other systems, the development of transformation systems is necessary for these analyses. The recent transformation by Forbes et al. (1998) of *Gigaspora rosea* suggests that manipulation of gene expression in AMF may also be possible.

Acknowledgements
The authors are grateful to Christine Arnould and Isabel Garcia for preparing the photographic material.

References

Bago B, Donaire J P, Azcón-Aguilar C (1997) ATPase activities of root microsomes from mycorrhizal sunflower (*Helianthus annuus*) and onion (*Allium cepa*) plants. New Phytol 136:305-311

Benabdellah K, Azcón-Aguilar C, Ferrol N (1999) Plasma membrane ATPase and H^+ transport activities in microsomal membranes from mycorrhizal tomato roots. J. Exp. Bot. 337:1343-1349

Blee K A, Anderson A J (1996) Defense-related transcript accumulation in *Phaseolus vulgaris* L. colonized by the arbuscular mycorrhizal fungus *Glomus intraradices* Schenck and Smith. Plant Physiol. 110:675-688

Bonfante P, Perotto S (1995) Strategies of arbuscular mycorrhizal fungi when infecting host plants. New Phytol. 130:3-21

Ewing N N, Bennett A B (1994) Assessment of the number and expression of P-type H^+-ATPase genes in tomato. Plant Physiol. 106:547-557

Ferrol N, Barea J M, Azcón-Aguilar C (2000a) The plasma membrane H^+-ATPase gene family in the arbuscular mycorrhizal fungus *Glomus mosseae*. Curr. Genet. 37:112-118

Ferrol N, Barea J M, Azcón-Aguilar C (2000b) Molecular approaches to study plasma membrane H^+-ATPases in arbuscular mycorrhizas. Plant Soil 226:219-225

Ferrol N, Pozo M J, Barea J M, Azcón-Aguilar C (2000c) Expression of genes encoding plasma membrane H^+-ATPases in wild-type and mycorrhiza-defective tomato mutants. In: Weber H C, Imhof S, Zeuske D (eds) Third International Congress on Symbiosis. Philipps University of Marburg, Germany, 13-19 August p 72

Forbes P J, Millam S, Hooker J E, Harrier L (1998) Transformation of the arbuscular mycorrhiza *Gigaspora rosea* by particle bombardment. Mycol Res. 102:497-501

Garril A (1995) Transport. In: Gow N A R, Gadd G M (eds) The growing fungus. Chapman and Hall, London, UK pp 161-181

Ghislain M, Goffeau A (1991) The *pma1* and *pma2* H^+-ATPases from *Schizosaccharomyces pombe* are functionally interchangeable. J. Biol. Chem. 266:18276-18279

Gianinazzi-Pearson V (1996) Plant cell responses to arbuscular mycorrhizal fungi: getting to the roots of the symbiosis. Plant Cell 8:1871-1883

Gianinazzi-Pearson V, Arnould C, Oufattole M, Arango M, Gianinazzi S (2000) Differential activation of H^+-ATPase genes by an arbuscular mycorrhizal fungus in root cells of transgenic tobacco. Planta 211:609-613

Gianinazzi-Pearson V, Dumas-Gaudot E, Gollotte A, Tahiri-Alaoui A, Gianinazzi S (1996) Cellular and molecular defence-related root responses to invasion by arbuscular mycorrhizal fungi. New Phytol. 133:45-57

Gianinazzi-Pearson V, Gianinazzi S (1988) Morphological integration and functional compatibility between symbionts in vesicular arbuscular endomycorrhizal associations. In: Scannerini S, Smith D C, Bonfante-Fasolo P, Gianinazzi-Pearson V (eds) Cell to cell signals in plant, animal and microbial symbiosis. NATO ASI, Series H : Cell Biology, Springer Verlag, Berlin, 17:73-84

Gianinazzi-Pearson V, Gollotte A, Lherminier J, Tisserant B, Franken P, Dumas-Gaudot E, Lemoine M C, van Tuinen D, Gianinazzi S (1995) Cellular and molecular approaches in the characterization of symbiotic events in functional arbuscular mycorrhizal associations. Can J Bot 73:S526-S532

Gianinazzi-Pearson V, Smith S E, Gianinazzi S, Smith F A (1991) Enzymatic studies on the metabolism of vesicular-arbuscular mycorrhizas. V: Is H^+-ATPase a component of ATP-hydrolysing enzyme activities in plant-fungus interfaces? New Phytol. 117:61-79

Harper J F, Manney L, Sussman M R (1994) The plasma membrane H^+-ATPase gene family in *Arabidopsis*: genomic sequence of AHA10 which is expressed primarly in developing seeds. Mol. Gen Genet. 244:572-587

Harrison M J (1999) Biotrophic interfaces and nutrient transport in plant/fungal symbioses. J. Exp. Bot. 50:1013-1022

Harrison M J, Dixon R A (1994) Spatial patterns of expression of flavonoid/isoflavonoid pathway genes during interactions between roots of *Medicago truncatula* and the mycorrhizal fungus *Glomus versiforme*. Plant J. 6:9-20

Harrison M J, van Buuren M L (1995) A phosphate transporter from the mycorrhizal fungus *Glomus versiforme*. Nature 378:625-629

Jakobsen I (1999) Transport of phosphorus and carbon in arbuscular mycorrhizas. In: Varma A, Hock B (eds) Mycorrhiza. structure, function, molecular biology and biotechnology. Springer, Berlin Heidelberg New York pp 305-332

Katz D B, Sussman M R, Mierzwa R J, Evert R F (1988) Cytochemical localization of ATPase activity in oat roots localizes a plasma membrane-associated soluble phosphatase, not the proton pump. Plant Physiol. 86:841-847

Krajinski F, Gianinazzi-Pearson V, Franken P (2000) Analysis of mycorrhiza-specific expression of a H^+-ATPase gene using transgenic approaches in *Medicago truncatula*. In: Weber H C, Imhof S, Zeuske D (eds) Third International Congress on Symbiosis. Philipps University of Marburg, Germany p 104

Lagunas R (1993) Sugar transport in *Saccharomyces cerevisiae*. FEMS Microbiol. Rev. 104:229-242

Lei J, Bécard G, Piché Y (1991) Root factors stimulate ^{32}P uptake and plasmalemma ATPase activity in the vesicular-arbuscular mycorrhizal fungus *Gigaspora margarita*. New Phytol. 118:289-294

Liu H, Treu A T, Blaylock L A, Harrison M J (1998) Cloning and characterization of two phosphate transporters from *Medicago truncatula* roots: regulation in response to phosphate and colonization by arbuscular mycorrhizal (AM) fungi. Mol. Plant Microb. Interact. 11:14-22

Maathuis J F, Sanders D (1999) Plasma membrane transport in context making sense out of complexity. Curr. Opin. Plant Biol. 2:236-243

Marger M D, Saier M H (1993) A major superfamily of transmembrane facilitators that catalyze uniport, symport and antiport. Trends Biochem. Sci. 18:13-20

Marx C, Dexheimer J, Gianinazzi-Pearson V, Gianinazzi S (1982) Enzymatic studies on the metabolism of vesicular-arbuscular mycorrhiza. IV. Ultra-cytoenzymological evidence (ATPase) for active transfer processes in the host-arbuscule. New Phytol. 90:37-43

McArthur D A J, Knowles N R (1993) Influence of vesicular-arbuscular mycorrhizal fungi on the response of potato to phosphorus deficiency. Plant Physiol. 101:147-160

Michelet B, Boutry M (1995) The plasma membrane H^+-ATPase: a highly regulated enzyme with multiple physiological functions. Plant Physiol. 108:1-6

Michelet B, Lukaszewicz M, Dupriez V, Boutry M (1994) A plant plasma membrane proton-ATPase gene is regulated by development and environment and shows signs of translational regulation. Plant Cell 6:1375-1389

Moriau L, Michelet B, Bogaerts P, Lambert L, Michel A, Ouffatole M, Boutry M (1999) Expression analysis of two gene subfamilies encoding the plasma membrane H^+-ATPase in *Nicotiana plumbaginifolia* reveals the major transport functions of this enzyme. Plant J 19:31-41

Murphy P J, Langridge P, Smith S E (1997) Cloning plant genes differentially expresed during colonization of roots of *Hordeum vulgare* by the vesicular-arbuscular mycorrhizal fungus *Glomus intraradices*. New Phytol. 135:291-301

Nehls U, Wiese J, Guttengerger M, Hampp R (1998) Carbon allocation in ectomycorrhizas: identification and expression analysis of an *Amanita muscaria* monosaccharide transporter. Mol Plant-Microbe Interact 11:167-176

Ouffatole M, Arango M, Boutry M (2000) Identification and expression of three new *Nicotiana plumbaginifolia* genes which encode isoforms of a plasma-membrane H^+-ATPase, and one which is induced by mechanical stress. Planta 210:715-722

Palmgren M G, Christensen G (1994) Functional comparison between plant plasma membrane H^+-ATPase isoforms expressed in yeast. J. Biol. Chem. 269:3027-3033

Palmgren M G, Harper J F (1999) Pumping with plant P-type ATPases. J. Exp. Bot. 50:883-893

Rao R, Slayman C W (1996) Plasma-membrane and related ATPases. In: Brambl R, Marzluf G A (eds) The mycota III. Biochemistry and molecular biology. Springer-Verlag Berlin Heidelberg pp 29-56

Saier M H Jr (1999) Eukaryotic transmembrane solute transport systems. Int. Rev. Cytol. 190:61-136

Schaller A, Oecking C (1999) Modulation of plasma membrane H^+-ATPase activity differentially activates wound and pathogen defense responses in tomato plants. Plant Cell 11:263-272

Schlesser A, Ulaszewski S, Ghislain M, Goffeau A (1988) A second transport ATPase gene in *Saccharomyces cerevisiae*. J. Biol. Chem. 263:19480-19487.

Serrano R (1989) Structure and function of plasma membrane ATPase. Annu. Rev. Plant Physiol. Plant Mol Biol. 40:61-94

Serrano R, Kielland-Brandt M C, Fink G R (1986) Yeast plasma membrane ATPase is essential for growth and has homology with (Na^+/K^+), K^+ and Ca^{2+}-ATPases. Nature 319:689-693

Smith S E, Smith F A (1990) Structure and function of the interfaces in biotrophic symbioses as they relate to nutrient transport. New Phytol. 114:1-38

Smith S E, Read D J (eds) (1997) Mycorrhizal symbiosis. Academic Press, London.

Supply P, Wach A, Goffeau A (1993) Enzymatic properties of the PMA2 plasma membrane-bound H^+-ATPase of *Saccharomyces cerevisiae*. J Biol Chem 268:19753-19759

Tanner W, Caspari T (1996) Membrane transport carriers. Annu. Rev. Plant Physiol. Plant Mol. Biol. 47:595-626

Wiese J, Kleber R, Hampo R, Nehls U (2000) Functional characterization of the *Amanita muscaria* monosaccharide transporter, AmMst1. Plant Biol 2:278-282

Woolhouse H W (1975) Membrane structure and transport problems considered in relation to phosphorus and carbohydrate movements and regulation of endotrophic mycorrhizal association. In: Sanders F E, Mosse B, Tinkers P B (eds) Endomycorrhizas. Academic Press, London, pp 209-230

Xing T, Higgins V J, Blumwald E (1996) Regulation of plant defense response to fungal pathogens: two type of protein kinases in the reversible phosphorylation of the host plasma membrane H^+-ATPase. Plant Cell 8:555-564

Mycorrhizal Technology in Agriculture
ed. by S. Gianinazzi, H. Schüepp, J.M. Barea and K. Haselwandter

Arbuscular mycorrhizal fungi nitrate assimilation: Genes and ecophysiological aspects

H. Bothe and U. Hildebrandt

Botanical Institute, The University of Cologne, Gyrhofstr.15, D-50923 Köln, Germany

Keywords: Nitrogen cycle, nitrate assimilation, nitrate reductase, differential gene expression, metabolite transfer in AMF symbiosis, bacteria and AMF

The involvement of arbuscular mycorrhizal fungi (AMF) in the biological nitrogen cycle

Plants and many microorganisms are N-autotrophs. They meet their demand for organic nitrogen by the conversion of nitrate to ammonia. The ammonia thus formed or directly taken up from the soil is incorporated into amino acids and heterocyclic N-compounds. Fungi can assimilate nitrate. *Neurospora crassa*, e.g. has been one of the model organisms for studying the process physiologically, biochemically and genetically (Marzluf 1997). The other reactions of the biological nitrogen cycle (nitrification, denitrification, N_2-fixation) are assumed to be performed only by prokaryotes. Any occasional claim in the past that fungi can fix N_2 has not been confirmed by critical re-examinations. However, bacteria associated with fungi may perform N_2-fixation (Bianciotto et al. 1996; 2000). *Burkholderia* spp, e. g. reside inside of AMF-spores and are known to fix N_2. No detailed investigations on whether *Burkholderia* significantly contributes to the N-supply of fungi were presented as yet.

Like bacteria, also fungi can perform denitrification (dissimilatory nitrate reduction, nitrate respiration). In the absence of O_2, many bacteria utilise nitrate and reduce it via nitrite, nitric oxide and nitrous oxide to molecular nitrogen. The pathogenic fungus *Fusarium oxysporum* has been described to perform denitrification in mitochondria (Kobayashi et al. 1996). The dissimilatory nitric oxide reductase of this organism has even been crystallized and characterized by X-ray analysis (Park et al. 1997). Thus fungi could well contribute to the global N_2-release and the formation of NO and N_2O (which are detrimental gases in several ways (Kloos et al. 1998; Bothe et al. 2000). This aspect has not been studied for AMF. As the fungi themselves, bacteria associated with the spores and other fungal structures could well perform denitrification reactions. Fungi (both parasitic and saprophytic species) significantly contribute to the decomposition of organic nitrogen. No information is available on to which extent AMF are involved in this process.

The role of AMF in the acquisition of nitrogen by plants

The relative contribution of AMF is not so clear as in the case of phosphorus. The elegant work of M. Harrison and her colleagues (Harrison and van Buuren 1995; Liu et al. 1998) indicated that a specific fungal phosphate translocator is expressed upon mycorrhizal colonization, whereas the plant counterpart is repressed. Many experiments with ^{32}P labelled phosphate clearly showed that hyphae of mycorrhizal fungi more efficiently exploit soils from phosphorus than roots and effectively transfer it to the plant partner (see e.g. Pearson and Jakobsen 1993). For the element nitrogen, an easily amenable radioactive isotope is not available as in the case of ^{32}P, therefore N-uptake and transfer measurements are restricted to mass spectrometric determinations with ^{15}N. Such studies are generally confronted with methodological problems and thus often do not provide clear results. However, measurements of the $^{15}NO_3^-$ or $^{15}NH_4^+$-uptake as well as enzyme activity determinations have been interpreted to indicate a significant contribution of AMF to the N-budget of plants (Johansen et al. 1992; George et al. 1992; Frey and Schüepp 1993; Tobar et al. 1994; Bago et al. 1996; Subramanian and Charest 1998). It has, however, also been reported that the hyphal N supply was not sufficiently high to sustain adequate N-nutrition of the plants supplied with very low amounts of N to the roots (Hawkins and George 1999). Any transfer of N in the opposite direction, from the root tissues to the AMF, seems to be insignificant (Johansen et al. 1994; 1996).

The enzymes of nitrate assimilation and their expression in the AMF plant symbiosis

Nitrate is taken up from the soil into the cells by a symport with H^+. In plants, several low and high affinity nitrate uptake carriers exist which reside at the plasmalemma or the tonoplast (Forde 2000). A computer analysis of the nitrate carrier sequences deposited into the databanks allowed to develop specific primers by which segments of two low and two high affinity nitrate carriers from tomato could be amplified by PCR (Hildebrandt and Bothe in preparation, Hildebrandt 2000). Sequencing of these PCR-products and hybridisation with DNA from tomato identified them as different putative carriers from this plant. The expression of one low and two high nitrate carriers was enhanced by the addition of nitrate to the tomato plants as shown by Northern analysis. This approach and also *in situ* hybridisations indicated their main expression in roots. One putative high affinity nitrate carrier was particularly interesting as its sequence has not yet been described and as its expression is negatively controlled by ammonia but remarkably not by glutamine. The specific transcript level of this carrier is significantly higher in tomato roots colonized with AMF than in non-colonized controls. We assume that this particular high affinity nitrate carrier plays a crucial role in the transfer of nitrate to the plant cells. No information is available for any AMF nitrate carrier as yet.

In the cytoplasm nitrate is reduced to nitrite catalysed by nitrate reductase. Plant species vary with respect to the organs in which nitrate reductase (and nitrite reductase catalysing the subsequent reduction of nitrite to ammonia) is localised. Some plants (conifers, e.g.) mainly reduce nitrate in roots, whereas others (maize, tomato) perform this reduction preferentially in leaves. AMF also possess an assimilatory nitrate reductase which was shown by activity measurements with isolated spores (Ho and Trappe 1975). Specific oligonucleotide primers for nitrate reductase allowed to amplify, clone and sequence an about 1 kb large gene segment of nitrate reductase from several fungi, namely from *Aspergillus nidulans, Pythium intermedium, Phytophthora infestans, Phytophthora megasperma* and also from the *Glomus* isolate D13 (Kaldorf et al. 1994). Sequence comparison and/or DNA-DNA hybridisations confirmed that the amplificate belonged to the organisms from which the DNA had been isolated. A hybridization experiment with the digoxigenin labelled PCR segment and a DNA preparation from about 0.5 million of *Glomus* spores confirmed the occurrence of nitrate reductase in AMF. It is not surprising that parasitic fungi (*Phytophthora, Pythium*) possess nitrate reductase because this had already been shown by activity measurements (Kaldorf et al. 1994).

Recently performed Northern hybridisations as well as activity measurements with hyphae grown in the compartment system (thus separated from the roots) showed that the nitrate reductase gene is expressed in the fungal mycelium separately from the plant partner (Hildebrandt 2000).

Tomato plants were grown for 12 weeks with *Glomus intraradices* Sy167 (= M) and without an AMF (= NM). Total RNA from the roots was then isolated by standard techniques (Hildebrandt 2000). Northern hybridisations were performed with the digoxigenin-labelled riboprobe of *Glomus* nitrate reductase. The signal intensity was determined densitometrically, the filter was stripped and subsequently hybridized with the digoxigenin labelled 18S-rRNA probe from tomato. The densitometrically determined signal intensity for the 18S-rRNA probe was used to calibrate the signal strength obtained with the nitrate reductase probe. Thus for nitrate reductase, AMF colonized tomato roots exhibited an almost five times more intensive signal than the controls. The weak signal obtained with the non mycorrhizal probe are due to unspecific cross–hybridisations of the *Glomus* nitrate reductase probe with tomato RNA. Mean signal intensities and variations are given for the two lanes.

The gene probe available for nitrate reductase from both *Glomus* and maize allowed to study the differential expression of this enzyme in maize in dependence of the AMF colonization (Kaldorf et al. 1998). Northern analysis showed that the transcript level of the *Glomus* nitrate reductase gene is enhanced upon mycorrhizal colonization, whereas that of the maize enzyme is reduced. Similar conclusions came from the *in situ* hybridisations and activity measurements. Transcription of nitrite reductase from maize is also lower in AMF colonized maize as compared with the non-colonized controls. Similar exper-

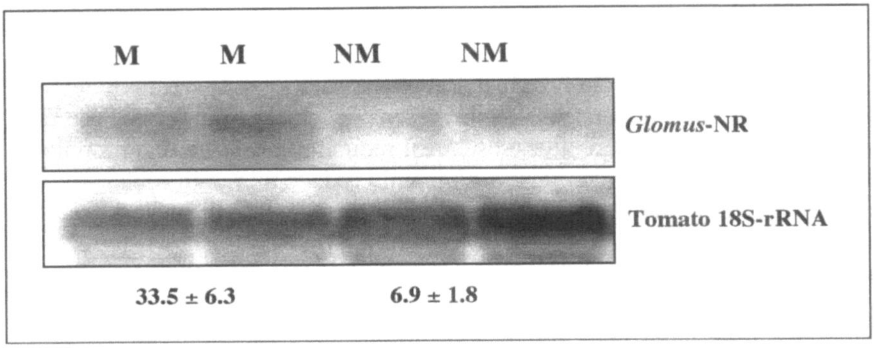

Figure 1. The transcription of the fungal nitrate reductase gene in AMF-colonized plants

iments with tomato colonised by AMF (Hildebrandt 2000) confirmed that the fungal nitrate reductase transcript level is enhanced upon mycorrhizal colonisation (Fig. 1). In the AMF-tomato symbiosis, the nitrate reductase gene of the plant is, however, not significantly downregulated (Hildebrandt 2000) as in the AMF-maize system (Kaldorf 1998).

Summing up the data, it is concluded that the transcript level of at least one specific high affinity nitrate carrier of tomato is enhanced upon mycorrhizal colonization. Nitrate uptake appears to be increased in AMF colonized plants. In addition, the fungal partner, partly, but not exclusively, overtakes the reduction of nitrate to ammonia which is reflected by the increase of *Glomus* nitrate reductase transcripts in the maize and tomato symbioses. Thus nitrogen might be transferred from the fungal partner to the plants both as nitrate and in a reduced, not yet identified form.

Concluding remarks

It is clear from the current experimental state that the fungal partner is involved in nitrate reduction of the AMF-plant symbiosis. However, it is not yet possible to predict the percentage to be assigned to each partner. This is mainly due to the fact that fungal specific probes could not yet be developed, however, with the exception of the nitrate reductase. With the other enzymes, the plant and fungal sequences are apparently so different that oligonucleotide primers developed from the plant sequences consistently failed to provide PCR-products with DNA from *Glomus*. Thus differential display approaches have to be tried to identify fungal specific genes and to develop gene probes for fungal nitrate uptake carriers, for nitrite reductase, glutamine synthetase, GOGAT and other enzymes. It may also be rewarding to sequence the ORFs upstream and downstream of the *Glomus* nitrate reductase, since the genes involved in nitrate assimilation and uptake cluster in some organisms. Finally it should be noted that an involvement of the fungal partner in nitrate uptake makes sense. In aerobic soils, the main form of nitrogen is nitrate due to the activity of nitrifying bacteria. Nitrate is easily washed off from the soil particles in contrast to ammonia. Thus an effective mobilization and binding of nitrate by the fungal partner may well be a key factor in the functioning of the fungal-plant symbiosis.

Acknowledgements
The authors are indebted to the Deutsche Forschungsgemeinschaft for continuous financial support of their own work on mycorrhiza. The diverse extensive discussion meetings of the COST Action 838 on arbuscular mycorrhiza significantly contributed for obtaining the own results cited in the text.

References

Bago B, Vierheilig H, Piché Y, Azcón-Aguilar C (1996) Nitrate depletion and pH-changes induced by the extraradical mycelium of the arbuscular mycorrhizal fungus *Glomus intraradices* grown in monoxenic culture. New Phytol. 133:273-280

Bianciotto V, Bandi C, Minerdi D, Sironi M, Tichy H V, Bonfante P (1996) An obligately endosymbiotic mycorrhizal fungus itself harbors obligately intracellular bacteria. Appl. Environ. Microbiol. 62:3005-3010

Bianciotto V, Lumini E, Lanfranco L, Minerdi D, Bonfante P, Perotti S (2000) Detection and identification of bacterial endosymbionts in arbuscular mycorrhizal fungi belonging to the family Gigasporaceae. Appl. Environ. Microbiol. 66:4505-4509

Bothe H, Jost G, Schloter M, Ward B B, Witzel K-P (2000) Molecular analysis of ammonia oxidation and denitrification in natural environments. FEMS Microbiol. Rev. 24:673-690

Forde B G (2000) Nitrate transporters in plants: structure, function and regulation. Biochim. Biophys. Acta 1465:219-235

Frey B, Schüepp H (1993) Acquisition of nitrogen by external hyphae of arbuscular mycorrhizal fungi associated with *Zea mays* L. New Phytol. 124:221-230

George E, Häussler K, Vetterlein D, Gorgus E, Marschner H (1992) Water and nutrient translocation by hyphae of *Glomus mosseae*. Can. J. Bot. 70:2130-2137

Harrison M J, van Buuren M (1995) A phosphate transporter from the mycorrhizal fungus *Glomus versiforme*. Nature 378:626-629

Hawkins H-J, George E (1999) Effect of plant nitrogen status on the contribution of arbuscular mycorrhizal hyphae to plant nitrogen uptake. Physiol. Plant. 105:694-700

Hildebrandt U (2000) Untersuchungen zur molekularbiologischen Charakterisierung der Nitratassimilation in der Symbiose zwischen dem arbuskulären Mykorrhizapilz *Glomus* und Tomate. Thesis. The University of Cologne

Ho I, Trappe J M (1975) Nitrate reducing capacity of two vesicular-arbuscular mycorrhizal fungi. Mycologia 67:886-888

Johansen A., Finlay R D, Olsson P A (1996) Nitrogen metabolism of external hyphae of the arbuscular mycorrhizal fungus *Glomus intraradices*. New Phytol. 133:705-712

Johansen A, Jakobsen I, Jensen E S (1994) Hyphal N-transport by a vesicular arbuscular mycorrhizal fungus associated with cucumber grown at three nitrogen levels. Plant and Soil 160:1-9

Kaldorf M, Schmelzer E, Bothe H (1998) Expression of maize and fungal nitrate reductase genes in arbuscular mycorrhiza. Molecular Plant Microbes Interactions 11:438-448

Kaldorf M, Zimmer W, Bothe H (1994) Genetic evidence for the occurrence of assimilatory nitrate reductase in arbuscular mycorrhizal and other fungi. Mycorrhiza 5:23-28

Kloos K, Hüsgen M, Bothe H (1998) DNA probing for genes coding for denitrification, N_2-fixation and nitrification in bacteria isolated from different soils. Z. Naturf. 53c:69-81

Kobayashi M, Matsuo Y, Takimoto A, Suzuki S, Maruo F, Shoun H (1996) Denitrification, a novel type of respiratory metabolism in fungal mitochondria. J. Biol. Chem. 271:16263-16267

Johansen A., Jakobsen I, Jensen E S (1992) Hyphal transport of ^{15}N-labelled nitrogen by a vesicular-arbuscular mycorrhizal fungus and its effect on depletion of inorganic soil N. New Phytol. 122:281-288

Liu H, Trieu A T, Blaylock L A, Harrison M J (1998) Cloning and characterization of two phosphate transporters from *Medicago truncatula* roots: Regulation in response to phosphate and to colonization by arbuscular mycorrhizal (AM) fungi. Mol. Plant-Microbe Interact. 11:14-22

Marzluf G A (1997) Genetic regulation of nitrogen metabolism in the fungi. Microbiol. Mol. Biol. Rev. 61:17-32

Park S Y, Shimizu H, Adachi S, Nakagawa A, Tanaka I, Nakahara K, Shoun H, Obayashi E, Nakamura H, Iizuka T, Shiro Y (1997) Crystal structure of nitric oxide reductase from the denitrifying fungus *Fusarium oxysporum*. Nature Struct. Biol. 4:827-832

Pearson J N, Jakobsen I (1993) Symbiotic exchange of carbon and phosphorus between cucumber and three arbuscular mycorrhizal fungi. New Phytol. 124:481-488

Subramanian K S, Charest C (1998) Arbuscular mycorrhizae and nitrogen assimilation in maize after drought and recovery. Physiol. Plant. 102:285-296

Tobar R, Azcón R, Barea J M (1994) Improved nitrogen uptake and transport from ^{15}N-labelled nitrate by external hyphae of arbuscular mycorrhiza under water-stressed conditions. New Phytol. 126:119-122

Mycorrhizal Technology in Agriculture
ed. by S. Gianinazzi, H. Schüepp, J.M. Barea and K. Haselwandter

An ecological point of view on arbuscular mycorrhiza research

D.J. Read

Department of Animal and Plant Sciences, University of Sheffield, Sheffield S10 2TN, UK

Keywords: Wide functional potential, diversity and versatility, plant life cycles and plant community, intra-specific selection over generations, population structure and dynamics of natural plant communities, role in complex ecosystem processes, ecologically realistic conditions

Introduction

It is now around one hundred years since those who pioneered the description of what became known as the arbuscular mycorrhizal (AM) symbiosis made their extensive analyses of its distribution through the plant kingdom. What was striking about their studies, in addition to the fact that so many autotrophic plant species were shown to be colonised by arbuscular mycorrhizal fungi (AMF), was the extraordinary diversity of climates and soil types in which the symbiosis occurred. These were shown to range from tropical rain forest (Janse 1897) to temperate woodlands (Gallaud 1905) and from saline to acidic soils (Stahl 1900). While plant ecologists would recognise that each of these distinctive habitats has selected for its own particular functional groups of plants, there has been little or no acknowledgement that the similar selective forces can be expected to have shaped the functional attributes of the fungal symbionts occurring in these habitats. On the contrary, the vast majority of studies of AM function have been based upon a very small number of fungal 'species', mostly selected from within the single genus *Glomus* and isolated from agricultural soils. This is not necessarily a criticism, and, in view of the fact that studies of the ectomycorrhizal symbiosis have likewise concentrated on a small number of fungal 'species' which conveniently grow in culture, it is certainly not a stricture that can be directed solely at those working with the AM system. Rather, the point is made to serve as a reminder, for those who may have ecological or landscape restoration interests, that the genotypic variability of AMF may be far greater than is normally recognised.

One of the major challenges to those working on the AM symbiosis in the next hundred years will be to identify and unlock the capabilities which have enabled these organisms to flourish in a vast range of terrestrial environments since they first appeared in the fossil record around 460 million years ago (Redecker et al. 2000). Some of the possible constraints upon our ability so far to realise this potential as well as possible avenues for advance are discussed below.

Constraints upon progress towards understanding AM function

Taxonomic factors

The dependence of the taxonomy and identification of AMF upon morphological features of their spores has had a restricting influence upon appreciation of their probable functional versatility. Their mode of spore formation has enabled the identification of only three families with six genera and around 150 'species' worldwide (Morton and Benny 1990, Walker and Trappe 1993). However, identification based upon morphology is of little value in applied or ecological contexts not only because the extent of spore production is highly dependent upon environmental factors such as disturbance and plant species composition (Merryweather and Fitter 1998a), but also, perhaps more importantly, because a single taxon may encompass a wide range of functional types. The fact that the spores are produced asexually has led widely to the perception that there may be less genetic and functional diversity in AMF than in other groups of mycorrhizal fungi. However, it is increasingly recognised that individual AMF spores contain many, perhaps thousands, of nuclei (Sanders et al. 1996). Each such spore, if mutations occurring over time were stable, would be expected to be a major repository of genetic diversity. In addition, it has recently been demonstrated (Giovanetti et al. 1999, 2000) that fusions occur frequently between hyphae of some AMF 'species'. The potential which such events provide for genetic exchange is clear. In future, considerable advance in our ability to recognise genotypic diversity should arise from the use of specific PCR primers to amplify glomalean rDNA directly from colonised roots (Simon et al. 1992; Helgason et al. 1998; van Tuinen et al. 1998). Application of these techniques will reveal much about the spatio-temporal structures of glomalean communities in nature while providing a tool with which to characterise genotypes of identified function.

Growth conditions

While we ultimately require answers to questions concerning the functions of AM symbioses in the diverse range of habitats in which they were originally described, it is a fact that most experimentalists, when confronted with the complexities of the natural environment, have sought to greatly simplify the conditions under which their studies were carried out. Thus the now classical 'big plant-little plant' pot experiments often using one 'species' of fungus and individual plants have enabled a blinkered view of AM function to become established. Some useful insights have clearly been obtained by these approaches. Thus, the sheer number of experiments carried out under such conditions enable us to make firm predictions about the processes whereby phosphorus supplies to individual roots, as well as to single plants, albeit usually grown for short periods of time, are augmented. However, the results produced by such experiments reveal only those attributes of the symbiosis which can be expressed under ecologically unrealistic conditions.

As we advance to ask questions about the role of mycorrhiza in ecosystem processes, the complexity required of our experimental designs will, of necessity, become progressively greater. Some, the 'reductionists', would say that they become so great as to render the questions unanswerable but such counsels of despair may derive as much from fear of the logistics of ecosystem scale experiments as from problems inherent in the science itself. Those who latterly have ventured into the new and challenging domain have done so by two different routes which can be designated 'microcosm' (Grime et al. 1987; van der Heijden et al. 1998a, b) and 'field' (Fitter 1986). The information gained from both approaches can be assimilated to produce broader pictures. What is evident is that if we are to investigate complex ecosystem processes by either route we should do so in such a way as to maximise relevance and to minimise the number of confounding factors.

The soil biota

Because emphasis has always been placed upon plant response to AMF colonisation, we still know very little about the structure of soil microbial and mesofaunal communities and even less about the role played by mycorrhizal mycelial networks in their trophic activities. Studies of some of these aspects have now been undertaken using ectomycorrhizal systems (Setala et al. 1999) and there is increasing awareness of the need for realistic assessment of these trophic levels in AM systems (Gange 2000). This is an area in which progress can be achieved through collaborative investigation involving specialists in other groups of the soil microbiota. The need to remove AMF inoculum in order to obtain non mycorrhizal 'control' treatments introduces a series of difficulties which are often underestimated. Since selective fungicides are not yet available (Paul et al. 1989) most 'control' treatments are confounded by wider effects upon the soil food web. It seems unlikely in view of the complexity of these systems that addition to sterile soil of a microbial 'filtrate' or its recolonisation by air borne propagules will reproduce a balanced natural community.

Restricted perceptions of possible impacts upon plant life cycles and plant community structure

Not only has there been emphasis upon the role of AM in the phosphorus nutrition of plants but observations of the processes of P acquisition have normally been restricted to relatively short periods of the plant life cycle. As a result we are unable to say a great deal about the possible impacts of the symbiosis upon those stages of the plant life cycle, for example seedling establishment and seed production which have a direct bearing on fitness (Read 1999). Since these aspects are likely to be at least as important as short-term productivity in determination of the composition of plant communities it is vital that attention be paid to them. Those studies that have been made of the impacts of AMF colonisation upon such features as fecundity (Shumway and Koide 1994) and seedling establishment (Francis and Read 1995) indicate that the impacts on plant fitness may be profound. The changing perception of the possible impacts of the AM symbiosis upon plant nutrition has been paralleled by an appreciation of other potential attributes. Amongst these its role in providing resistance to pathogens has achieved prominence (Carey et al. 1992; Newsham et al. 1995a), but possible contributions to resistance to climatic stresses such as drought (Allen and Allen 1986) and to metal toxicity (Gildon and Tinker 1981) should also be taken into account.

Outlook into the future

Greater awareness of the wider functional potentials of the symbiosis is timely because it should enable them to be considered if ecosystem scale experiments are being designed and analysed. It leads to emphasis on the need to examine responsiveness of the mycorrhizal partners over their full life cycles or at least over critical parts thereof, in the presence of the environmental variables likely to be imposed upon the systems in nature. Since these variables themselves differ at global (e.g. latitudinal) and regional (e.g. edaphic) scales it can be predicted that selection will have favoured distinctive attributes in the plant and fungal partners under each environmental circumstance. These selective forces have been recognised by plant ecologists as 'drivers' which have produced clearly defined biomes (Odum 1971) or communities (Ellenberg 1988) which are characterised by distinctive suites of largely phenological characteristics seen above ground.

That powerful selective forces will also have moulded mycorrhizal fungal communities below ground is indicated by the changes observed at the global scale in the extent of occurrence of arbuscular, ericoid and ecto mycorrhizal categories of the symbiosis in different biomes (Read 1991). While these large-scale patterns are relatively easy to detect there is every likelihood that effects at the more local scales will influence selection of genotypes within each category of the symbiosis but, to date, we have been less aware of this possibility. Small scale changes of community structure, for example of the kind enabling identification of particular phyto-sociological units within broader groupings, *sensu* Braun-Blanquet (1928) and Rodwell (1991) may well be linked to, even driven by, distinctive genotypes of mycorrhizal fungi.

Soil as a selective determinant of AM fungal population structure

Amongst the soil factors which can cause shifts in AMF population structure at the species level are water content (Anderson et al. 1984) pH, (Porter et al. 1987;, Wang et al. 1993), temperature (Koske 1987) and organic content (Johnson et al. 1991). In combination such effects are likely to select for populations that are genotypically adjusted to a given soil type. In a study of the selective effects imposed by soil upon AMF community structure Johnson et al. (1992) showed that of 12 abundant species, the occurrence of 6 was strongly influenced by soil type. There is thus much to suggest that the population of microbial symbionts will be moulded by local environmental conditions in the same way as is the plant population seen above ground and that, particularly in stable natural and semi-natural environments, particular genotypic traits will have been selected at the intra-specific level over many generations. The fact that disturbance appears to lead on the one hand to severe losses of genetic diversity and on the other to selection of distinctive, perhaps ruderal, genotypes (Helgason et al. 1998), indirectly lends support to the view that AMF populations are specially adapted to particular local conditions.

Recognition of the likelihood that AMF populations may be edaphically selected is important. It runs counter to the prevailing view, perhaps largely based upon the low levels of speciation seen in the Glomales, that genetic diversity is low. While it may be appropriate, indeed ideal, for studies of molecular or physiological processes to use 'model' organisms, the trend in ecosystem studies should be away from use of ecologically irrelevant genotypes obtained from culture collections, towards selection of those isolated from the ecosystem being investigated. Since some of these may be poorly represented in the local spore population, the use, as initial inoculum, of root segments freshly cut from the plant species in question has much to recommend it. Parallel taxonomic characterisation of these AMF communities, using conventional or molecular methods as appropriate, is clearly desirable. Genetic characterisation will in turn provide the potential for determination of the extent of intra-specific specialisation within AMF communities. Access to these populations will enable experimental investigation of the relative importance of soil type on the one hand, and plant genotype on the other, in selecting for genetic differentiation in populations of the fungal symbionts.

'Host' compatibility as a factor determining AM fungal population structure

While sensitivity towards particular soil conditions is likely to be one factor exerting selective effects upon AM fungal populations, another might be host-specificity. However, there is much empirical evidence that a given AMF can readily penetrate the roots of most host species. Indeed it is this assumption that has formed the basis of experimental designs involving inoculation with 'model' fungi such as *G. mosseae*. Such views are strengthened by observations based upon both morphological (Merryweather and Fitter 1998b) and molecular (Helgason et al. 1999) analyses, that roots of a single plant, in this case of bluebell (*Hyacinthoides non-scripta*), may simultaneously support up to 7 taxa of AMF. Some of these fungi have been shown to sporulate infrequently, if at all, and so would have been overlooked by conventional approaches involving, for example, use of trap cultures. What emerges from such observations is that the plants appear to exert little influence upon infectivity of the fungus. Recent studies by van der Heijden et al. (1998) have indicated that, despite this apparent lack of specificity, a given cohort of fungal taxa occupying a root system will contain some genotypes which stimulate plant growth more effectively than others. If this situation led to a positive feedback to that fungus from the plant it is predictable that it would be selected, eventually to the exclusion of other inhabitants of the root. The presence of multiple infections even under these conditions appears to suggest that the host is exerting little influence upon the composition of its fungal symbiont population. The long evolutionary history of interdependence of plant and fungus in this symbiosis may indeed have produced selection against resistance to colonisation. As pointed out by Law and Lewis (1983), the internal environment of the root is far more uniform than that of the soil in which the external mycelial network proliferates. For this reason alone, selective forces exerted by inherently complex soil environments may have considerably greater impacts than any arising from host plants upon AMF populations.

Conclusion

The constraints imposed upon AM research by a simplistic view of fungal diversity, and reductionist approaches to experimental design are at last being overcome. Earlier preoccupations with the involvement of the fungal symbiont in the phosphorus nutrition of single plants are being replaced by more holistic analyses which recognise the potential for functional diversity in the symbiosis and emphasise the need for analyses of plant response over entire lifecycles as well as in real communities. In addition to providing new perspectives on the structure and dynamics of natural plant communities these approaches will increase our scope to manipulate the symbiosis in conservation and landscape restoration schemes.

References

Allen E B, Allen M F (1986) Water relations of xeric grasses in the field: interactions of mycorrhizae and competition. New Phytologist 104:559-71

Anderson R C, Liberta A E, Dickman L A (1984) Interaction of vascular plants and vesicular-arbuscular mycorrhizal fungi across a soil moisture-nutrient gradient. Oecologia (Berlin) 64:111-117

Braun-Blanquet J (1928) Pflanzensoziologie. Grundzüge der Vegetationskunde Springer, Berlin, Germany

Carey P D, Fitter A H, Watkinson A R (1992) A field study using the fungicide benomyl to investigate the effect of mycorrhizal fungi on plant fitness. Oecologia 90:550-555

Ellenberg H (1988) Vegetation Ecology of Central Europe. Cambridge University Press, Cambridge, UK

Fitter A H (1986) Effect of benomyl on leaf phosphorus concentration in alpine grassland: a test of mycorrhizal benefit. New Phytologist 103:767-776

Francis R, Read D J (1995) Mutualism and antagonism in the mycorrhizal symbiosis, with special reference to impacts on plant community structure. Canadian Journal of Botany 73 (Suppl. 1):1301-1309

Gallaud I (1905) Etudes sur les mycorrhizes endotrophs. Revue Générale de Botanique 17:5-48, 66-83, 123-135, 223-239, 313-325, 425-433, 479-500

Gange A C (2000) Arbuscular mycorrhizal fungi by Collembola and plant growth. Trends in Ecology and Evolution 15:369-372

Gildon A, Tinker P B (1981) A heavy metal tolerant strain of a mycorrhizal fungus. Transactions of the British Mycological Society 77:648-649

Giovannetti M, Azzolini D, Citernesi A S (1999) Anastomosis formation and nuclear and protoplasmic exchange in arbuscular mycorrhizal fungi. Applied and Environmental Microbiology 65:5571-5575

Giovanetti M, Fortuna P, Citernesi S, Morini S, Nuti M P (2001) At the root of the wood wide web: anastomosis formation and nuclear exchange in intact arbuscular mycorrhizal extraradical networks. New Phytologist (in press)

Grime J P, Mackey J M L, Hillier S H, Read D J (1987) Floristic diversity in a model system using experimental microcosms. Nature 328:420-422

Helgason T, Daniell T J, Husband R, Fitter A H, Young J P Y (1998) Ploughing up the wood-wide web? Nature 392:431

Helgason T, Fitter A H, Young J P W (1999) Molecular diversity of colonising *Hyacinthoides non-scripta* (bluebell) in a semi-natural woodland. Molecular Ecology 8:659-666

Janse J M (1897) Les endophytes radicaux de quelques plantes Javanaise. Annales du Jardin Botanique de Buitenzorg 14:53-212

Johnson N C, Pfleger F L, Crookston R K, Simmons S R, Copeland P J (1991) Vesicular-arbuscular mycorrhizae respond to corn and soybean cropping history. New Phytologist 117:657-663

Johnson N C, Tilman D, Wedin D (1992) Plant and soil controls on mycorrhizal fungal communities. Ecology 73:2034-2042

Koske R E (1987) Distribution of VA mycorrhizal fungi along a latitudinal temperature gradient. Mycologia 79:55-68

Law R, Lewis D H (1983) Biotic environments and the maintenance of sex - some evidence from mutualistic symbioses. Biological Journal of the Linnean Society 20:249-276

Merryweather J, Fitter A H (1998a) The arbuscular mycorrhizal fungi of *Hyacinthoides non-scripta* II. Seasonal and spatial patterns of fungal populations. New Phytologist 138:131-142

Merryweather J, Fitter A H (1998b) The arbuscular mycorrhizal fungi of *Hyacinthoides non-scripta* I. Diversity of fungal taxa. New Phytologist 138:117-129

Morton J B, Benny G L (1990) Revised classification of arbuscular mycorrhizal fungi (Zygomycetes): a new order, Glomales, two new suborders, Glomineae and Gigasporineae, and two new families, Acaulosporaceae and Gigasporaceae, with an emendation of Glomaceae. Mycotaxon 37:471-491

Newsham K K, Fitter A H, Merryweather J W (1995a) Multifunctionality and biodiversity in arbuscular mycorrhizas. Tree 10:407-411

Odum E P (1971) Fundamentals of Ecology. Saunders. Philadelphia, USA, p 574

Paul N D, Aryes P G, Wyness L E (1989) On the use of fungicides for experimentation in natural vegetation. Functional Ecology 3:759-769

Porter W M, Robson A D, Abbott L K (1987) Field survey of the distribution of vesicular-arbuscular mycorrhizal fungi in relation to soil pH. Journal of Applied Ecology 24:659-662

Read D J (1991) Mycorrhizas in ecosystems. Experientia 47:376-391

Read D J (1999) The ecophysiology of mycorrhizal symbioses with special reference to impacts upon plant fitness. In: Press M C, Scholes J D, Barker M G (eds) Physiological plant ecology. Blackwell Science, Oxford, UK

Redecker D, Kodner R, Graham L E (2000) Glomalean fungi from the Ordovician. Science 289:1920-1921

Rodwell J S (1991) British Plant Communities Vols 1-5 Cambridge University Press, Cambridge, UK

Sanders I R, Clapp J P, Wiemken A (1996) The genetic diversity of arbuscular mycorrhizal fungi in natural ecosystems: A key to understanding the ecology and functioning of the mycorrhizal symbiosis New Phytologist 133:123-134

Setälä H, Kulmala P, Mikola J, Markkola A M (1999) Influence of ectomycorrhiza on the structure of detrital food webs in pine rhizosphere. Oikos 87:113-122

Shumway D L, Koide R T (1994) Reproductive responses to mycorrhizal colonisation of *Abutilon theophrasti* Medic. Plants grown for two generations in the field. New Phytologist 128:219-224

Simon L, Lalonde M, Bruns T D (1992) Specific amplification of 18S fungal ribosomal genes from vesicular-arbuscular endomycorrhizal fungi colonizing roots. Applied and Environmental Microbiology 58:291-295

Stahl E (1900) Der Sinn der Mycorhizenbildung. Jahrbuch für wissenschaftliche Botanik 34:539-668

van der Heijden M G A, Boller T, Wiemken A, Sanders I R (1988a) Different arbuscular mycorrhizal fungal species are potential determinants of plant community structure. Ecology 79:2082-2091

van der Heijden M G A, Klironomos J N, Ursic M, Moutoglis P, Streitwolf-Engel R, Boller T, Wiemken A, Sanders I R (1998b) Mycorrhizal fungal diversity determines plant biodiversity, ecosystem variability and productivity. Nature 396:69-72

van Tuinen D, Jacquot E, Zhao B, Gollotte A, Gianinazzi-Pearson V (1998) Characterization of root colonization profiles by a microcosm community of arbuscular mycorrhizal fungi using 25 rDNA-targeted nested PCR. Molecular Evolution 7:879-887

Walker C, Trappe J M (1993) Names and epithets in the Glomales and Endogonales. Mycological Research 97:339-344

Wang G M, Stribley D P, Tinker P B, Walker C (1993) Effects of pH on arbuscular mycorrhiza. I. Field observations on the long-term liming experiments at Rothamsted and Woburn. New Phytologist 124:465-472

Mycorrhizal Technology in Agriculture
ed. by S. Gianinazzi, H. Schüepp, J.M. Barea and K. Haselwandter

Arbuscular mycorrhizal fungi, an essential component of soil microflora in ecosystem restoration

K. Turnau[1] and K. Haselwandter[2]

[1] *Institute of Botany of the Jagiellonian University, PL-31512 Krakow, ul. Lubicz 46, Poland*
[2] *Institut für Mikrobiologie, Universität Innsbruck, Technikerstrasse 25, A-6020 Innsbruck, Austria*

Keywords: biodiversity, drought, mine spoils, salinity, alpine habitats, endangered plant species, rhizosphere

Introduction

A wide range of natural factors such as lightning-caused fires and geomorphic or palaeotectonic processes may affect the stability of natural ecosystems (Herrera et al. 1993). Additionally, human activities causing pollution of air, water and soil, and overuse of resources like grasslands or clear-cutting of forests have a strong impact on a wide range of ecosystems. They may become degraded to such an extent that spontaneous recovery is strongly limited, especially if the damaging agent is continuously present. In general, successful restoration requires the reconstruction of adequate biological, physico-chemical, hydrological and morphological conditions. Moreover, the presence of hazardous substances can necessitate chemical or bioremediated clean-up. A common reason for the failure of many restoration attempts is the neglect of the fact that the plant root systems are associated with a diverse community of active soil micro-organisms. It is well known that a functioning association between plants and rhizosphere micro-organisms can modify the substratum, facilitate plant establishment under hostile conditions, and counteract the stagnation of the succession.

Among favourable plant-microbe interactions, the mycorrhizal symbiosis is now known to be an essential component of any plant community (Smith and Read 1997). So far, the potential contribution of arbuscular mycorrhizal fungi (AMF) to ecosystem functioning has not been adequately taken into consideration in a number of restoration programmes although plant-microbe interactions must be considered a pre-requisite of the successful establishment of stress tolerant, self-sustaining soil-plant systems. Hence, this chapter focuses on AMF as a potential tool in ecosystem restoration, with special emphasis on problems inherent to temperate regions in central Europe, and thus complementary to the chapter on mediterranean areas (Jeffries et al. in this book).

Why are AMF essential for ecosystem restoration?

Mycorrhizal symbiosis affects a wide range of aspects of the biology of a plant at all levels, from that of a functioning ecosystem to that of individual plants. AMF exhibit a very wide distribution and are able to form a symbiosis with about 80% of all plant species on Earth including angiosperms, gymnosperms, pteridophytes and hepatics. This kind of mycorrhiza dominates in ecosystems inhabited by herbaceous plants that are among the first organisms recolonizing previously damaged ecosystems (Smith and Read 1997).

The activity and diversity of AMF may be strongly affected by ecosystem damages (Jha et al. 1992). Not only the intensity of mycorrhizal colonisation, but also the arbuscule richness was found to reflect mycorrhizal functionality. In case of serious damage it might be necessary to introduce inoculum containing AMF spores or mycorrhizal roots. Recently AMF inoculum has become commercially available (see Feldmann and Grotkass in this book, von Alten et al. in this book). It is, however, important to remember that native autochtonous fungal strains may be best adapted to actual soil and climatic conditions; in some cases inoculation with inoculum developed especially for a given area could be more successful. If AMF are still present in the soil to be restored, the most valuable technique would rely on the management of the indigenous microbiota promoting their proliferation. This aim could be achieved by the application of experimentally estimated levels of selected fertilizers, the introduction of mycorrhizal and the avoidance of non-mycorrhizal plants, which may have a deleterious influence on symbiotic fungi. During soil restoration the evaluation of mycorrhiza development could be used as an important indicator of ecosystem changes and could serve as a tool for biomonitoring the soil quality and success of restoration practices (Lovera and Cuenca 1996; Haselwandter 1997).

Rhizosphere microorganisms as allied colonisers for restoration

Rhizosphere microorganisms such as nitrogen fixing bacteria, plant growth promoting rhizobacteria (PGPR) and mycorrhizal helper bacteria (MHB) are also components of the mycorhizosphere and have been shown to stimulate the development of the mycorrhizal symbiosis (Barea et al. in this book). Together with AMF all these organisms create unique beneficial systems that determine the structure of the plant community (Francis and Read 1995; Gryndler and Vosatka 1996). Among the rhizosphere microorganisms, only inoculation with nitrogen fixing organisms has received comparatively much attention. Reciprocal stimulation of nodule bacteria (*Azorhizobium*, *Bradyrhizobium*, *Rhizobium*, *Sinorhizobium*) and AMF has been known for quite a long time, but contrary to AMF that form symbiosis with a wide range of plants, nodulating bacteria are almost completely restricted to leguminous plants. Rhizobia studied in microcosm experiments have been shown to increase the number of

AMF entry points in plant roots and increase the development of mycelium from axenically germinated spores. Both kinds of stimulation are probably due to increased root exudation induced by *Rhizobium* polysaccharides and to increased production of phytohormones by rhizobia or other bacteria (Azcón-Aguilar and Barea 1992). The efficiency of such a triple symbiosis (bacteria, fungi and plants) resulting in increased plant biomass, N and P uptake depends on the selection of bacterial and fungal strains involved (Azcón et al. 1991).

Managing the rhizosphere by using a combination of various microorganisms including PGPR, MHB, rhizobia and AMF with multiple purposes offer an interesting possibility and could provide plants with benefits crucial for restoration. Some of these organisms may be applied in the form of seed inoculants or soil amendments. Matching appropriate ecotypes of plants and microbes is an important task. Additionally, some species of *Pseudomonas* proved to be able to degrade a vast range of compounds deposited in soil due to pollution (Crowley et al. 1996). Some of these organisms have been genetically modified and possess an increased biological control activity or an improved detoxification effect (Dowling and O'Gara 1994). Recently these genetically modified bacteria were demonstrated to have no detrimental effect on mycorrhiza (Edwards and Abivardi 1997). This obviously opens new perspectives, especially in combination with the potential for bioremediation, which AMF possess *per se* (Leyval et al. in this book).

AM as a tool for re-establishment of endangered plant species and population diversity

Natural processes and anthropogenic impact strongly affect plant populations and their viability, and finally can lead to the extinction of more vulnerable taxa. The multiplication of endangered plant species is the main goal of special protection projects such as those developed for *Pulsatilla silvatica* or *Senecio umbrosus* in the Polish Tatra Mountains (Mirek and Piekos-Mirkowa 2000). Active protection involves the cultivation of the plant *ex situ*. So far the technique employed in the project mentioned above does not include the management of mycorrhizosphere organisms and AMF. In general, knowledge of the mycorrhization of endangered plant species is lacking. On the other hand, information of this kind must be considered a pre-requisite for making plant re-establishment programmes successful (van der Heijden et al. 1998a). Thus, a study of the mycorrhizal status e.g. of a range of wetland plant species is bound to yield data on root colonisation as well as on spore numbers and types present in the habitat, which may well prove crucial for any attempts of re-establishment of such endangered plants (Fuchs 2001). As plants and microbial strains differ in the effectiveness of the symbiosis, the selection of adequate symbiotic consortia is bound to be of particular value, ensuring better chances for survival.

In cases of very small plant populations, micropropagation techniques have been recently introduced into protection practices. *In vitro* cultures have been used to multiply ferns such as *Osmunda regalis*, *Phyllitis scolopendrion* (Zenkteler 1993) and monocotyledons such as *Lilium martagon* (Bach and Kodra 1997) which are under strict protection in most European countries. According to Harley and Harley (1987) all these plants form arbuscular mycorrhiza. During the *post vitro* stage the plants are seriously endangered by pathogens (Mc Rae 1998). The risk could be attenuated by the introduction of AMF either *in vitro* or during the *post vitro* stage. The methods used, depending on the host species, may be adopted from those developed for commercially important plants (Hooker et al. 1994; Vosatka et al. 1999).

The structure of the plant community and the species composition appear to be driven by the biodiversity of AMF (van der Heijden et al. 1998b). Arbuscular mycorrhizal associations significantly influence the dynamics of plant succession (Janos 1980; Gange et al. 1990, 1993). Sites lacking AMF are colonised by facultatively mycorrhizal or non-mycorrhizal plant species (Reeves et al. 1979). If non-mycorrhizal vegetation becomes established, successional processes stagnate. On the contrary, the presence of AMF enables mycorrhizal plants to colonise the site and to explore the benefits of increased water and nutrient uptake, faster growth and production of higher biomass, thus making them more competitive for light or space. All of that is due to the association with symbiotic fungi that also confer a higher resistance to extreme conditions to mycorrhizal plants (Hetrick et al. 1989). Since there are large differences among plant species in growth response to mycorrhization, AMF significantly affect plant competition and community structure (Fitter 1977; Harnett et al. 1993; Francis and Read 1994; Sanders and Koide 1994). Changes in nutrient status or fitness due to AMF colonisation may alter the relationship between the host and other plants within the community.

Restoration of ecosystems subjected to water deficit in temperate areas

Diminution of the water supply is a general phenomenon and results in particularly serious changes in natural vegetation. In about 40% of the country area of Poland there is a deficit in surface water. Moreover, improper management of sandy soils resulted in the degradation of over 16% of the territory, of which 100,000 hectares need to be reclaimed (Grodzinska et al. 1995).

The functioning of the communities subjected to drought stress and the role of symbiotic organisms may be easily demonstrated on coastal or inland dunes, which are especially fragile. In Poland research focussed on active mobile dunes and deflation hollows of the Slowinski National Park, the Hel Penisula and the Bledowska Desert, which is an inland sandy area located in the eastern part of the Silesian Upland (Blaszkowski 1994; Blaszkowski et al. 2000; Tadych and Blaszkowski 2000). The Hel Peninsula, a narrow strip of land created by sand brought by waves of the Baltic Sea, is progressively disappearing due to changes of the course of sea currents. Restoration of the disturbed area is accomplished by the transfer of sand from the Puck Gulf. This sand, however, is devoid of AMF which may account for the delayed establishment of mycorrhizal plants and thus for the delay in sand stabilisation. The three studies quoted above give precise information on AMF species abundant in sandy areas, sizes of spore populations developed in the root zone of individual plant species and intensity of mycorrhiza development. Fungal strains are kept in pot cultures and could be a good source of inoculum, possibly useful for areas similar to the Bledowska Desert which is endangered by diminution of water resources as well as water contamination by industrial sludges from a neighbouring cellulose factory.

Restoration of alpine ecosystems

Alpine habitats belong to those ecosystems which are most susceptible to natural and anthropogenic impact causing ecosystem degeneration, mainly due to poor soil stability and the threat of massive landslides. Previously, the construction, use and maintenance of ski runs often led to damages in high-alpine areas (Urbanska 1995). Without suitable countermeasures almost bare surfaces are created, devoid of a reasonably high number of plant and fungal propagules. Hence the establishment of self-sustaining plant communities exhibiting an appropriate level of species diversity is an important task. This can be facilitated through transplantation of topsoil (Densmore 1994) containing pieces of roots as well as AMF mycelium and spores, which are all together good sources of mycorrhizal inoculum. The stabilising effect of these organisms is more important than the improvement of the nutrient status by extensive use of fertilisers (Urbanska 1995).

In natural high-alpine vegetation systems AMF infection is a widespread feature of dominant and sub-dominant plants (Read and Haselwandter 1981). A fine endophyte of the type of *Glomus tenuis* is common at higher elevation, and progressively becoming dominant over the coarse endophytes with altitude (Haselwandter and Read 1980). It is interesting to note that in comparison to a mineral fertiliser, at least one slow-release organic fertiliser appeared to enhance the development of arbuscular mycorrhiza (Haselwandter 1997). Thus, where the mycorrhizal inoculum potential is low, the few mycorrhizal propagules present could be stimulated through an adequate fertilizer treatment, ultimately leading to the build-up of a functioning self-sustaining AMF – plant – soil system. Mycorrhizal infection can be anticipated to be of paramount importance for plants to tolerate the climatic and nutritional stress they are exposed to in an alpine environment (Haselwander 1987).

Restoration of surface-mined areas, industrial wastes and polluted soils

Strip mining, used to recover coal or metal ores, destroys vegetation, alters microbial communities, disrupts soils, and can cause soil compaction and reduced water infiltration (Grunwald et al. 1988). Strip mine reclamation can ameliorate these sites' conditions. It includes grading and levelling of spoil piles, topsoiling, adding soil amendments and introduction of plants such as grasses which can slow down erosion, stabilise the soil and successfully outcompete local weedy species (Helm 1995). Unfortunately, some of the reclamation practices such as topsoil stockpiling may lead to further compacting of the soil (Liberta 1981), and if sewage sludge is used as amendment, heavy metals may be introduced (Rodgers and Anderson 1989). The practices mentioned above, together with liming which is often used to increase soil pH may cause a further decline in microbial population including AMF (Sylvia 1990; Cuenca and Lovira 1992). The presence of heavy metals or polyaromatic hydrocarbons may strongly affect the quantity, diversity and activity of soil microbial populations including AMF (Ganesan et al. 1991; Del Val et al. 1999). In such cases physico-chemical extraction methods are the main technique employed, although they are extremely expensive and additionally damage the microbial community.

Restoration of surface-mined areas or industrial wastes is essential for reducing erosion and dust problems. Such areas are often located close to large agglomerations and constitute a serious health hazard. Hence, in general it is uttermost urgent to establish a plant cover on the surface of the abandoned mines or industrial wastes, as delayed revegetation may result in the transfer of toxic substances into the surroundings. The rate of natural revegetation processes is usually slow due to low levels of nutrients such as N and P and, especially at the initial stages, the lack of organic matter. The application of appropriate levels and kinds of organic amendments may be a valid choice (Martens and Frankenberger 1992). Spontaneous establishment of mycorrhizal inoculum is a long process. The first arbuscular mycorrhizal (AM) plants appearing on zinc wastes investigated in Southern Poland take two or more decades to become established after the deposition of industrial dusts, and there are several wastes which are over 40 years old where only few AM plants have so far been observed (Turnau 1998; Turnau et al. 2001).

Introduction of tree seedlings (*Populus, Pinus, Betula* spp.) and seed mixtures of herbaceous plants is a common, but not always effective practice. The establishment of AMF may speed up the process. The role of AMF in the revegetation of mine wastes is comparatively well known (Daft and Nickolson 1974; Daft and Hacskaylo 1976; Allen 1989 a, b; Allen and Allen 1990; Miller and Jastrow 1992). The rate of restoration and land rehabilitation may be increased by the stimulation of the mycorrhizal fungal population or the application of inoculation techniques (Reeves et al. 1979; Janos 1980; Haselwandter and Bowen 1996).

The usefulness of indigenous AMF used as inoculants to reclaim mine spoils (Khan 1981; Stahl et al. 1988) and oil polluted soil (Call and McKell 1982; Cabello 1995, 1997, 1999) has been shown by several authors. The number of AMF propagules depends on many factors such as soil nutritional status, host plant, AMF propagule density, effectiveness of AMF species and competition between them and other soil microorganisms (Cabello 1999). AMF isolated from polluted areas were more effective than those originating from unpolluted sites, suggesting the adaptation of fungi to persistent soil toxicants. Fungi are capable of genetic differentiation at the intraspecific level (Daniels and Duff 1978; Stahl and Smith 1984).

Phytoremediation will give better results when plants are associated with efficient AMF isolates that can tolerate pollution. In places where native AMF are present the restoration techniques should enhance mycorrhiza development assuring survival of the AMF until vegetation is reasonably stabilised (Jasper 1994). However, surface-mined areas and industrial wastes are often totally devoid of AMF. In such cases surface soil (topsoil) may be reapplied after mining. It is essential for the re-establishment of mycorrhizas, as it represents a primary source of mycorrhizal inoculum (Waaland and Allen 1987; Lambert and Cole 1980).

AMF isolates differ in their effect on heavy metal uptake by plants (Leyval et al. 1997). They should be not only resistant to toxic conditions but also well adapted to particular waste and climatic conditions. AMF have a great potential for bioremediation (Leyval et al. 2001). The selection of appropriate isolates for a given phytoremediation strategy (Khan et al. 2000) depends on whether the main goal is phytoextraction or whether a plant cover should be established which does not accumulate heavy metals and does not create a health hazard, for example to animals. In the case of phytoextraction the idea is to employ fast growing plants with high accumulation properties towards heavy metals. *Ricinus communis* and *Sonchus oleraceus* have been proposed as mycorrhizal plants suitable for phytoextraction of Pb and Cd. Both plant species can be colonised by *Glomus* and *Gigaspora* strains (Brooks et al. 1998). Other research carried out on industrial wastes has shown that phytoremediation processes may strongly influence the number of AMF spores. The growth of nonmycorrhizal plants such as *Silene vulgaris* and *Thlaspi caerulescens* had a more negative effect on the number of AMF spores in the substratum than if mycorrhizal *Zea mays* was present. Sulphur amendments had also a negative effect on spore density in maize rhizosphere (Pawlowska et al. 2000).

Fungal strains can be isolated from areas that are either naturally enriched in heavy metals or originate from old mine/industry wastes. One isolate has recently been obtained from *Viola calaminaria*, molecularly characterised (RFLP) and shown to confer heavy metal tolerance to plants including maize, alfalfa, barley and others (Hildebrandt et al. 1999). However, there are usually several AMF types which may colonise the root systems in a given plant ecosystem. Some may form abundant spores but have little effect on plant growth while others may behave in the opposite way. The selection of effective isolates should therefore include investigations on their colonisation abilities as well as on their growth stimulation effectivity in the sites where inoculum is to be introduced.

Further research is needed on AMF ecotypes isolated from polluted soils and inoculated onto plants under field conditions. Caution should be paid to the use of fertilizers and soil amendments, which can contain heavy metals additionally polluting the soil. The use of AM symbiosis may play an important role in diminishing the level of fertilizer application.

Restoration of salino-sodified areas

Soil salinity, natural or anthropogenic, may cause problems for spontaneous revegetation, as high concentrations of sodium and chloride ions can affect organisms directly or indirectly by increasing the pH value of the soil, decreasing the nutrient availability and damaging the structure of the soil (Sucoff 1975; Kelsey and Hootman 1990). Certain plants that are known to be halophilic or tolerate high salinity levels are colonized by AMF (Sylvia and Williams 1992). This was recently confirmed by Hildebrandt et al. (2001). As reviewed by Juniper and Abbott (1994), increased salt concentrations affect AMF spore germination, hyphal growth and effectiveness of mycorrhiza formation. Saline areas are either originally devoid of AMF (as it happens in the case of sedimentation tanks of soda factories) or the number of AMF propagules is considerably decreased. Stahl and Williams (1986) demonstrated that increasing the salinity of the soil solution markedly changes the species distribution among AMF. The presence of AMF may, however, strongly affect the growth of plants under salt stress (Ojala et al. 1983; Rozema et al. 1986; Rosendahl and Rosendahl 1991), although the mechanism is not yet known. Tsang and Maun (1999) have shown that arbuscular mycorrhizal specimens of *Strophostyles helvola* were characterised by higher chlorophyll content, shoot dry weight and number of nodules than nonmycorrhizal ones.

An interesting case of succession was observed on wastes from an ammonia-soda processing factory in Krakow (Poland). Liquid wastes were dumped into sedimentation tanks. The material dried out in about six months and fissured into polygons. When the first organisms appeared the material was characterised by a pH value of about 12, a very high $CaCO_3$ and a very low N, P and organic matter content (Turnau and Rybka 1991). Research on the mycorrhization of plants developing in this area (Pawlowska 1991) has shown that in early stages of succession non-mycorrhizal or facultatively mycorrhizal plants dominated. In later stages an increasing proportion of obligatory mycorrhizal plants was observed. However, even on sediments over 30 years old the substratum was still a rather harsh environment for plants. In this case again the restoration of the disturbed land clearly depends upon the formation of the mycorrhizal symbiosis. Fungal isolates differ in their plant growth stimulation properties under salinity stress (Pond et al. 1984). This suggests that it should be possible to select the most effective strains to be used in the restoration of saline areas. Such AMF ecotypes could be obtained from natural saline habitats.

Conclusion

More research is needed to select those AMF strains which are most efficient in a specific restoration programme, and to check whether they survive the competition with spontaneously appearing or indigenous fungi. This aim requires the identification of the fungus within the roots. Molecular tools seem to be most promising. Recently the PCR discriminating probes based on the variability of the 5' end, including the variable domains D1 and D2, of the large ribosomal subunit have been developed. They can be subsequently used as primers in a nested PCR reaction for the identification of different AMF in trypan blue stained root fragments (van Tuinen et al. 1998a, b). This method has been successfully used to monitor a mixed population of AMF in substrates containing sewage sludge enriched or not with heavy metals or organic pollutants (Jacquot-Plumey et al. 1999). It allows the detection of fungal hyphae present in roots as well as in soil (van Tuinen et al. 1998a, b; Jacquot-Plumey et al. 1999) and opens new possibilities for investigating mycorrhizal community structure and competition between different fungal species in roots and in soil. Recently the method was successfully used to study AMF colonisation of *Fragaria vesca* roots on zinc wastes in southern Poland (Turnau et al. 2001). Taxon-specific primers as described e.g. by Redecker (2000) could also be valuable tools for the assessment of the success of individual fungal species in mycorrhizal colonisation of restored sites.

AMF as well as other soil microorganisms and their activities can be used as bioindicators of soil quality and the efficacy of restoration techniques (Haselwandter 1997). Understanding the relationships in below ground systems can be used to develop new restoration strategies ensuring the biodiversity which is required for ecosystem functioning. However, estimation of the costs and benefits of individual restoration techniques appears necessary (Edwards and Abivardi 1997).

References

Allen E B (1989a) The restoration of disturbed arid landscapes with special reference to mycorrhizal fungi. J Arid Environm 17:279-286

Allen M F (1989b) Mycorrhizae and rehabilitation of disturbed arid soils: Processes and practices. Arid Soil Res 3:229-241

Allen E B, Allen M F (1990) The mediation of competition by mycorrhizal fungi in successional and patchy environments. In: Grace J R, Tilman G D (eds) Perspectives in plant competition. Academic Press, New York pp 367-389

Azcón-Aguilar C, Barea J M (1992) Interactions between mycorrhizal fungi and other rhizosphere microorganisms. In: Allen M J (ed) Mycorrhizal functioning: an integrative plant-fungal process. Chapman and Hall, New York pp 163-198

Azcón R, Rubio R, Barea J M (1991) Selective interactions between different species of mycorrhizal fungi and *Rhizobium meliloti* strains and their effects on growth, N_2-fixation (^{15}N) and nutrition of *Medicago sativa* L. New Phytol 117:399-404

Bach A, Kodra M (1997) Somatic embryogenesis of *Lilium martagon* L. seedling culture. 22nd Conference of Embryology. Plants, Animals, Humans. Osieczany. Acta Biol Cracov 39 (Suppl 1):47
Blaszkowski J (1994) Arbuscular fungi and mycorrhizae (Glomales) of the Hel Peninsula, Poland. Mycorrhiza 5:71-88
Blaszkowski J, Tadych M, Madej T (2000) Arbuscular mycorrhizal fungi (Glomales, Zygomycota) of the Bledowska Desert, Poland. Acta Soc Bot Pol (in press)
Brooks R R, Chambers M F, Nicks L J, Robinson B H (1998) Phytomining. Water Air & Soil Pollut 104:389-402
Cabello M N (1995) Effect of hydrocarbon pollution on vesicular-arbuscular mycorrhizal fungi (VAM). Bol Micol (Chile) 10:77-83
Cabello M N (1997) Hydrocarbon pollution: its effect on native arbuscular mycorrhizal fungi (AMF). FEMS Microbiol Ecol 22:233-236
Cabello M N (1999) Effectiveness of indigenous arbuscular mycorrhizal fungi (AMF) isolated from hydrocarbon polluted soils. J Basic Microbiol 39:89-95
Call C A, McKell C M (1982) Vesicular-arbuscular mycorrhizae - a natural revegetation strategy for disposed processed oil shale. Reclam Revegn Res 1:337-347
Crowley D E, Brennerova M V, Irwin C, Brenner V, Focht D D (1996) Rhizosphere effects on biodegradation of 2,5-dichlorobenzoate by a bioluminescent strain of root-colonizing *Pseudomonas fluorescens*. FEMS Microbiol Ecol 20:79-89
Cuenca G, Lovira M (1992) Vesicular-arbuscular mycorrhizae in disturbed and revegetated sites from La Gran Sabana, Venezuela. Can J Bot 70:73-79
Daft M J, Hacskaylo E (1976) Arbuscular mycorrhizas in the anthracite and bituminous coal wastes of Pensylvania. J Appl Ecol 13:523-531
Daft M J, Nicolson T H (1974) Arbuscular mycorrhizas colonizing coal wastes in Scotland. New Phytol 74:1129-1132
Daniels B A, Duff D M (1978) Variation in germination and spore morphology among four isolates of *Glomus mossae*. Mycologia 70:1261-1267
Del Val C, Barea J M, Azcon-Aguilar C (1999) Diversity of arbuscular mycorrhizal fungus populations in heavy-metal-contaminated soils. Appl Environm Microbiol 65:718-723
Densmore R (1994) Succession on regraded placer mine spoil in Alaska, U.S.A., in relation to initial site characteristics. Arctic Alpine Res 26:354-363
Dowling D N, O'Gara F (1994) Metabolites of *Pseudomonas* involved in the biocontrol of plant disease. Trends Biotechnol 12:133-141
Edwards P J, Abivardi C (1997) Ecological engineering and sustainable development. In: Urbanska K, Webb N R, Edwards P J (eds) Restoration ecology and sustainable development. Cambridge Univ. Press, Cambridge pp 325-352
Edwards S G, Young J P W, Fitter A H (1988) Interactions between *Pseudomonas fluorescens* biocontrol agents and *Glomus mossae*, an arbuscular mycorrhizal fungus, within the rhizosphere. FEMS Microbiol Lett 166:297-303
Fitter A H (1977) Influence of mycorrhizal infection on competition for phosphorus and potassium by two grasses. New Phytol 70:119-125
Francis R, Read D J (1994) The contribution of mycorrhizal fungi to the determination of plant community structure. Plant & Soil 159:11-25
Francis R, Read D J (1995) Mutualism and antagonism in the mycorrhizal symbiosis, with special reference to impacts on plant community structure. Can J Bot 73 (Suppl 1):1301-1309
Fuchs B (2001) Mykorrhizierung ausgewählter und gefährdeter Pflanzenarten zweier Feuchtstandorte im Bundesland Salzburg. MSc Thesis, University of Salzburg, Austria
Ganesan V, Ragupathy S, Parthipan B, Raijini Rani D B, Mahadevan A (1991) Distribution of vesicular-arbuscular mycorrhizal fungi in coal, lignite and calcite mine spoils of India. Biol Fertil Soils 12:131-136
Ganesan V, Ragupathy S, Parthipan B, Raijini Rani Ganesan V, Ragupathy S, Parthipan B, Raijini Rani D B, Mahadevan A (1991) Distribution of vesicular-arbuscular mycorrhizal fungi in coal, lignite and calcite mine spoils of India. Biol Fertil Soils 12:131-136
Gange A C, Brown V K, Foster L M (1990) A test of mycorrhizal benefit in an early succession plant community. New Phytol 115:85-91
Gange A C, Brown V K, Sinclair G S (1993) Vesicular-arbuscular mycorrhizal fungi: a determinant of plant community structure in early succession. Funct Ecol 7:616-622
Grodzinska K, Mirek Z, Szarek G (1995) Environmental damage and restoration attempts in Poland -

an Overview. In: Urbanska K M, Grodzinska K (eds) Restoration ecology in Europe. Geobotanical Institute SFIT, Zürich pp 113-130

Grunwald C, Iverson L R, Szafoni D B (1988) Abandoned mines in Illinois and North Dakota: towards an understanding of revegetation problems. In: Cairns J (ed) Rehabilitating damaged ecosystems, vol. 1. CRC Press, Boca Raton, Florida pp 40-56

Gryndler M, Vosatka M (1996) The response of the arbuscular mycorrhizal fungus *Glomus fistulosum* to treatments with culture fractions from *Pseudomonas putida*. Mycorrhiza 6:207-211

Harley J L, Harley E L (1987) A check-list of mycorrhiza in the British flora. New Phytol (Suppl) 105:1-102

Harnett D C, Hetrick B A D, Wilson G W T, Gibson D J (1993) Mycorrhizal influence on intra-and inter-specific neighbour interactions among co-occurring praire grasses. J Ecol 81:787-795

Haselwandter K (1987) Mycorrhizal infection and its possible ecological significance in climatically and nutritionally stressed alpine plant communities. Angew Bot 61:107-114

Haselwandter K (1997) Soil micro-organisms, mycorrhiza, and restoration ecology. In: Urbanska K., Webb N R, Edwards P J (eds) Restoration ecology and sustainable development. Cambridge Univ. Press, Cambridge pp 65-80

Haselwandter K, Bowen G D (1996) Mycorrhizal relations in trees for agroforestry and land rehabilitation. Forest Ecol Managem 81:1-17

Haselwandter K, Read D J (1980) Fungal associations of roots of dominant and sub-dominant plants in high-alpine vegetation systems with special reference to mycorrhiza. Oecologia 45:57-62

Helm D J (1995) Native grass cultivars for multiple restoration goals on a proposed mine site in south-central Alaska. Restoration Ecol 3:111-122.

Herrera M A, Salamanca C P, Barea J M (1993) Inoculation of woody legumes with selected arbuscular mycorrhizal fungi and rhizobia to recover desertified mediterranean ecosystems. Appl Environm Microbiol 59:129-133

Hetrick B A D, Wilson G W T, Harnett D C (1989) Relationship between mycorrhizal dependence and competitive ability of two tallgrass prairie grasses. Can J Bot. 67:2608-2615

Hildebrandt U, Janetta K, Ouziad F, Renne B, Nawrath K, Bothe H (2001) Arbuscular mycorrhizal colonization of halophytes in Central European salt marshes. Mycorrhiza 10:175-183

Hildebrandt U, Kaldorf M, Bothe H (1999) The zinc violet and its colonization by arbuscular mycorrhizal fungi. J Plant Physiol 154:709-717

Hooker J E, Gianinazzi S, Vestberg M, Barea J M, Atkinson D (1994) The application of arbuscular mycorrhizal fungi to micropropagation systems: an opportunity to reduce chemical input. Agric Sci (Finland) 3:227-232

Jacquot-Plumey E, van Tuinen D, Gianinazzi S, Gianinazzi-Pearson V (1999) Monitoring species of arbuscular mycorrhizal fungi *in planta* and in soil by nested PCR: application to the study of the impact of sewage sludge. Plant & Soil (in press)

Janos D P (1980) Mycorrhizae influence tropical succession. Biotropica 12:56-64

Jasper D A (1994) Management of mycorrhizas in revegetation. In: Robson A D, Abbott L K, Malajczuk N (eds) Management of mycorrhizas in agriculture, horticulture and forestry. Dordrecht, Kluwer pp. 211-219

Jha D K, Sharma G D, Mishra R R (1992) Ecology of soil microflora and mycorrhizal symbionts in degraded forests at two altitudes. Biol Fertil Soils 12:272-278

Juniper S, Abbott L (1993) Vesicular arbuscular mycorrhizas and soil salinity. Mycorrhiza 4:45-57

Kelsey P, Hootman R (1990) Soil recource evaluation for a group of sidewalk street tree planters. J Arboric 16:113-117

Khan A G (1981) Growth responses of endomycorrhizal onions in unsterilized coal wasts. New Phytol 87:363-370

Khan A G, Kuek C, Chaudhry T M, Khoo C S, Hayes W J (2000) Role of plants, mycorrhizae and phytochelators in heavy metal contaminated land remediation. Chemosphere 41:197-207

Lambert D H, Cole H (1980) Effects of mycorrhizae on establishment and performance of forage species in mine spoil. Agron J 72:257-260

Leyval C, Turnau K, Haselwandter K (1997) Effect of heavy metal pollution on mycorrhizal colonization and function: physiological, ecological and applied aspects. Mycorrhiza 7:139-153

Liberta A E (1981) Effects of topsoil-storage duration on inoculum potential of vesicular-arbuscular mycorrhizae. In Graves DH (ed) Symposium on Surface Mining, Hydrology, Sedientation and Reclamation. Office of Engineering Services, College of Engineering, University of Kentucky, Lexington, Kentucky pp 45-48

Lovera M, Cuenca G (1996) Arbuscular mycorrhizal infection in Cyperaceae and Gramineae from natural, disturbed and restored savannas in La Gran Sabana, Venezuela. Mycorrhiza 6:111-118
Martens D A, Frankenberger W T (1992) Modification of infiltration rates in an organic-amended irrigated soil. Agron J 84:707-717
Mc Rae E A (1998). Lilies. Timber Press, Portland, Oregon (USA)
Miller R M, Jastrow J D (1992) The application of VA mycorrhizae to ecosystem restoration and reclamation. In: Allen M F (ed) Mycorrhizal functioning: an integrative plant-fungal process. Chapman and Hall, New York pp 438-467
Mirek Z, Piekos-Mirkowa H (2000) Operat ochrony gatunkowej roslin jako narzedzie ochrony flory na obszarze parku narodowego. In: Woloszyn B, Postawa T (eds) Drugie forum dyskusyjne: Parki Narodowe i ich funkcja w czasie i przestrzeni. Komitet Ochrony Przyrody PAN, Kraków pp 33-52
Ojala J C, Jarrell W M, Menge J A, Johnson E L V (1983) Influence of mycorrhizal fungi on the mineral nutrition and yield of onion in saline soil. Agron J 75:255-259
Pawlowska T (1991) Plant mycorrhizae in the sedimentation tanks of the Cracow soda factory. Zesz Nauk UJ, Prace Bot. 22:163-170
Pawlowska T, Chaney R L, Chin L, Charvat I (2000) Effects of metal phytoextraction practices on the indigenous community of arbuscular mycorrhizal fungi at a metal-contaminated landfill. Appl Environm Microbiol 66/6:2526-2530
Pond E C, Menge J A, Jarrell W M (1984) Improved growth of tomato in salinized soil by vesicular-arbuscular mycorrhizal fungi collected from saline soils. Mycologia 76:78-84
Read D J, Haselwandter K (1981) Observations on the mycorrhizal status of some alpine plant communities. New Phytol 88:341-362.
Redecker D (2000) Specific PCR primers to identify arbuscular mycorrhizal fungi within colonized roots. Mycorrhiza 10:73-80
Reeves F B, Wagner D, Moorman T, Keil J (1979) The role of endomycorrhizae in revegetation practices in the semi-arid west. I. A comparison of incidence of mycorrhizae in severely disturbed vs. natural environments. Amer J Bot 66:6-13
Rodgers C S, Anderson R C (1989) Establishment of grasses on sewage-sludge amended strip mine spoils. In: Bragg T B, Stubbendieck J (eds) Proceedings of the 11th North American Praire Conference. University of Nebrasca Press, Lincoln, Nebrasca pp 103-108
Rosendahl C N, Rosendahl S (1991) Influence of vesicular-arbuscular mycorrhizal fungi (*Glomus* spp.) on the response of cucumber (*Cucumis sativus* L.) to salt stress. Environm Exp Bot 31:313-318
Rozema C N, Arp W, van Diggelen, van Esbroek M, Brokeman R, Punte H (1986) Occurrence and ecological significance of vesicular-arbuscular mycorrhiza in the salt marsh environment. Acta Bot Neerl 35:457-467
Sanders I R, Koide R T (1994) Nutrient acquisition and community structure in co-occurring mycotrophic and non-mycotrophic old field annuals. Funct Ecol 18:77-84
Smith S E, Read D J (1997) Mycorrhizal symbiosis. Academic Press, San Diego
Stahl P D, Smith W K (1984) Effects of different geographic isolates of *Glomus* on the water relations of *Agropyron smithii* 76:261-267
Stahl P D, Williams S E (1986) Oil shale process water affects activity of vesicular-arbuscular fungi and *Rhizobium* 4 years after application to soil. Soil Biol Biochem 18:451-455
Stahl P D, Williams S E, Christensen M (1988) Efficacy of native vesicular-arbuscular mycorrhizal fungi after severe soil disturbance. New Phytol 110:347-354
Sucoff E (1975) Effect of deicing salts on woody vegetation along Minnesota roads. Tech Bull Minn Agric Exp: 303-309
Sylvia D M (1990) Beneficial root-associated fungi improve establishment of sea oats on beach nourishment sand. In: Proceedings of the Third Annual National Conference on Beach Preservation Technology (St. Petersburg, FL, 1990) pp 100-111
Sylvia D M, Williams S E (1992) Vesicular-arbuscular mycorrhizae and environmental stress. In: Mycorrhizae in sustainable agriculture. ASA Special Publication no. 54, ASA, CSSA, SSSA, Madison, WI pp 101-124
Tadych M, Blaszkowski J (2000) Arbuscular fungi and mycorrhizae (Glomales) of the Slowinski National Park, Poland. Mycotaxon 74:463-482
Tsang A, Maun M A (1999) Mycorrhizal fungi increase salt tolerance of *Strophostyles helvola* in coastal foredunes. Pl Ecol 144:159-166

Turnau K (1998) Heavy metal uptake and arbuscular mycorrhiza development of *Euphorbia cyparissias* on zinc wastes in South Poland. Acta Soc Bot Poloniae 67:105-113

Turnau K, Rybka R (1991) Succession of microorganisms on the spoil mounds of the Cracow soda factory. Zesz. Nauk. UJ, Prace Bot. 22:155-161

Turnau K, Ryszka P, Gianinazzi-Pearson V, van Tuinen D, (2001) Identification of arbuscular mycorrhizal fungi in soils and roots of plants colonizing zinc wastes in southern Poland. Mycorrhiza 10:169-174

Urbanska K M (1995) Ecological restoration above the timberline and its demographic assessment. In: Urbanska K M, Grodzinska K (eds) Restoration ecology in Europe. Geobotanical Institute SFIT, Zürich pp 113-130

Van der Heijden M G A, Klironomos J N, Ursic M, Moutoglis P, Streitwolf-Engel R, Boller T, Wiemken A, Sanders I R (1998a) Mycorrhizal fungal diversity determines plant biodiversity, ecosystem variability and productivity. Nature 396:69-72

Van der Heijden M G A, Boller T, Wiemken A, Sanders I R (1998b) Different arbuscular mycorrhizal fungal species are potential determinants of plant community structure. Ecology 79:2082-2091

van Tuinen D, Jacquot E, Zhao B, Gallotte A, Gianinazzi-Pearson V (1998a) Charaterization of root colonization profiles by a microcosm community of arbuscular mycorrhizal fungi using 25S rDNA-targeted nested PCR. Mol Ecol 7:879-887

van Tuinen D, Zhao B, Gianinazzi-Pearson V (1998b) PCR in studies of AM fungi: from primers to application. In: Varma A K (ed) Mycorrhizal Manual. Springer-Verlag, Heidelberg pp 387-399

Vosatka M, Jansa J, Regvar M, Sramek F, Malcova R (1999) Inoculation with mycorrhizal fungi – a feasible biotechnology for horticulture. Phyton 39:219-224

Waaland M E, Allen E B (1987) Relationships between VA mycorrhizal fungi and plant cover following surface mining in Wyoming. J Range Managem 40:271-276.

Zenkteler E (1993) Homo- and heterophasis in the in vitro reproduction of *Phyllitis scolopendrium* L. Newm. Bull PAN, Biol Sci 41:267-261

Mycorrhizal Technology in Agriculture
ed. by S. Gianinazzi, H. Schüepp, J.M. Barea and K. Haselwandter

Application of arbuscular mycorrhizal fungi in the revegetation of desertified Mediterranean ecosystems

P. Jeffries[1], A. Craven-Griffiths[1], J. M. Barea[2], Y. Levy[3] and J. C. Dodd[4]

[1] *Research School of Biosciences, University of Kent, Canterbury, Kent CT2 6NJ, UK*
[2] *Departamento de Microbiología del Suelo y Sistemas Simbióticos, Estación Experimental del Zaidín, CSIC, Profesor Albareda 1, E-18008 Granada, Spain*
[3] *Department of Citriculture, Institute of Horticulture, Gilat Experiment Station, ARO, Mobile Post Negev 85280, Israel*
[4] *International Institute of Biotechnology, 1/13 Innovation Building 1000, Sittingbourne Research Centre, Sittingbourne, Kent ME9 8HL, UK*

Keywords: desertification, water stress, restoration, revegetation, rehabilitation

Characterisation of desertified ecosystems

Over 40 million km^2, or 35% of the total land surface, can be described as drylands with permanent, seasonal or periodic significant moisture deficiency. These ecosystems are found in tropical, subtropical, temperate and polar latitudes (Barrow 1991) and share common edaphic characteristics which are summarised as follows:

- Vegetation cover and plant community structure are easily damaged and difficult to recover
- Soils are often deficient in humus and/or nutrients especially P
- Soil nutrients and organic matter are often low and easily lost
- There are seasonal and/or diurnal extremes of temperature
- Precipitation falls in short periods, commonly intense and erosive
- Strong winds and bush fires are a risk

Such ecosystems are highly susceptible to degradation, which is usually referred to as 'desertification'. Barrow (1991) describes desertification as the 'degradation of ecosystems in semi-arid or arid regions, degeneration usually being measured in loss of primary productivity'. Desertification was initially used in 1949 to describe the process of vegetation changes leading to denuded and degraded soils in the Sahel region of Africa (Barrow 1991). Since then an array of definitions have been coined with an obvious trend over time to incorporate and emphasise the contribution of human activity as a cause (Perez-Trejo 1994). As a result 'desertization' is generally applied to degradation arising due to physical events and processes and 'desertification' used to denote the involvement of human activity (Jeffries and Barea 1994). Desertification of terrestrial ecosystems is claiming several million hectares annually (Warren et al. 1996). It often results from human impacts exacerbating the effects due to the natural agents. Disturbance of natural plant communities is the first symptom, but is often accompanied or preceded by loss of the key physico-chemical and biological soil properties which largely determine soil quality and fertility. Hence their degradation results in a loss of sustainability (Fig. 1). Since soil degradation limits the potential for re-establishment of native plants (Warren et al. 1996), erosion and desertification are accelerated. Desertification has a negative environmental impact, particularly in arid, semi-arid and subhumid areas of the world (Warren et al. 1996). In particular, desertification reduces the inoculum potential of mutualistic microbial symbionts (mycorrhizas and rhizobia) that are key ecological factors in governing the cycles of major plant nutrients, and hence in sustaining the vegetation cover in natural habitats (Jeffries and Barea 2000). A reduction in the potential to form these symbioses therefore hinders revegetation success (Allen 1989; Schreiner et al. 1997).

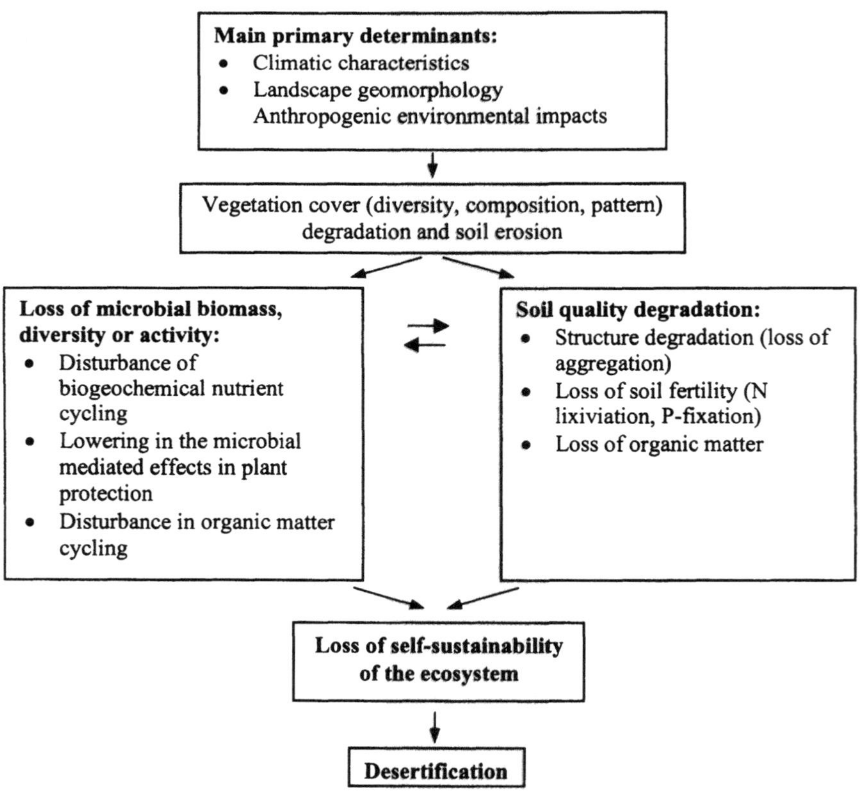

Figure 1. Various and diverse ecological determinants, which act either as causes or effects, create a downward spiral leading to the progressive degradation of both soil quality and vegetation and, ultimately, to ecosystem desertification.

Desertification in European Mediterranean ecosystems

Mediterranean regions are characterized by a set of climatic conditions which include a long dry and hot summer, with scarce, erratic but torrential winter rainfalls. This climate, together with anthropogenic activity, means that desertification in Mediterranean regions is increasing worldwide (Miller and Lodge 1997; Warren et al. 1996). The European Mediterranean contains extensive areas of degraded lands (UNEP 1992). Approximately 50% of Spain is arid or semi-arid and it has been estimated that around 60% of these drylands are at least moderately desertified (Barrow 1991). This constitutes the biggest environmental threat for the EU. Human activity has continually played a significant role in shaping the landscape across the European Mediterranean, with changes in and use of cultivation. Evidence suggests that severe erosion had occurred certainly by classical times with degradation evident as far back as the bronze age (Thornes 1993). Although some of the changes are reversible, recovery would take a long time, possibly tens of thousands of years (Thornes 1993). Such desertified regions are prone to flooding, sediment movement and poor ground water recharge due to inadequate vegetation cover, poor soil structure and low capacity for water infiltration in climates where precipitation is infrequent yet intense. Together with the renewed demands upon the land and greater public concern for the environment concerns, these factors have driven a new awareness in the EU. From this, programmes have arisen in recent years, which aim to improve the understanding of desertification in the Mediterranean and to redress the problem.

In order to combat the trend towards desertification, strategies are being employed to revegetate the degraded lands, thus restoring ground cover, reducing erosive loss and increasing soil fertility. The success of these schemes relies on functional below-ground microbial symbioses necessary to support the above-ground plant growth. In this review, we have chosen three particular case studies to illustrate how the arbuscular mycorrhizal symbiosis can play a crucial role in alleviating the stresses of living in desertified soils and how they can be managed to aid recovery of natural plant communities.

The role of arbuscular mycorrhizas in alleviating water stress

Arbuscular mycorrhizas can modify a host plant's response to environmental pressures in a wide variety of ways and are an essential component of sustainable plant growth in a natural ecosystem. In particular, arbuscular mycorrhizal fungi (AMF) can ameliorate the effects of unfavourable conditions. In desertified or semi-arid ecosystems plant populations are exposed to considerable environmental pressures, principally the shortage of adequate water resources for prolonged periods of the year. The majority of experiments on drought tolerance of mycorrhizal plants have been carried out using cultivated plants, such as maize, alfalfa and cowpea, but they demonstrate that arbuscular mycorrhizas have the potential to alter host response to water stress in a semi-arid ecosystem.

In semi-arid and arid environments the major limiting factor is water availability which results in limited availability of P (Gupta 1989) and AMF are likely to play an important role because of beneficial changes in water relations of host plants. Improved water status and relation to drought tolerance are well documented. Sánchez-Díaz and Honrubia (1994) provide a comprehensive review of the relationship between AM associations and drought stress. The role of AMF in alleviation of water stress is not fully understood or straightforward and there are a number of aspects to consider. Firstly, it is evident that behaviour of mycorrhizas under well-watered and water-stressed conditions is not the same (Fitter 1985; Hetrick et al. 1987; Sieverding 1991; Allen and Allen 1992; Tobar et al. 1994; Ebel et al. 1997; Goicoechea et al. 1997a, b). Secondly, improvements in water status may be due to improved nutritional status, particularly P (Nelson and Safir 1982; Fitter 1988). Experimentally it has been difficult to separate the effects of AMF from other physiological effects such as the relative size of plants being compared. It is very difficult to obtain non-mycorrhizal plants that have developed to a similar extent to mycorrhizal counterparts, and even if this is achieved their nutrient status (particularly P) will invariably differ. In the case study presented below these differences were minimised and, although the data highlight the difficulties in minimising these factors, they show that – if they are addressed – there are potential benefits accruing from the mycorrhizal status. It should always be remembered that the non-mycorrhizal status of many plants is an anomaly, except in high input systems. Besides the benefits of enhanced water relations, the multi-functional role played by AMF will probably mean that overall benefits will also depend on plant species, soil type and environment. Generalisations should be avoided!

Case study 1: Effect of AMF on water relations of cotton and pepper (Dodd and Levy, unpublished data)

Growth room studies were undertaken with cotton (*Gossypium herbaceum* L. cv SJ-2) and bell pepper (*Capsicum annuum* L., cv Ma'or) inoculated with an isolate of *Glomus coronatum* from the Ramat HaNegev region in Israel (originally described as *Glomus mosseae* by Dodd and Krikun 1984). Sterile 2.5 kg pots were filled with a γ-irradiated (2.5 Mrad) sandy soil (from Mivtahim) in the pepper experiment and a sandy loam (from Hamra) in the cotton experiment. These were amended to obtain 15 mg kg^{-1} bicarbonate extractable P by the addition of triple superphosphate. Chopped colonised roots of Sudan grass were placed 5 cm below the soil surface and non-mycorrhizal controls received washings from the root inoculum. Plants were grown in a controlled environment growth-room with a 12 h photoperiod with a light intensity (PAR 400–700 nm wavelength) of 320–400μmol m^{-2} s^{-1} and a temperature range of

21–31 °C. Whilst measurements of plant water relations were taken, pots were enclosed in polyethylene bags tied around the base of the stem, aerated constantly to provide saturated air.

In the cotton trial, plants were grown from seed for 3 months to flower bud initiation when leaves at the same position on several plants were checked to estimate %P content. Similar plant growth and P levels were found in both inoculated and non-inoculated plants (Tab. 1) at the end of the first water stress cycle. Whole plant transpiration, measured gravimetrically (Levy et al. 1980, 1983), in non-inoculated plants, was found to be significantly higher prior to the stress induction.

All pots were brought to maximum water holding capacity (WHC) and two groups of seven replicate pots selected from each of the inoculated and non-inoculated treatments. One set of plants was subjected to water stress and the other was maintained at maximum WHC by daily irrigation. Water stress was

Table 1. Transpiration rates prior to first stress cycle, P levels in leaves at the first and second cycles of stress and final growth parameters of cotton plants across stressed and non-stressed plants (n = 14)

Parameter (means)	Non-mycorrhizal	Mycorrhizal	LSD (P = 0.05)
Transpiration (μg cm^{-2} s^{-1}) – mean of 8 days prior to first cycle of stress	8.1	7.4	*
Height of plants (cm) at harvest	31.8	32.0	NS
Shoot DW (g) at harvest	4.3	4.1	NS
Root DW (g) at harvest	2.7	2.9	NS
Flower bud production per plant	2.4	3.1	*
%P in lower leaves following first stress cycle	0.21	0.22	NS
%P in lower leaves at final harvest	0.16	0.50	**

induced by withholding irrigation for 48 h before restoring maximum WHC. Water-stressed mycorrhizal plants were slower to recover from water stress as indicated by measurements of transpiration (Fig. 2) and conductance. The greater extent of stress levels of mycorrhizal plants was also indicated by comparing proline concentrations, which were twice as high in leaves of inoculated plants compared with non-inoculated plants during the drying cycle and subsequent recovery. Leaf water potentials were also significantly lower at peak stress (Fig. 3) prior to resumption of irrigation in inoculated compared with non-inoculated plants even where there was no significant difference in transpiration rates (Fig. 2). This indicated that mycorrhizal plants were able to endure greater stress but that recovery was slower. This greater stress, as indicated by lower leaf water potentials, was also seen in the second cycle in the

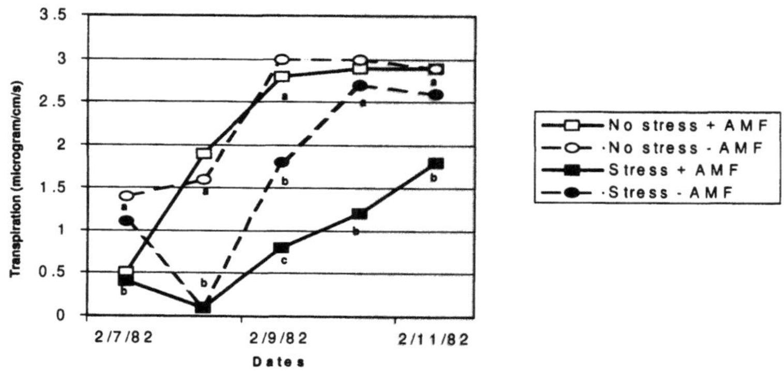

Figure 2. Transpiration rates of leaves, taken from the same leaf position, in stressed or non-stressed cotton plants during the first cycle of water stress. Differences between means are not significantly different when followed by the same letter ($P < 0.05$ Duncan's Multiple Range Test).

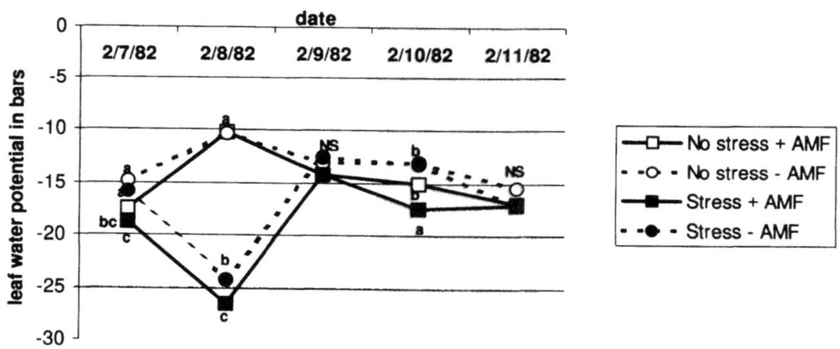

Figure 3. Leaf water potentials of leaves, taken from the same leaf position, in stressed or non-stressed cotton plants during the first cycle of water stress. Differences between means are not significantly different when followed by the same letter ($P < 0.05$ Duncan's Multiple Range Test).

pepper trial (see below) although recovery from stress was the same as that for non-mycorrhizal plants in that case.

A second cycle of water stress was instigated on the cotton plants a month later after excision of flower buds. Leaf P levels were again measured after the stress cycle (Tab. 1). Gravimetrically measured transpiration rates were significantly lower in mycorrhizal *versus* non-mycorrhizal plants, but only in the non-stressed plants as observed prior to the first cycle of stress for all plants (data not presented). Final analysis of leaf tissue however showed that by this time mycorrhizal plants contained 3 times the levels of P of non-mycorrhizal plants. This may explain the significantly greater flower bud production noted in mycorrhizal plants irrespective of stress (Tab. 1).

In the similar trial conducted with pepper, seedlings were again grown for 3 months, in conditions described above for cotton, before two consecutive water stress cycles were instigated. Leaf P levels were 0.13% for non-mycorrhizal and 0.11% for mycorrhizal plants immediately prior to the first water stress and plant growth was similar as indicated by height measurements. All non-stressed mycorrhizal plants had significantly lower leaf water potentials than non-mycorrhizal plants, as observed for cotton in the second stress cycle (Fig. 4). Over the two stress cycles, water-stressed mycorrhizal plants had lower leaf water potentials than non-mycorrhizal plants at maximum water stress but this was only significant in the second cycle. There were no discernible trends for the other parameters measured. At final harvest (1month later) P levels in leaves across stressed and non-stressed plants were 0.13% and 0.17% in non-mycorrhizal and mycorrhizal plants respectively, whilst the values in roots were 0.14% and 0.22%. Once again these results highlight the primary role of AMF in enhancing the P nutrition of plants in nutrient-limited soils.

As Auge (2000) states, in a thorough recent review of the stomatal behaviour of arbuscular mycorrhizal plants, the conclusions from many growth-room experiments seem to suggest that AMF do not alter stomatal physiology but rather postpone the onset of drought responses. Ultimately he concluded that this can be considered a form of drought avoidance, as highlighted by the

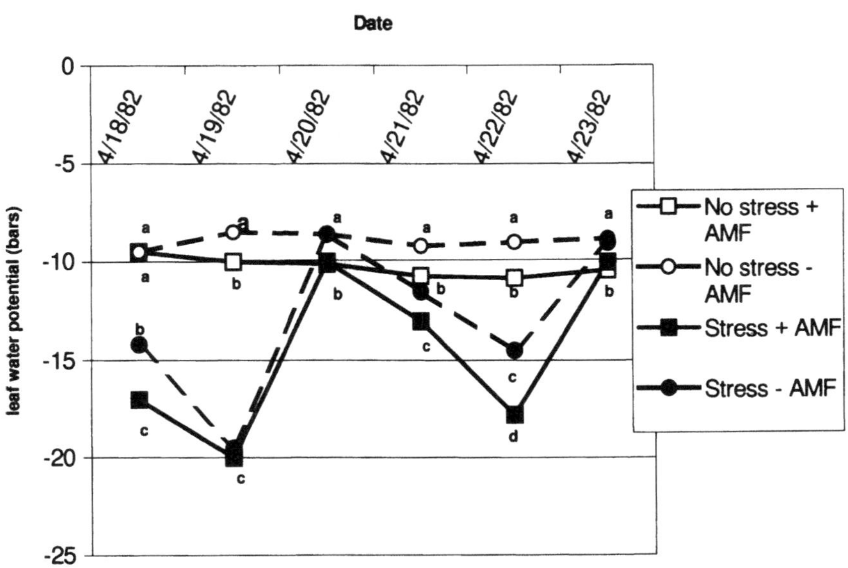

Figure 4. Leaf water potentials of leaves, taken from the same leaf position, in stressed or non-stressed pepper plants during two cycles of water stress. Differences between means are not significantly different when followed by the same letter (P < 0.05 Duncan's Multiple Range Test).

case study described above for the two crop plants. The explanation for stomata remaining open in AM plants longer than non-mycorrhizal plants is the result of more complete extraction of soil water (Auge 2000) which enables the plant to stay more hydrated at low soil water levels. Plants may benefit through the direct uptake of water through the soil mycelium (Faber et al. 1991; Sánchez-Díaz and Honrubia 1994) but George et al. (1992) contest that hyphae of AMF would have to be much larger for a significant increase in water uptake. That is not to say that the extraradical mycelium (ERM) does not play a role in water uptake. Davies et al. (1992, 1993) have also proposed that improvements in drought resistance of bell pepper inoculated with an isolate of *G. deserticola* were due to contributions of ERM in the soil that facilitated water uptake. The ERM, which was enhanced by drought acclimation, potentially helped to maintain hydraulic contact between the root and soil through aggregation and binding. This was not associated with plant size or nutrition but was a result of hyphal development. The debate will continue, but why are transpiration rates different in case study 1 between well-watered mycorrhizal and non-mycorrhizal plants? Auge (2000) suggests that any modification of stomatal behaviour by AMF may be the result of the greater carbon demand from leaves to roots to support the developing ERM. We suggest that the increased flowering and fruit production observed in mycorrhizal *versus* non-mycorrhizal plants in this case study and other studies (Dodd et al. 1983; Krikun et al. 1990) is probably the result of more efficient uptake of P following the greater tolerance of water stress.

The beneficial role of AMF in natural semi-arid ecosystems

Not all mycorrhizal benefits in desertified ecosystems are conferred through improvements in water relations. The hyphal network of AMF within the soil is a vital component of any soil ecosystem (Dodd 1994; Requena et al. 1996; Dodd et al. 2000) and its fragile nature means that it is especially vulnerable in semi-arid environments. In desertified ecosystems, topsoils are readily lost through erosion because the soils are poor in structure with low organic matter content, and plant cover may be sparse (Perez-Trojez 1994). This means that nutrients are often limited and are heterogenously dispersed. In general most significant plant growth benefits from arbuscular mycorrhizas are attributed to increases in phosphate absorption (Marschner and Dell 1994). The amount of available phosphate in desert soils is generally adequate for native vegetation (Bloss 1985) but arbuscular mycorrhizas in desert soils can improve the nutrient uptake of other minerals through the extensive hyphal network permeating through the soil. Such uptake is important where soils are poorly structured, low in fertility and water deficient (Tarafdar and Praveen-Kumar 1996). By its nature, ERM is able to exploit larger volumes of soil and scavenge nutrients which have limited mobility due to low water availability, or are patchy and in short supply (Jeffries and Barea 1994). In this way, nutrients are retained within the biomass and the soil-plant system relies more on mineralisation than on extraction from the soil parent materials. The hyphal network can enable sustainable plant productivity where nutrients are limited by negating heterogeneous distribution of nutrients within a soil. There is also evidence (Tommerup 1992) that hyphal connections can act as channels for nutrient transfer of C, P and N between host plants and that this may be significant in the field, certainly for N (Newman et al. 1992; Martins 1992).

Under desertified conditions, AMF can also significantly contribute to the improvement of soil stability, through the considerable binding properties of ERM. Mycorrhizal hyphae physically enmesh microaggregates to form more stable macroaggregates (Jastrow and Miller 1991; Miller and Jastrow 1992; Tisdall et al. 1997). This is complemented by hyphal residues and exudates. In particular, a recalcitrant soil glycoprotein, glomalin, is sloughed from hyphae of AMF, accumulates in the soil and may play a major role in soil stabilisation (Wright and Upadhyaya 1996, 1998, Wright et al. 1996, 1998). Hence, mycorrhizas can play a significant role in soil conservation, providing vital resistance to erosive forces through improved plant establishment and also ERM development throughout the soil. Improvement in plant nutrition confers additional benefits, which enhance development and survival in such a potentially hostile environment. For example, AMF increased seed yield of *Pisum sativum* without enhancing shoot growth (Schreiner and Bethlenfalvay 1996) and have also altered yield and nutrient composition of soybean seeds (Bethlenfalvay et al. 1997). This is also demonstrated clearly for indigenous legume shrubs, widespread throughout the Mediterranean landscape, which have an intrinsic requirement for P and dependence on mycorrhizas (Herrera et al. 1993; Lopez-Sanchez et al. 1992). The improved uptake of P and health status enables effective nodulation and nitrogen fixation (Osonubi et al. 1991; Herrera et al. 1993; Thiagarajan and Ahmad 1993; Roldán-Fajardo 1994). Other significant effects of AMF are mediated through interactions with other microorganisms in the rhizosphere (Azcón-Aguilar and Barea 1992). These interactions play an important role in natural ecosystems as they regulate arbuscular mycorrhiza formation and function, and can determine the growth and development of hosts. In addition to rhizobia, beneficial interactions are reported between AMF and other N-fixing microorganisms such as nodule forming actinomycetes, *Frankia* spp., and plant-growth-promoting rhizobacteria (PGPR) (Piccini and Azcón 1987). This group of bacteria can influence plant development as modifiers of soil fertility and have been implicated as facilitators of plant establishment (Linderman 1992).

In addition to these synergistic effects, antagonistic interactions of AMF with potential pathogenic fungi can have beneficial results. Fitter and Garbaye (1994) have proposed that a major role of arbuscular mycorrhiza symbioses in natural ecosystems is as protection against root pathogenic fungi. Studies have shown that AMF can enhance tolerance or resistance to root pathogens including pathogenic fungi such as *Pythium* spp., *Phytophthora* spp., *Rhizoctonia* spp. and *Fusarium* spp. and also nematodes and pathogenic bacteria. These are widespread throughout soils of natural ecosystems, including desertified soils where plants might be particularly susceptible due predisposition by the environmental stresses.

Population dynamics of AMF in desertified ecosystems

Whilst there are clear ecological advantages for plants in natural semi-arid ecosystems to form effective arbuscular mycorrhizas, there is a wide variation in the efficacy of host-fungus-climate-soil combinations. Thus populations of AMF must remain diverse to satisfy ecological demand. In Mediterranean ecosystems, annual plants die in the summer and re-establish with the onset of winter rains. Thus there needs to be sufficient inoculum potential within the soil to maximise colonisation of emerging seedlings in the autumn. There may also be small windows of opportunity at other times when unseasonal rain causes seed germination and seedling emergence. 'False breaks' may thus occur before soil moisture becomes limiting again and the seedlings die. The occurrence of false breaks does not seem to affect the overall levels of mycorrhizal colonisation, except for colonisation by the 'fine endophyte' (Braunberger et al. 1994). Soil temperatures at the time of false breaks can have a large impact on resultant colonisation of germinating seedlings by AMF. Fungi that remain quiescent in warm soil may avoid damage in a false break (Braunberger et al. 1997). The population dynamics of AMF are complicated as the spread of species and their propagules is heterogeneous, and temporally and spatially distributed within an ecosystem. Arbuscular mycorrhizas and propagules of AMF are most common in the top 10–30 cm of soil, decreasing with increasing soil depth (Schwab and Reeves 1981; McGee 1989; Brundrett et al. 1996). McGee (1989) determined the most probable number (MPN) of propagules in a semi-arid site of Australia and found significantly fewer spores below 30 cm. This was supported by Bellgard (1993) also in a semi-arid region of Australia, who found at least double the numbers of spores in the top 15 cm than at 15–30 cm depth.

The principal determinant of distribution of AMF is vegetation (Abbott and Robson 1991), particularly in semi-arid ecosystems where vegetation may be sparse, and host plants display seasonal variations in growth and development to survive long hot summers and unpredictable rainfall events. Hence, the development of mycorrhizas and propagules displays seasonal variation. In a survey of eroded soils from southern Spain López-Sánchez et al. (1992) found that numbers of spores reached a maximum in summer and a minimum in winter. The role of spores as prime propagules, however, of semi-arid soils is complicated. Dodd and Krikun (1984) isolated spores from 3 different isolates of *Glomus coronatum* in a semi-arid region of Israel, produced under the same growth conditions, that showed widely differing germination rates on various agar media. Similarly, in the same study (Dodd and Krikun 1984) a fungus originally designated as an unidentified *Glomus*, and found all over the

Mediterranean countries in Europe and North Africa has failed to germinate under many different environmental conditions (Dodd et al. personal communication). Other AMF have behaved similarly e.g. *Acaulospora longula* BEG52 from soils in SE Spain (Craven-Griffiths 1999).

In natural ecosystems it is now recognised that non-spore propagules are significant for the initiation of mycorrhizas for many species of AMF (Moorman and Reeves 1979; McGee 1989; Dodd 1994; Requena et al. 1996). Spores, in contrast, may represent survival units which preserve the inoculum potential during extended periods of unfavourable conditions. The viability of hyphae is limited and mycelial disturbance can seriously diminish the inoculum potential of certain taxa of AMF. For example, Jasper et al. (1993) showed that the infectivity of hyphae of *Acaulospora laevis* declined rapidly in dry field soils, dependent on the stage in the life cycle. In contrast, hyphae of *Scutellospora calospora* remained infective for at least 11 weeks. The potential effect of disturbance on fungal populations and propagules has been demonstrated using soils from semi-arid America. In a bioassay with corn (*Zea mays)* percentage colonisation of roots was only 2% in disturbed soils compared with over 70% in adjacent undisturbed soils containing the same indigenous AMF species (Moorman and Reeves 1979). The loss of vegetation cover is also destructive for the survival and perpetuation of AMF. If vegetation loss continues, the infective potential of denuded areas will deteriorate with time, relying on long-surviving spores. In this way, heterogeneity increases, the numbers and diversity of AMF populations are reduced and the potential for revegetation and effective relationships also deteriorates (Moorman and Reeves 1979; Allen et al. 1995). Hence, whilst AMF are an essential component of sustainability in desertified ecosystems they are susceptible to loss of propagules and the potential for formation of mycorrhizas must be taken into account where revegetation is planned.

Earlier studies based on identification of spores of AMF from soils collected directly from arid areas suggested that species diversity was low. Stutz and Morton (1996) showed that this method underestimated species richness and that a trap culture method was preferred. Species richness at 13 sampling sites in arid regions of the USA and Africa ranged from 6 to 12 (Stutz et al. 2000). The species were limited mostly to small-spored AMF from the Glomaceae and Acaulosporaceae. These results were paralleled by Craven-Griffiths (1999), who found that trap cultures in a desertified ecosystem in SE Spain revealed the presence of at least 12 species of AMF, with 10 species of *Glomus*, 1 species of *Acaulospora* and 1 species of *Scutellospora* being identified. Of these, *Glomus coronatum* BEG49, *G. microaggregatum* BEG51, *G. viscosum* BEG50, *G. constrictum* BEG130 and *Acaulospora longula* BEG52 were established in pure culture. Biodiversity studies in such semi-arid regions should undertake appropriate sampling techniques over prolonged periods to truly collect complete information on these stressed environments (Stutz and Morton 1996; Craven-Griffiths 1999; Dodd et al. 2001).

Work by Craven-Griffiths (1999) has highlighted how disturbance via sieving of the stored inoculum of an isolate of *A. longula* BEG52 reduced potential infective propagules to negligible levels (Fig. 5) possibly through disruption of the ERM. In contrast, the colonisation capacity of mixed and relatively undisturbed stored inoculum increased after 6 months. This complicates the potential use of an indigenous consortium of these AMF in restoring natural semi-arid vegetation. Stutz et al. (2000) also noted a preponderance of *Glomus* species in the desert soils in the US suggesting that this genus can exploit a range of propagule types and contains isolates adapted to fluctuating environmental conditions. This still masks the difficulty in managing these AMF for inoculum production as found by Craven-Griffiths (1999) on comparing 4 species of *Glomus* (5 isolates) from SE Spain. Negative effects on the subsequent capacity of the AMF to colonise leeks resulted from storing inoculum under low moisture conditions or from disturbance prior to storage. The uniqueness of the AMF biodiversity and their physiology found in such semi-arid environments clearly indicates why inocula should contain isolates from these ecosystems rather than exotic isolates from alien environments.

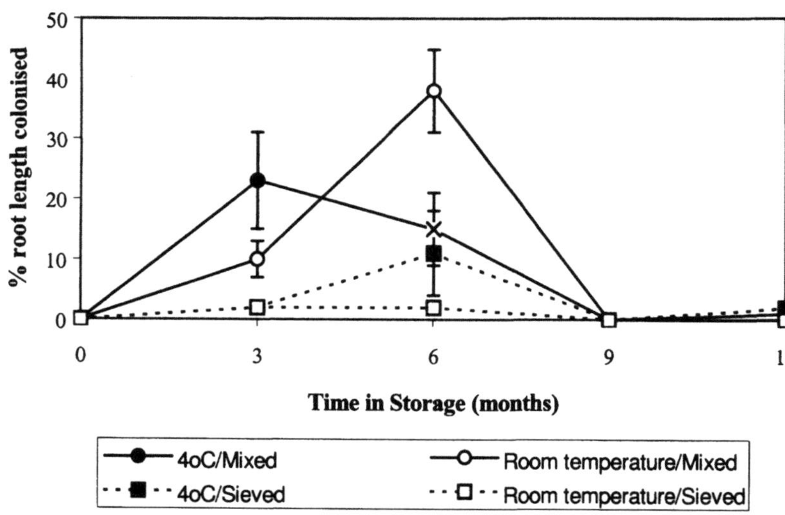

Figure 5. The effect of storage, storage temperature and inoculum composition on the % root length colonisation of leek roots inoculated with *A. longula* BEG52, after 6 weeks growth. Sieved inoculum comprised principally spores whilst mixed inoculum comprised spores, infected root fragments and ERM. Data are means ± SE of 4 plants.

AMF and the rehabilitation of desertified ecosystems

In order to combat desertification, rehabilitation programmes are being attempted in order to restore sustainable ecosystems. The method adopted to improve chemical, physical and biological properties of the soil is revegetation (Barea et al. 1990; Francis and Thornes 1990; Morgan et al. 1990; Roldan et al. 1997). The aim should be to re-establish a stable ecosystem with fully functioning nutrient cycling processes (Jasper 1994) for which AMF are important through the following mechanisms:

- Enhancing establishment and growth of plants by increasing nutrient uptake
- Maintaining diversity, by boosting the ability of host plants to compete for resources
- Contributing to efficient recycling of nutrients and thus to long-term stability
- Stabilising the soil

In particular, leguminous shrubs are recommended for revegetation of semi-arid areas because of their ability to develop tripartite symbioses with rhizobial bacteria and mycorrhizal fungi (Barea et al. 1990, 1996). The symbiosis enhances the ability of the host to become established and cope with stress in the ecosystem, namely nutrient deficiency, drought, trace element imbalance and soil disturbance. However, if soil erosion is pronounced, the scarcity of microbial propagules in such ecosystems may be a serious handicap to plant establishment and survival. In such cases it may be necessary to inoculate with indigenous fungal species (Jasper 1994; Vosatka and Dodd in this book) or augment the natural AMF already present within the rhizosphere of indigenous leguminous plants.

Most semi-arid Mediterranean ecosystems are characterised by shrub communities, associated with other small woody plants, with nitrogen-fixing legumes as key components of the natural succession (Azcón and Barea 1997; Francis and Thornes 1990; Puigdefábregas et al. 1996). Thus re-establishing this shrubland is a key step in revegetation strategies. In the tri-partite rhizobial-AMF-legume symbiosis, there is synergism between the partners in that the scarcity of available P in desertified ecosystems limits legume establishment and N_2 fixation (Barea et al. 1992) in the absence of AM formation. Two important indigenous legumes are *Anthyllis cytsoides* and *Retama sphaerocarpa*, both are mycotrophic (López-Sánchez et al. 1992; Roldán-Fajardo 1994).

Case Study 2: Effect of AMF on physiology of indigenous plants from desertified ecosystems

In studies of the effect of inoculation with AMF isolated from a desertified Mediterranean ecosystem, Craven-Griffiths (1999) found that the native legumes *Anthyllis cytisoides* and *Retama sphaerocarpa* both responded to inoculation with *G. coronatum* BEG49 and *G. microaggregatum* BEG56 in that both isolates of AMF increased shoot hydration after 12 weeks growth. In a second experiment, inoculation with *G. coronatum* BEG49 lowered the shoot fresh weight, shoot dry weight, leaf area, root fresh weight and root length of *A. cytisoides* plants, at 8 weeks, compared with non-mycorrhizal plants. However, this was a temporary effect and there was no difference between mycorrhizal treatments by 17 weeks. In both growth experiments inoculation with *G. microaggregatum* BEG56 altered the root architecture and development of *R. sphaerocarpa*. In the first experiment *G. microaggregatum* BEG56 increased the proportion of lateral roots formed in the 0–10 cm root zone by 18%. In the second growth experiment, *G. microaggregatum* BEG56 increased the proportion of lateral roots that were 2nd order or higher in the 0–10 cm root zone from 6.8% to 14.5% (Tab. 2).

In a third growth experiment, inoculation with *G. microaggregatum* BEG56 increased the proportion of roots in the upper 0–20 cm root zone to 22% compared with 15% in non-mycorrhizal plants, without affecting overall root biomass. In this latter experiment colonisation was largely restricted to the zone in which inoculum was originally placed (0–10 cm or 10–20 cm rooting zone). However, the placement of the inoculum did not affect the host response to inoculation with *G. microaggregatum* BEG56. The placement of inoculum and distribution of inoculum influenced the extent to which roots of *R. sphaerocarpa* became colonised by *G. microaggregatum* BEG56 and ERM developed, 12.5 weeks after planting. Thus the root architecture of *R. sphaerocarpa* was influenced by inoculation with *G. microaggregatum* BEG56, irrespective of the placement or distribution of inoculum.

The alteration of root development within the rooting zones indicates differences in resource allocation between treatments. The root system of *R. sphaerocarpa* consists of a dense layer of fine roots, mainly in the top 20 cm of soil, along with deeply penetrating, sparsely branched vertical roots (Haase et al. 1996). Thus the bulk of the lateral roots by *R. sphaerocarpa* was localised in the zone of inoculation. However, the zone of inoculation, either the 0–10 cm or the 10–20 cm rooting zone, did not influence the alterations in

Table 2. The effect of inoculation with *microaggregatum* BEG56 on the proportion of lateral root orders of *R. sphaerocarpa* in the 0–10 cm rooting zone

	Proportion of total lateral root length (%)	
AMF	1st order lateral roots	2nd order lateral roots
Non-mycorrhizal plants	93.1 (1.26)	6.8 (0.07)
G. microaggregatum BEG56	85.2 (1.11)	14.5 (0.14)
Statistical significance		
AMF	P < 0.05 (LSD = 0.14)	P < 0.05 (LSD = 0.06)

Data are combined means of 16 plants. Data was transformed (arcsin) for ANOVA. Transformed values are given in parentheses and LSD refers to transformed data.

root development that occurred in the upper 0–20 cm of the root system. This contrasts with the findings of Yano et al. (1996a, b), using pigeon pea (*Cajanus cajan*) and peanut (*Arachis hypogea*), in split root boxes, who proposed that resources were preferentially allocated to the mycorrhizal roots. However, it is evident from the work of Yano et al. (1996a, b) that increases in lateral root production on the mycorrhizal side of the split-root pots were in fact localised in zones as demonstrated in the experiments presented here. Hence, these experiments indicate that changes in root development reflect changes to resource allocation that are mediated by AMF, possibly in response to factors of the local substrate. Certainly, the absence of overall root length changes corresponds to 'correlative inhibition' that some species display, in which certain meristems can develop at the expense of others (Robinson 1996), for example, in exploitation of resource-rich soil niches.

In conclusion, if *R. sphaerocarpa* plants were used in a revegetation programme it would be wise to place inoculum in the region where most lateral roots are produced, i.e. throughout the top 20 cm of substrate. If plants were grown in a nursery initially, the location and distribution of inoculum of AMF would not affect colonisation levels in roots of *R. sphaerocarpa* since root density in a pot would exceed that in the field. However, given the importance of the phreatophytic root system of *R. sphaerocarpa* for survival in the native semi-arid ecosystem, it would be advisable to grow *R. sphaerocarpa* in pots that minimise potential damage to roots or alteration of normal root development. In this case, inappropriate distribution of inoculum relative to the lateral roots would result in poor and possibly inadequate colonisation. In practice, the most appropriate placement and distribution of inoculum will depend on the root system of a plant species and in particular the relative distribution of the finer lateral roots. These results demonstrate the complexity of the symbiosis between native plant species and AMF. They highlight the importance of the host:fungus:environment combination in determining the establishment of the symbiosis and response of a host species to colonisation.

Arbuscular mycorrhizal associations are commonly reported as providing growth benefits to host plants. However, in the experiments of Craven-Griffiths (1999), neither *A. cytisoides* or *R. sphaerocarpa* exhibited enhanced growth when inoculated with AMF under the prevailing environmental conditions in the greenhouse. Instead, as detailed above, host responses included increased shoot hydration and alterations in the distribution of lateral roots in *R. sphaerocarpa* plants. These results support the increasing evidence that benefits for plants in natural ecosystems are not necessarily exhibited as enhanced growth, but are manifested in other ways e.g. increased survival, recovery from short term stress, resistance to pathogens (Fitter and Garbaye 1994; Allsop and Stock 1993; Roldán-Fajardo 1994; Berta et al. 1995). Responses such as enhanced survival or short term tolerance of abiotic stress, such as drought, are not unexpected when the climatic conditions in semi-arid ecosystems are taken into consideration i.e. unpredictable and frequently limited water availability with high temperatures, sometimes for long periods of the year

(Puigdefábregas et al. 1996). Native plant species have evolved strategies to enable them to endure or avoid these unfavourable conditions. For example, in their natural environment *A. cytisoides* is summer deciduous whilst *R. sphaerocarpa* maintains a minimal open canopy (Puigdefábregas et al. 1995).

Enhanced growth of shoots would, therefore, not necessarily provide any survival advantage to native plant species and excessive shoot development would increase the water and/or heat stress on a plant during these periods. Indeed nursery production of inoculated plants should avoid overwatering and favourable conditions which will ultimately lead to high mortality rates at outplanting into fierce environmental conditions in semi-deserts. The outplanting of plants raised in a greenhouse with luxuriant foliar growth could lead to high mortality rates as plants would not be able to support such growth under the prevailing drought stress. Figure 6 shows an example of the difference between similar plants raised in a greenhouse compared to those raised under field conditions. Those raised under field conditions appear smaller and less luxuriant yet are better equipped to withstand the trauma of outplanting.

However, improvements in water relations may alleviate water stress imposed by short-term drought in the field. The phreatophytic root system of *R. sphaerocarpa* comprises extensive lateral roots in the upper soil ones and very deep vertical roots which enable exploitation of water in the infiltration zone and access of deep waters, respectively (Haase et al. 1996). *G. microaggregatum* BEG56 consistently influenced the distribution and topography of roots of *R. sphaerocarpa*, increasing lateral root development in the upper infiltration zone of the soil. This may have a considerable impact on functioning in the field, since *R. sphaerocarpa* plants grow in soils with the lowest water holding capacities. Hence, increased lateral roots in the infiltration zone may enable *R. sphaerocarpa* to access greater volumes of water before it is lost to the deeper layers. It must also be remembered that the effects of AMF may

Figure 6. Growth of *Anthyllis cytisoides* inoculated with AMF in soil from Almeria, SE Spain. Those on the left were raised in the field for 6 months, whilst those on the right were grown in a greenhouse.

not necessarily be realised within early growth. Once pioneer legumes are established, such as *R. sphaerocarpa,* they facilitate the growth of other species underneath the canopy through the creation of an 'island of fertility' (Pugnaire et al. 1996). Although, the mycorrhizal status of neighbouring plants has not been considered, it is likely that AMF associated with pioneer legumes would colonise and contribute to the development of other plants.

In summary, this work has shown that, within a revegetation programme, AMF can contribute to the long-term benefits for a desertified region other than through the initial establishment of the target species. A primary consideration for revegetation is the quantity and quality of indigenous inoculum, which can be significantly depleted in degraded ecosystems (Dodd and Thomson 1994; Requena et al. 1996). If there is inadequate or a low diversity of inoculum, it is recommended to transplant inoculated plants into the target ecosystem using isolates of AMF that are suited to or native to the target ecosystem (Dodd and Thomson 1994; Monzon and Azcón 1996; Vosatka and Dodd in this book). Likewise, it is important to use an appropriate plant species for the target ecosystem. Both *A. cytisoides* and *R. sphaerocarpa* are successful pioneer species of the semi-arid ecosystems in Mediterranean ecosystems but they differ in their growth strategies and the niches they occupy. Inoculation of transplanted legumes has been shown to improve the success of establishment in the field and the final case study will illustrate the potential.

Case Study 3: Integration of a mycorrhizal technology into revegetation schemes

Requena et al. (2001) conducted two long-term experiments in a desertified Mediterranean ecosystem, that demonstrated that inoculation with indigenous AMF, and with rhizobial nitrogen-fixing bacteria, not only enhanced establishment of key plant species but also increased soil fertility and quality. The dual symbiosis increased soil nitrogen (N) content, organic matter and hydrostable soil aggregates, and enhanced N-transfer from N-fixing to non-fixing species associated within the natural succession. It was concluded that the introduction of target indigenous species of plants associated with a managed community of microbial symbionts was a successful tool to aid the recovery of desertified ecosystems. The results of a 5-year trial showed significant improvements in the performance of *Anthyllis* plants inoculated with a mixture of native AMF and rhizobial ecosymbionts. These were compared with non-inoculated plants or those inoculated with a non-indigenous isolate of *G. intraradices.* Survival rates were higher in AMF-inoculated plants, but at least 25 plants in any treatment, out of the 40 transplanted, survived the dry and adverse conditions during the first year after outplanting, with no more losses afterward. In year 1, the plants inoculated with the exotic AMF *G. intraradices* were larger than those of the other two treatments, but by year 3 the plants inoculated with the mixed indigenous AMF were the largest. By year 5, the plants inoculated with *G. intraradices* were not significantly larger than plants not inoculated with AMF, whereas plants inoculated with a mixed inoculum were almost twice as large as those of the other two treatments.

Inoculation with the microbial ecosymbionts resulted in an increase in the number of AMF propagules able to develop colonization units on plant roots in the soil around the *Anthyllis* plants. *Scutellospora calospora, Glomus coronatum, G. constrictum, Acaulospora* sp. and an undescribed hyaline species frequent in Mediterranean soils (Dodd and Krikun 1984) were all present in the rhizosphere of all plants in year 5, whilst spores of the introduced exotic AMF were scarce. There were also significant improvements in years 3 and 5 in the physico-chemical properties in the soil around the *Anthyllis* plants inoculated with the mixed AMF inoculum, including N content, amount of organic matter (OM) and aggregation (Requena et al. 2001). Bioaugmentation of the soil with inoculum of a mixed, native AMF inoculum thus increased plant productivity. This correlates with other studies which show that native AMF are important contributors to plant biodiversity and ecosystem productivity (Allen et al. 1995; Herrera et al. 1993; Van der Heijden et al. 1998).

An increase in soil levels of both OM and N stimulates plant development (Francis and Thornes 1990; Puigdefábregas et al. 1996). Organic matter increases mainly through leaf and branch fall, but it has been also related to the AM status of the root (Carillo-Garciá et al. 1999). The increase in N content in the rhizosphere of the legume can be accounted for by an improvement in nodulation and N-fixation capacity resulting from inoculation with AMF. The improvement of soil aggregation contributes to the ability of a soil to maintain good water infiltration rates, good tilth and adequate aeration for plant growth (Wright and Upadhyaya 1998).

In a second field trial over 12 months, ^{15}N isotope dilution techniques were used to study N fixation in *Anthyllis* and N transfer from this plant to *Lavandula*, a non-leguminous woody species commonly associated with *Anthyllis* in the natural plant succession in the target area (Azcón and Barea 1997). The results showed: (i) *Lavandula* plants benefited from growing with the N-fixing legume, with regard to both biomass accumulation and N acquisition, (ii) inoculation with native AMF benefited plant growth, N fixation and N-transfer in both plants, despite the fact that indigenous fungi had colonised high levels of the roots of non-inoculated plants after 10 months in the field, and (iii) inoculation of *Anthyllis* with AMF also enhanced the mycorrhization of uninoculated *Lavandula* plants growing nearby. It is clear from these results that inoculation with native AMF benefited plant growth, N fixation and N-transfer. Improved N status of non-leguminous plants grown in association with legumes has previously been described for agricultural crops (Azcón-Aguilar et al. 1979), but this is the first demonstration of this phenomenon for natural plant communities in a semi-arid ecosystem. The results emphasize the important role of shrub legumes as a source of AMF inoculum for the surrounding area and in improving N nutrition for non-N-fixing vegetation, and support the general conclusion that the introduction of target indigenous species of plants associated with a managed community of microbial symbionts is a successful technological tool to aid the recovery of desertified ecosystems

Conclusions

Overall, it is clear that consideration of AMF must be included in the development of biological stategies to combat desertification (Fig. 7). However, the work reviewed here demonstrates the complexity of the biological interactions

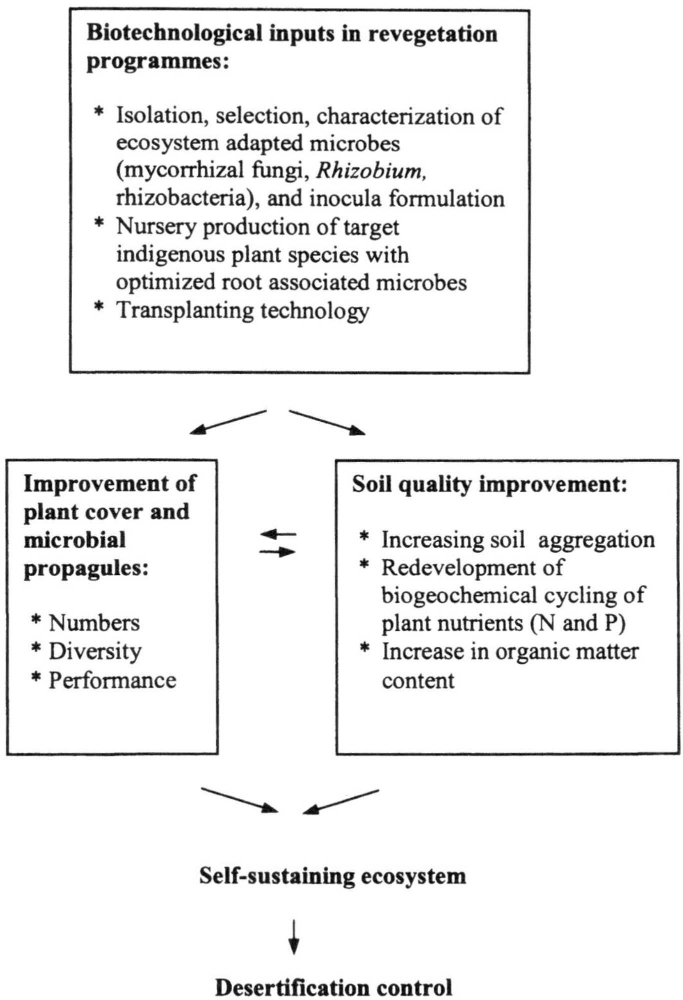

Figure 7. Proposed approaches to help plant establishment and to improve physical, chemical and biological soil properties essential to redevelop a self-sustaining ecosystem and to combat desertification.

which involve AMF, as well as the importance of host:fungus:environmental combinations in determining the response of a host to colonisation by AMF. The importance of AMF for plant survival and development in natural ecosystems, particularly those which are environmental under stress, is undeniable given the increasing wealth of evidence that AMF contribute to, if not determine, the fitness of native plants and communities. In fact, it is now believed that it is the co-evolution of associations with AMF that enable native plant species to grow successfully in nutrient poor, water limited environments (Azcón and Barea 1997).

Futhermore, the benefits of AMF extend beyond individual host responses to include improvements in nutrient cycling and stabilisation of natural ecosystems (Jeffries and Barea 1994). However, the results described here clearly demonstrate that it is not possible to generalise about the interactions between a specific host and specific isolates of AMF in desertified ecosystems. This is reinforced by the realisation that environmental influences on mycorrhizal functioning include not only physicochemical characteristics of soil, but also the relationships with other soil micro-organisms i.e. PGPR and rhizobia (Requena et al. 1997). Whilst the nature of effects of AMF on host development varied under different environmental conditions, semi-arid climates are notoriously unpredictable and it is possible that all host responses may potentially occur. The work reviewed here also illustrates the potential significance of biodiversity of AMF in semi-arid ecosystems. As well as recognising the potential benefits of AMF for agriculture and revegetation, it highlights the need to protect the diversity of AMF populations and vegetation, to ensure a robust plant community in abandoned agricultural lands in semi-arid zones.

Acknowledgements
The authors acknowledge several funding bodies which have financed this work including BARD, CICYT, OECD, and the Wain Fund. Significant amounts of the experimental work was carried out under the EU Environment Programme project 'REDEEM' (EV5V-CT94-0488). We also thank our numerous colleagues in COST Action 838 for their interesting discussions and suggestions.

References

Abbott L K, Robson A D (1991) Factors affecting the occurrence of vesicular-arbuscular mycorrhizas. Agriculture, Ecosystems and Environment 35:39-64

Allen E B (1989) The restoration of disturbed arid landscapes with special reference to mycorrhizal fungi. Journal of Arid Environments 17:279-320

Allen E B, Allen M F, Helm D J, Trappe J M, Molina R, Rincon E (1995) Patterns and regulation of mycorrhizal plant and fungal diversity. Plant and Soil 170:47-62

Allen M F, Allen E B (1992) Development of mycorrhizal patches in a successional arid ecosystem. In: Read D J, Lewis D H, Fitter A H, Alexander I J (eds) Mycorrhizas in ecosystems CAB International, Oxon. pp 164-170

Allsop A, Stock W D (1993) Mycorrhizas and seedling growth of slow growing sclerophylls from nutrient-poor environments. Acta Oecologica 14:577-587

Augé R M (2000) Stomatal behaviour of arbuscular mycorrhizal plants. In: Kapulnik Y, Douds D (eds) Arbuscular mycorrhizas: Physiology and function. Kluwer Academic Publishers, The Netherlands pp 201-238

Azcón R, Barea J M (1997) Mycorrhizal dependency of a representative plant species in mediterranean shrublands (*Lavandula spica* L.) as a key factor to its use for revegetation strategies in desertification-threatened areas. Applied Soil Ecology 7:83-92

Azcón-Aguilar C, Azcón R, Barea J M (1979) Endomycorrhizal fungi and *Rhizobium* as biological fertilizers for *Medicago sativa* in normal cultivation. Nature 27:235-237

Azcón-Aguilar C, Barea J M (1992) Interactions between mycorrhizal fungi and other rhizosphere microorganisms. In: Allen M (ed) Mycorrhizal functioning: an integrated plant fungus process. Chapman and Hall, London pp 163-198

Barea J M, Azcón R, Azcón-Aguilar C (1992) The use ^{15}N to assess the role of VA mycorrhiza in plant N nutrition and its application to evaluate the role of mycorrhiza in restoring Mediterranean ecosystems. In: Read D J, Lewis D H, Fitter A H, Alexander I J (eds) Mycorrhizas in ecosystems CAB International, Oxon pp 190-197

Barea J M, Requena N, Jimenez I (1996) A revegetation strategy based on the management of arbuscular mycorrhizas, *Rhizobium* and rhizobacteria for the reclamation of desertified mediterranean shrubland ecosystems. In: Lasram M (ed) Mycorrhization of forest plants under arid and semi-arid conditions and sesertification control in the Mediterranean Basin. CIHEAM, IAMZ, Zaragoza pp 75-86

Barea J M, Salamanca C P, Herrera M, Roldán-Fajardo B E (1990) Microorganisms-plant symbioses in the establishment of a vegetal cover on degraded lands. In: Albaladejo J, Stocking M, Diaz E (eds) Soil degradation and rehabilitation in Mediterranean environmental conditions. Consejo Superior do Investgaciones Cientificas, Murcia pp139-158

Barrow C J (1991) Land Degradation. Cambridge University Press, Cambridge

Bellgard S E (1993) The topsoil as the major store of the propagules of the vesicular-arbuscular mycorrhizal fungi in SE Australian sandstone soils. Mycorrhiza 3:19-24

Berta G, Trotta A, Fusconi A, Hooker J E, Munro M, Atkinson D, Giovannetti M, Morini S, Fortuna P, Tisserant B, Gianinazzi-Pearson V, Gianinazzi S (1995) Arbuscular mycorrhizal induced changes to plant growth and root system morphology in *Prunus cerasifera*. Tree Physiology 15:281-293

Bethlenfalvay G J, Schreiner R P, Mihara K L (1997) Mycorrhizal fungi effects on nutrient composition and yield of soybean seeds. Journal of Plant Nutrition 20:581-591

Bloss H E (1985) Studies of symbiotic microflora and their role in the ecology of desert plants. Desert Plants 7:119-127

Braunberger P G, Abbott L K, Robson A D (1994) The effect of rain in the dry-season on the formation of vesicular-arbuscular mycorrhizas in the growing season in annual clover-based pastures. New Phytologist 127:107-117

Braunberger P G, Abbott L K, Robson A D (1997) Early vesicular-arbuscular mycorrhizal colonisation in soil collected from an annual clover-based pasture in a Mediterranean environment: soil temperature and the timing of autumn rains. Australian Journal of Agricultural Research 48:103-110

Brundrett M C, Bougher N, Dell B, Grove T, Malajczuk, N (1996) Working with Mycorrhizas in Forestry and Agriculture. ACIAR Monograph 32

Carrillo-García A, Leon de la Luz J L, Bashan Y, Bethlenfalvay G J (1999) Nurse plants, mycorrhizae, and plant establishment in a disturbed area of the Sonoran Desert. Restoration Ecology 7:321-335

Craven-Griffiths A (1999) Ecology of arbuscular mycorrhizal fungi from a desertified ecosystem. PhD thesis, University of Kent, UK

Davies F T, Potter J R, Linderman R G (1992) Mycorrhiza and repeated drought exposure affect drought resistance and extraradical hyphae development of pepper plants independent of plant size and nutrient content. Journal of Plant Physiology 139:289-294

Davies F T, Potter J R, Linderman R G (1993) Drought resistance of mycorrhizal pepper plants independent of leaf P concentration-response in gas exchange and water relation. Physiologia Plantarum 87:45-53

Dodd J C (1994) Approaches to the study of the extraradical mycelium of arbuscular mycorrhizal fungi. In: Gianinazzi S, Schüepp H (eds) Impact of arbuscular mycorrhizas on sustainable agriculture and natural ecosystems. Birkhäuser Verlag, Basel pp 147-166

Dodd J C, Boddington C L, Rodriguez A, Gonzalez-Chavez C, Mansur I (2000) Mycelium of arbuscular mycorrhizal fungi (AMF) from different genera: form, function and detection. Plant and Soil 226:131-151

Dodd J C, Dougall T A G, Clapp J P, Jeffries P (2001) The role of arbuscular mycorrhizal fungi in plant community establishment at Samphire Hoe, Kent, UK – The reclamation platform created

during the building of the Channel Tunnel. Biodiversity and Conservation (In Press)
Dodd J C, Krikun J (1984) Observations on endogonaceous spores in the Negev desert. Transactions of the British Mycological Society 82:536-540
Dodd J C, Krikun J, Haas J (1983) Relative effectiveness of indigenous populations of vesicular-arbuscular mycorrhizal fungi from four sites in the Negev (Israel). Israel Journal of Botany 32: 10-21
Dodd J C, Thomson B (1994) The screening and selection of inoculant arbuscular-mycorrhizal and ectomycorrhizal fungi. Plant and Soil 59: 149-158
Ebel R C, Duan X, Still W, Augé R M (1997) Xylem sap abscisic acid concentration and stomatal conductance of mycorrhizal *Vigna unguiculata* in drying soil. New Phytologist 135:755-761
Faber B A, Zasoski R J, Munns D N, Shackel K (1991) A method for measuring hyphal nutrient and water uptake in mycorrhizal plants. Canadian Journal of Botany 69:87-94
Fitter A H (1985) Functioning of vesicular-arbuscular mycorrhizas under field conditions. New Phytologist 99:257-265
Fitter A H (1988) Water relations of red-clover *Trifolium pratense* as affected by VA mycorrhizal infection and phosphorus supply before and during drought. Journal of Experimental Botany 39:595-603
Fitter A H, Garbaye J (1994) Interactions between mycorrhizal fungi and other soil organisms. In: Robson A D, Abbott L K, Malajczuk N (eds). Management of mycorrhizas in agriculture, horticulture and forestry. Kluwer Academic Publishers, Dordrecht, Netherlands pp123-132
Francis C F, Thornes J B (1990) Matorral: Erosion and Reclamation. In: Albaladejo J, Stocking M A, Díaz E (eds) Soil degradation and rehabilitation in Mediterranean environmental conditions. CSIC, Spain pp 87-115
George E, Häussler K-U, Vetterlein D, Gorgus E, Marschner H (1992) Water and nutrient translocation by hyphae of *Glomus mosseae*. Canadian Journal of Botany 70:2130-2137
Goicoechea N, Antolín M C, Sánchez-Díaz M (1997a) Influence of arbuscular mycorrhizae and *Rhizobium* on nutrient content and water relations in drought stressed alfalfa. Plant and Soil 192:261-268
Goicoechea N, Antolín M C, Sánchez-Díaz M (1997b) Gas exchange is related to the hormone balance in mycorrhizal nitrogen-fixing alfalfa subjected to drought. Physiologia Plantarum 100:989-997
Gupta R K (1989) Drought response in fungi and mycorrhizal plants. In: Arora D K, Rai B, Mukerji K G, Knudsen G R (eds) Handbook of Applied Mycology. Volume I: Soil and Plants. Dekker, New York pp 55-76
Haase P, Pugnaire F I, Fernández E M, Puigdefábregas J, Clark S C, Incoll L D (1996) An investigation of rooting depth of the semiarid shrub *Retama sphaerocarpa* (L.) Boiss. by labelling of ground water with a chemical tracer. Journal of Hydrology 177:23-31
Hetrick B A D, Kitt D, Gerschefske K, Wilson G T (1987) Effects of drought stress on growth response in corn, sudan grass and big bluestem. New Phytologist 105:403-410
Herrera M A, Salamanca C P, Barea J M (1993) Inoculation of woody legumes with selected arbuscular mycorrhizal fungi and rhizobia to recover desertified mediterranean ecosystems. Applied and Environmental Microbiology 59:129-133
Jasper D A (1994) Management of mycorrhizas in revegetation. In: Robson A D, Abbott L K, Malajczuk N (eds) Management of mycorrhizas in agriculture, horticulture and forestry. Kluwer Academic Publishers, Dordrecht, Netherlands pp 211-219
Jasper D A, Abbott L K, Robson A D (1993) The survival of infective hyphae in dry soil: an interaction with sporulation. New Phytologist 124:473-479
Jastrow J D, Miller R M (1991) Methods for assessing the effects of biota on soil structure. Agriculture, Ecosystems and Environment 34:279-303
Jeffries P, Barea J M (1994) Biogeochemical cycling and arbuscular mycorrhizzas in the sustainability of plant-soil systems. In: Gianinazzi S, Schüepp H (eds) Impact of arbuscular mycorrhizas on sustainable agriculture and natural ecosystems. Birkhäuser Verlag, Basel pp 101-115
Jeffries P, Barea J M (2000) Arbuscular Mycorrhiza – a key component of sustainable plant-soil ecosystems. In: Hock B (ed) The mycota, Volume IX. Fungal Associations. Springer-Verlag, Heidelberg pp 95-113
Krikun J, Haas J H, Dodd J C, Kinsbursky R (1990) Mycorrhizal dependence of four crops in a P-sorbing soil. Plant and Soil 122: 213-217
López-Sánchez M E, Díaz G, Honrubia M (1992) Influence of vesicular-arbuscular mycorrhizal infection and P addition on growth and P nutrition of *Anthyllis cytisoides* L. and *Brachypodium retusum* (Pers.) Beauv. Mycorrhiza 2:41-45

Levy Y, Krikun J 1980. Effect of vesicular arbuscular mycorrhiza on Citrus jambhiri water relations. New Phytol. 85:25-31.

Levy Y, Syvertsen J P, Nemec S 1983. Effect of drought stress and vesicular arbuscular mycorrhiza on citrus transpiration and hydraulic conductivity of roots. New Phytol. 93:61-66.

Linderman R G (1994) Role of VAM in Biocontrol. In: Pfleger F L, Linderman R G (eds) Mycorrhizae and plant health. APS Press, St. Paul, USA pp 1-25

Marschner H, Dell B (1994) Nutrient uptake in mycorrhizal symbiosis. Plant and Soil 159:89-102

Martins M A (1992) The role of the external mycelial network of VAMF: a study of carbon transfer between plants interconnected by a common mycelium. Mycorrhiza 2:66-73

McGee P A (1989) Variation in propagule numbers of vesicular-arbuscular mycorrhizal fungi in a semi-arid soil. Mycological Research 92:28-33

Miller R M, Jastrow J D (1992) The role of mycorrhizal fungi in soil conservation. In: Bethlanfalvay G J, Linderman R G (eds) Mycorrhizae in sustainable agriculture. American Society of Agronomy, Crop Science Society of America and Soil Science Society of America, Madison,WI, USA pp 29-44

Miller R M, Lodge D J (1997) Fungal responses to disturbance: agriculture and forestry. In: Wicklow D W, Söderström, B (eds) The mycota. IV. Environmental and Microbial Relationships. Springer Verlag, Heidelberg pp 65-80

Monzon A, Azcón R (1996) Relevance of mycorrhizal fungal origin and host plant genotype to inducing growth and nutrient uptake in *Medicago* species. Agriculture, Ecosystems and Environment 60:9-15

Moorman T, Reeves F B (1979) The role of endomycorrhizae in revegetation practices in the semi-arid west II: a bioassay to determine the effect of land disturbance on endomycorrhizal populations. American Journal of Botany 66:14-18

Morgan R P C, Rickson R J, Wright E (1990) Regeneration of degraded soils. In: Albaladejo J, Stocking M A, Díaz E (eds) Soil degradation and rehabilitation in Mediterranean environmental conditions. CSIC, Spain pp 87-115

Nelson C E, Safir G R (1982) Increased drought tolerance of mycorrhizal onion plants caused by improved phosphorus nutrition. Planta 154:407-413

Newman E I, Eason W R, Eissenstat D M, Ramos M I R F (1992) Interactions between plants: The role of mycorrhizae. Mycorrhiza 1:47-53

Osonubi O, Mulongoy K, Awotoye O O, Atayese M O, Okali D U U (1991) Effects of ectomycorrhizal and vesicular-arbuscular mycorrhizal fungi on drought tolerance of four leguminous woody seedlings. Plant and Soil 136:131-143

Perez-Trejo F (1994) Desertification and Land Degradation in the European Mediterranean. European Commission, Luxembourg

Piccini D, Azcón R (1987) Effect of phosphate solubilizing bacteria and vesicular arbuscular mycorrhizal fungi on the utilization of Bayovar rock phosphate by alfalfa plants using a sand-vermiculite medium. Plant and Soil 101:45-50

Pugnaire F I, Haase, P, Puigdefábregas J (1996) Facilitation between higher plant species in a semi-arid environment. Ecology 77:1420-1426

Puigdefábregas J, Alonso J M, Delgado L, Domingo F, Cueto M, Gutiérrez L, Lázaro R, Nicolau J M, Sánchez G, Solé A, Torrentó J R, Vidal S, Aguilera C, Brenner A J, Clark S C, Incoll L D (1996) The Rambla Honda Field Site: Interactions of soil and vegetation along a catena in semi-arid SE Spain. In: Thornes J B, Brandt J (eds) Mediterranean desertification and land use. John Wiley, New York pp137-168

Requena N, Jeffries P, Barea J M (1996) Assessment of natural mycorrhizal potential in a desertified semiarid ecosystem. Applied and Environmental Microbiology 62:842-847

Requena N, Jimenez I, Toro M, Barea J M (1997) Interactions between plant-growth-promoting rhizobacteria (PGPR), arbuscular mycorrhizal fungi and *Rhizobium* spp. in the rhizosphere of *Anthyllis cytisoides*, a model legume for revegetation in Mediterranean semi-arid ecosystems. New Phytologist 136:667-677

Requena N, Perez-Solis E, Azcón-Aguilar C, Jeffries P, Barea J M (2001) Management of indigenous plant-microbe symbioses aids restoration of desertified ecosystems. Applied and Environmental Microbiology 67:495-498

Robinson D (1996) Variation, co-ordination and compensation in root systems in relation to soil variability. Plant and Soil 187:57-66

Roldan A, Garcia C, Albaladejo J (1997) AM fungal abundance and activity in a chronosequence of abandoned fields in a semi-arid mediterranean site. Arid Soil Research and Rehabilitation 11:211-220

Roldán-Fajardo B E (1994) Effect of indigenous arbuscular mycorrhizal endophytes in the development of six wild plants colonising a semi-arid area in south-east Spain. New Phytologist 127:115-121

Sánchez-Díaz M, Honrubia M (1994) Water relations and alleviation of drought stress in mycorrhizal plants. In: Gianinazzi S, Schüepp H (eds) Impact of arbuscular mycorrhizas on sustainable agriculture and natural ecosystems. Birkhäuser Verlag, Basel pp 167-177

Schreiner R P, Bethlenfalvay G J (1996) Mycorrhizae, biocides, and biocontrol. 4. Response of a mixed culture of arbuscular mycorrhizal fungi and host plant to three fungicides. Biology and Fertility of Soils 23:189-195

Schreiner R P, Milhara K L, McDaniel H, Bethlenfalvay G J (1997) Mycorrhizal fungi influence plant growth and soil functions and interactions. Plant and Soil 188:199-209

Schwab S, Reeves F B (1981) The role of endomycorrhizae in revegetation practices in the semi-arid west. III. Vertical distribution of vesicular-arbuscular (VA) mycorrhiza inoculum. American Journal of Botany 68:1293-1297

Sieverding E (1991) Vesicular-Arbuscular Mycorrhizal Management in Tropical Agrosystems, GTZ Publishers, Germany, p 371

Stutz J C, Morton J B (1996) Successive pot cultures reveal high species richness of arbuscular endomycorrhizal fungi in arid ecosystems. Canadian Journal of Botany 74:1883-1889

Stutz J C, Copeman R, Martin C A, Morton J B (2000) Patterns of species composition and distribution of arbuscular mycorrhizal fungi in arid regions of southwestern North America and Namibia, Africa. Canadian Journal of Botany 78:237-245

Tarafdar J C, Praveen-Kumar (1996) The role of vesicular-arbuscular mycorrhizal fungi on crop, tree and grasses green in an arid environment. Journal of Arid Environments 34:197-203

Thiagarajan T R, Ahmad M H (1993) Influences of a vesicular-arbuscular mycorrhizal fungus on the competitive ability of *Bradyrhizobium* spp. for nodulation of cowpea *Vigna unguiculata* (L.) Walp in non-sterilised soil. Biology and Fertility of Soils 15:294-296

Thornes J B (1993) MEDALUS Mediterranean Desertification and Land Use. The MEDALUS Office,Bristol, UK

Tisdall J M, Smith SE, Rengasamy P (1997) Aggregation of soil by fungal hyphae. Australian Journal of Soil Research 35:55-60

Tobar R M, Azcón R, Barea J M (1994) The improvement of plant N acquisition from an ammonium-treated, drought-stressed soil by the fungal symbiont in arbuscular mycorrhizae. Mycorrhiza 4:105-108

Tommerup I C (1992) The role of mycorrhiza in plant populations and communities. Mycorrhiza 1:123-125

UNEP (1992) World Atlas of Desertification. Edward Arnold, London

Van der Heijden M G A, Klironomos J N, Ursic M, Moutoglis P, Streitwolf-Engel R, Boller T, Wiemken A, Sanders I R (1998) Mycorrhizal fungal diversity determines plant biodiversity, ecosystem variability and productivity. Nature 396:69-72

Vosatka M, Dodd J C (2001) This volume

Warren A, Sud Y C, Rozanov B (1996) The future of deserts. Journal of Arid Environments 32:75-89

Wright S F, Franke-Snyder M, Morton J B, Upadhyaya A (1996) Time-course study and partial characterisation of a protein on hyphae of arbuscular mycorrhizal fungi during active colonisation of roots. Plant and Soil 181:193-203

Wright S F, Upadhyaya A (1996) Extraction of an abundant and unusual protein from soils and comparison with hyphal protein of arbuscular mycorrhizal fungi. Soil Science 161:575-586

Wright S F, Upadhyaya A (1998) A survey of soils for aggregate stability and glomalin, a glycoprotein produced by hyphae of arbuscular mycorrhizal fungi. Plant and Soil 198:97-107

Wright S F, Upadhyaya A, Buyer J S (1998) Comparison of N-linked oligosaccharides of glomalin from arbuscular mycorrhizal fungi and soils by capillary electrophoresis. Soil Biology and Biochemistry 30:1853-1857

Yano K, Yamauchi A, Kono Y (1996a) Localised alteration in lateral root development in roots colonised by an arbuscular mycorrhizal fungus. Mycorrhiza 6:409-415

Yano K, Yamauchi A, Kono Y (1996b) Modification of root system morphology in a peanut seedling inoculated with arbuscular mycorrhizal fungus, *Gigaspora margarita* Becker and Hall. Japanese Journal of Crop Science 65:361-367

Mycorrhizal Technology in Agriculture
ed. by S. Gianinazzi, H. Schüepp, J.M. Barea and K. Haselwandter

Potential of arbuscular mycorrhizal fungi for bioremediation

C. Leyval[1], E. J. Joner[1], C. del Val[2] and K. Haselwandter[3]

[1] *Laboratoire des Interactions Microorganismes-Minéraux-Matière Organique dans les Sols, CNRS, F-54501 Vandoeuvre-les-Nancy, France*
[2] *Depto. Microbiologia del Suelo y Sistemas Simbioticos, Estacion Experimental del Zaidin, CSIC, Profesor Albareda 1, E-18008 Granada, Spain*
[3] *Universität Innsbruck, Institut für Mikrobiologie, Technikerstr. 25, A-6020 Innsbruck, Austria*

Keywords: heavy metals, polycylic aromatic hydrocarbons, radionuclides, stress tolerance, phytoremediation

Introduction

Bioremediation is the use of organisms for the treatment of soil pollution. Root colonizing symbiotic microorganisms such as arbuscular mycorrhizal fungi (AMF) are mainly involved in phytoremediation, that uses plants for soil remediation. Phytoremediation comprises a set of technologies that use various plants as a containment, destruction or extraction technique (EPA 2000). These techniques have received considerable interest in recent years because of potential cost savings compared to conventional non biological techniques. Different strategies of phytoremediation can be applied depending on the kind of pollutants. In all cases, vegetation reduces infiltration of water and erosion. Heavy metals cannot be degraded and can only be extracted (phytoextraction) from the soil or immobilized in a non toxic form (phytostabilization). AMF can help alleviate metal toxicity to plants by reducing metal translocation from root to shoot (Leyval et al. 1997). Therefore they may contribute to plant establishment and survival in heavy metal polluted sites and could be used as a complement to immobilization strategies. Phytoextraction mainly uses plants accumulating high concentrations of heavy metals, which can be harvested, discarded and even extracted to recover metals. For this purpose plants with various capacities for metal accumulation are used, like members of the *Brassicaceae*, which are generally considered non mycorrhizal, but also other accumulators producing higher biomass, which can be mycotrophic. Organic pollutants such as polycylic aromatic hydrocarbons (PAH) can be transformed or degraded through microbial activity, which is commonly enhanced in the root zone (rhizodegradation). It is not known whether this enhanced degradation in the rhizosphere is due to plant exudates including enzymes, surfactants, and other physical/chemical effects, and/or to increased microbial activity. Another possible mechanism for degradation of organic pollutants could be directly dependent on plant metabolism. However, this is not a quantitatively important route for PAH (Binet et al. 2000a), and will not be discussed here. AMF may be beneficial for PAH rhizodegradation because they affect root exudation and root associated microbial populations and because, in some ways, they act as an extension of the roots outside the rhizosphere. They may potentially also have a direct effect on PAH degradation. Finally, AMF can be used in bioassays of soil quality or soil toxicity, due to their sensitivity towards a range of soil pollutants. In this way, they could be used to show that adequate soil quality has been re-established after remediation.

The objective of this chapter is to review recent findings concerning the potential of AMF for bioremediation of soils polluted with heavy metals, radionuclides and PAH as representatives of organic pollutants. In each case, it will underline obstacles, and outline the perspectives for a successful application of AMF in phytoremediation strategies.

Potential of AMF for phytoremediation of heavy metal polluted soils

Soil microorganisms are known to play a key role in the mobilization and immobilization of metal cations, thereby changing their availability to plants (Birch and Bachofen 1990). AMF are among the most common soil microorganisms and constitute an important functional component of the soil-plant system occurring in almost all habitats and climates (Barea et al. 1997), including disturbed soils (McGonigle et al. 1996; Bundrett et al. 1996). Degraded soils do, however, suffer from changes in diversity and abundance of AM fungal populations (Koomen et al. 1990; Jasper et al. 1991; Loth 1996).

More specifically, it has been shown that AMF can be affected by heavy metal toxicity, but in many cases mycotrophic plants growing in soils contaminated with heavy metals are colonized by AMF (see e.g. review by Leyval et al. 1997). Many reports concerning this have quantified spores and estimated root colonization *in situ*. Others have gone further and described metal tolerant AMF in heavy metal polluted soils (Gildon and Tinker 1983; Weissenhorn et al. 1995; del Val 1999a; Hildebrandt et al. 1999).

In the last few years, research interest has focused on the diversity and tolerance of AMF in heavy metal contaminated soils trying to understand the basis underlying adaptation and tolerance of AMF to heavy metals in soils, since this could facilitate the management of these soil microorganisms, for restoration/bioremediation programs. Vandenkoornhuyse (1998) showed that AMF species diversity associated with maize plants in a long term field experiment did not differ between three plots that had received different levels of heavy metals-containing sewage sludge. However, the number of spores of each species was lower in the soil with the highest concentration of heavy metals. Using the same long-term field experiment and the same plant variety, but more acidic soils, del Val et al. (1999) found a reduction of number, but also of diversity of AMF spores in the soil receiving the highest rate of sludge.

On a highly polluted soil in northern France where only adapted plants can grow, *Arrhenaterum elatius* roots were collected along a gradient of heavy metal concentration. Up to three different *Glomus* species were identified inside *Arrhenaterum* roots, which differed along the gradient of metals (Leyval et al. unpublished data). The contribution of these AMF to plant tolerance to heavy metals or heavy metal accumulation by plants has not been established. Four *Glomus* species were also found in the rhizosphere of another metal tolerant plant, *Viola calaminaria*, growing on a soil highly contaminated with heavy metals (20,961 and 41 mg kg^{-1} Zn and Cd, respectively) (Tonin et al. 2001). Only one of these fungi colonized clover roots growing in pots supplemented with Cd and Zn salts. This *Glomus* sp. increased Cd and Zn concentrations in clover roots, but not in shoots, and did not affect plant growth. On the contrary, a *Glomus* isolate from the rhizosphere of *Viola calaminaria* increased the growth of maize and lucerne in heavy metal polluted soils and reduced Zn concentration in roots and shoots (Hildebrandt et al. 1999; Kaldorf et al. 1999).

Although AMF have been recovered from numerous metal enriched habitats, their role in plant interaction with toxic metals is not well understood. At high metal concentrations reports show variations in metal accumulation and inter-plant translocation depending on the fungi, host-plant, root density, soil characteristics, metals and their availability (El-Kerbawy et al. 1989; Leyval et al. 1997; Joner and Leyval 2001b). Large variations have also been found between AM fungal species due to differences in hyphal growth outside the rhizosphere (Bürkert and Robson 1994; Joner and Leyval 1997). Metal-tolerant AMF isolates can decrease metal concentration in shoots or in roots, or decrease translocation from root to shoots (Diaz et al. 1996; Joner and Leyval 1997; Joner and Leyval 2001b; del Val et al. unpublished data). The latter could be due to the high metal sorption capacity of these fungi, which could 'filter' metal ions during uptake (Joner et al. 2000). The high concentrations of heavy metals in the intracellular hyphae of a heavy metal tolerant AMF colonizing maize roots (Kaldorf et al. 1999) and in phosphate rich material in hyphal vacuoles of mycorrhizal roots of *Pteridium aquilinum* (Turnau et al. 1993) strengthen the hypothesis of a sequestration of metals by AMF structures. However, the competitivity of such metal tolerant AMF in the field is often unknown and should be investigated. Further, the potential benefit of a consortium of AMF, which corresponds to the situation in the roots, to improve phytoremediation, should be considered.

Phytoextraction studies often use hyperaccumulators (plants accumulating high concentrations of heavy metals, e.g. 1% Zn in their dry matter), which are in most cases non-mycotrophic plants belonging to the *Brassicaceae*. One objective is to use plants with high concentrations of heavy metals in shoots, which may limit the potential use of AM plants. However, many of these hyperaccumulating plants, such *Thlaspi caerulescens,* are small and grow slowly, which limits phytoextraction rates. Other accumulators producing a higher biomass, such as sunflower and willow, are now receiving attention and these are mycorrhizal plants. Highly productive crops associated with metal-tolerant AMF may therefore be considered for decontamination of slightly contaminated soils (Ernst 2000).

The use of non-mycorrhizal plants for phytoextraction such as *Thlaspi caerulescens* can also change the glomalean community and reduce the propagule number in soil (Pawlowska et al. 2000). In this situation, re-inoculation with AMF after phytoextraction would be a possible option. Using metal-accumulating plants able to form mycorrhizas instead of non-host-plants should also improve the establishment of a metal-intolerant vegetation.

Phytostabilization refers to promoting plant growth to reduce or eliminate the bioavailability of heavy metals and various grasses such as *Festuca rubra* and *Agrostis tenuis* have been used commercially (Salt et al. 1995; van Tichelen et al. 1996). Physical immobilization of heavy metals in soil is often performed using various amendments such as beringite combined with planting of metal tolerant species. Within this context, AMF can play an important role by improving plant establishment, promoting plant growth and reducing metal translocation to shoots. The beneficial effect of the mycorrhizal symbiosis has been shown in the facultative mycotroph *Festuca arundinacea* when grown in mine spoils (Hetrick et al. 1994). The possible contribution of AMF to the success of restoration programmes developed for polluted soils is also outlined by Turnau and Haselwandter (in this book).

Further studies should be carried out with plants susceptible of mycorrhizal colonization and already known for their capacity to translocate heavy metals. There are some reports which show that plant species, including maize and *Sonchus solearaceus*, can extract considerable amounts of lead from soil (Huang et al. 1996). Other species, like barley, which are also potentially mycorrhizal, are able to extract Zn as efficiently as can the non-mycorrhizal *Brassica juncea* (Ebbs and Kochian 1998).

Potential of AMF for bioremediation of radionuclides

It is well known that mycorrhizal infection affects the mineral nutrition of plants (George et al. 1994; Smith and Read 1997). Nevertheless, remarkably little research has been carried out on the effect of AMF on plant uptake of (radio)cesium. The few results obtained so far appear controversial (Haselwandter et al. 1994).

McGraw et al. (1979) reported that root infection of *Paspalum notatum* by two of ten species of AMF resulted in a twofold increase of the radioactivity in leaf tissue 48 h after injection of ^{134}Cs into soils. A study carried out by Rogers and Williams (1986) also indicated that some AMF may enhance the radiocesium uptake. They found that the ^{137}Cs concentration in mycorrhizal *Melilotus officinalis* was 2.0 and 1.7 times that in the non-AM clover at the first (65 d) and second (93 d) harvest, respectively. Additionally, in the case of

Sorghum sudanense, the radiocesium concentration was greater in the AM than in the non-AM grass, albeit not significantly.

In case the biosphere is subjected to contamination with radiocesium, one possible strategy involves remediation of contaminated areas by using plants that accumulate large amounts of Cs. Obviously, at least some AMF species have the capacity to enhance the radiocesium uptake by plants and, hence, the mycorrhizal symbiosis can be very useful. Based on the use of a plant species considered to be most efficient in ^{137}Cs phytoextraction and not mycorrhizal, Ebbs et al. (2000) have estimated to achieve a 75% decrease of ^{137}Cs contamination in approximately 15 years. It appears essential to compare such a potential with that of a mycorrhizal plant species.

The alternative strategy aims at preventing radioactive Cs isotopes from entering the food chain, and relies on crops that do not accumulate substantial amounts of radiocesium. With regard to this strategy it is of interest to note that AMF can also lead to a significant decrease in ^{137}Cs content e.g. of *Festuca ovina*. After three and five weeks of growth, shoot tissue radioactivity in the mycorrhizal grass was about half that of the non-mycorrhizal controls (Haselwandter and Berreck 1994). Similar results were obtained in pot experiments with *Agrostis tenuis*, in which AM colonisation by *Glomus mosseae* led to a significant decrease in the Cs uptake from the first (4 weeks) to the third (8 weeks) harvest (Berreck and Haselwandter 2001). The reduction in Cs content of mycorrhizal *versus* non-mycorrhizal plants was between 18 and 23%. In the study quoted above, the effect of K fertilization on Cs uptake by AMF was monitored over a growth period of 10 weeks. Potassium was applied at a rate recommended as one of the countermeasures after contamination of grassland with radiocesium (Konoplev et al. 1993; Segal 1993). Potassium addition resulted in a significant decrease of Cs uptake of both AMF and non-AMF grass. The fertilized non-AMF plants contained 56 to 78% of the Cs concentration of the non-fertilized plants, whereas fertilized AM plants contained 61 to 78% of the non-fertilized AM plants (Berreck and Haselwandter 2001). Thus the findings summarized above corroborate the hypothesis that soil fungi represent a potential pool of Cs immobilization (Dighton et al. 1991).

It is clear that AMF play a key role with regard to both of the strategies, be it soil remediation on basis of plants accumulating large amounts of radiocesium or the prevention of Cs from entry into the food chain. Hence, AMF deserve particular interest with regard to the development of countermeasures for controlling the radionuclide uptake by plants. As Cs^+ transport across the plasma membrane of root cells determines the potential of a plant either to accumulate or exclude radiocesium it was envisaged that genetic modifications would allow for the development of means to control the uptake rates (White and Broadley 2000). However, as almost all plants are normally associated with mycorrhizal fungi which, obviously, are at least as crucial for cesium uptake as the plant uptake system *per se*, it seems feasible to suggest exploitation of mycorrhizal potential first. Ernst (2000) expressed a similar view with regard to metal (hyper)accumulation by plants.

Role of AMF in PAH-polluted soils

Polycyclic aromatic hydrocarbons (PAH) are hydrophobic organic molecules consisting of two or more fused benzene rings. A selection of 16 PAH are commonly quantified for characterization and monitoring of these pollutants, while in fact 200–300 PAH compounds and their derivatives are commonly extracted and may putatively be identified in polluted soil samples. The concern for PAH pollution derives from their ubiquitous distribution, their recalcitrance towards degradation and their proven or suspected mutagenic properties (WHO 1983). The origin may partially be natural (organic residues after fire) or anthropogenic (mainly processing and incomplete combustion of fossile fuels). Thus, oil spills and industrial sites, e.g. for coke distillation etc., commonly give rise to extreme pollution events for which physical, chemical and biological remediation strategies are employed for clean-up. These include, among others, bio-venting, land-farming, bio-augmentation and phytoremediation. The latter is only applicable when pollution levels and physical conditions of the polluted matrix permit the establishment of plants, but offers a cost-effective and efficient treatment that simultaneously restores an ecosystem, limits erosion and improves the esthetical impression of a polluted site (Cunningham et al. 1996; Schwab and Banks 1994; Wilson and Jones 1993).

Arbuscular mycorrhiza fungi may play a role in two aspects of bioremediation of PAH: the establishment of a plant cover on polluted soil and modification of PAH degradation rates or pathways.

Improved plant establishment on fallowed, degraded or polluted soils, waste materials or soil-waste mixtures in the presence of AMF is well known (Reddell and Milnes 1992; Shetty et al. 1994,Thompson 1994; Leyval and Binet 1998). The major mechanism behind the success of AM under such conditions is an improvement of plant nutrient acquisition, with additional benefits due to improved water relations, pollutant tolerance and sequestration. The impact of the latter factors are often difficult to distinguish due to the confounding effect of plant nutrition on plant size, but, in the case of enhanced survival in the absence of competitors/predators, the role of AM may be clear. Identification of the symbiotic features that have permitted continued subsistence is however difficult, and such investigations are still in their infancy. One mechanism that may be involved is the oxidation of the contaminant by activated oxygen species and concomitant enhancement of oxidoreductases to protect the plant from oxidative stress. Indeed, enhanced levels of hydrogen peroxide in AM roots (Salzer et al. 1999) as well as enhanced levels of peroxidase activity in mycorrhizal roots and the rhizosphere (Criquet et al. 2000) has been demonstrated.

One peculiarity of PAH polluted soil that may be overcome by AM plants is the hydrophobicity and resulting limitations in uptake of water and water-dissolved inorganic nutrients (Leyval and Binet 1998). Again confounding effects of improved mineral nutrition prevents the distinction of mycorrhizal effects on uptake of water and mineral nutrients, but even if the effect is limited to mineral uptake it is of no less importance.

Effects of AMF on PAH degradation in the rhizosphere may be direct or indirect. As PAH are not absorbed by plants (Binet et al. 2000b; Schwab and Banks 1994) and are metabolized intracellularly, all degrading activity would take place in soil or inside soil organisms other than AM. Furthermore, AMF have poor saprophytic capacities, so the only probable direct effect of AMF on PAH degradation would be through enhanced production of extracellular peroxidases. Indirect effects would be due to changes in the microbial community e.g. due to stronger competition for mineral nutrients, direct antagonistic or synergistic effects of AMF, or changes in root exudation patterns, phenomena that are well documented.

Binet et al. (2000b) found no effect of AM on the dissipation on light compounds like anthracene (3 rings) in spiked soil on the short term (<40 days). Continued phytoremediation treatments of spiked soil resulted in an almost complete (98%) dissipation of anthracene after only 56 days in the rhizosphere of clover/ryegrass (Joner et al. unpublished data). Heavier compounds (4–6 rings) are more persistent and often rely on degradation by co-metabolism rather than direct metabolism (Cutright and Lee 1994; Perry 1979). The rhizospheric effect on dissipation of these compounds is thus more pronounced than for the lighter PAH (Binet et al. 2000a) as the root exudates may drive co-metablism. In addition, we have observed an enhanced dissipation of 4–5 ring PAH in mycorrhizal treatments compared to non-mycorrhizal treatments after prolonged periods (>100 days) of phytoremediation (Joner et al. submitted). Mycorrhizal effects on PAH dissipation was in this case accompanied by a modification in biomass partitioning (but not total plant biomass) between two co-occurring plant species (clover and ryegrass) (Joner and Leyval 2001a), as well as changes in soil microbial community structure. The effect of colonization by mycorrhizal fungi on the competitive ability of clover on PAH-polluted soil (Joner and Leyval 2001a) is perhaps similar to that observed in non-polluted soil (Bolan et al. 1987), i.e. a phenomenon linked to the improved nutrient acquisition of the mycotrophic legume. The modification of the soil microbial community may on the other hand either be due to inherent qualitative (Germida et al. 1998) or quantitative (Griffiths et al. 1999) differences in root exudation between the two plant species coupled with a proportional change in plant biomass (indirectly mediated by the mycorrhizal effect on competitive ability/plant size of the two plants), or more direct effects of AMF, i.e. alterations in root exudation as a consequence of AMF colonization (Graham et al. 1981; Laheurte et al. 1990) and direct antagonistic/synergistic effects between the AMF hyphae and the soil microflora (Linderman 1988; Linderman 1991). The exploitation of root-free soil by AMF hyphae has a potential for modifying microbial composition (Olsson et al. 1996; Ravnskov et al. 1999) and activity. Ephemeral hyphae serving as a source of C outside the rhizosphere (see e.g. Schreiner and Bethlenfalvay 1995) may ultimately result in a microbial community with an improved capacity for PAH degradation (Joner et al. submitted).

A limited number of AMF was used in the above cited experiments on phytoremediation of PAH. The selection of fungi as well as of associated host-plants for improving the dissipation of PAH should be performed. Since the half-life of most of the PAH is high, longer-term and field experiments should also be carried out.

Potential use of AMF in bioassays for soil pollution

Arbuscular mycorrhizas are not only an aid to ecosystem remediation. They should also be considered as a key indicator for soil pollution or soil quality because:

(i) mycorrhizal fungi are ubiquitous microorganisms,
(ii) many plants are highly dependant on mycorrhizas for their growth,
(iii) they provide a direct link between soil and roots, and
(iv) are involved in the transfer of elements including pollutants from soil to plants.

Furthermore, AMF can be affected by pollutants in soil and can be more sensitive to pollutants than plants (Weissenhorn and Leyval 1995).

The toxicity of compounds such as xenobiotics, PAH and heavy metals on AMF has been studied using techniques based on estimation of spore germination (Weissenhorn et Leyval 1996), mycorrhizal colonisation of roots in pot cultures using nested PCR (Jacquot et al. 2000), mycorrhizal infectivity (Leyval et al. 1995) and mycorrhizal colonization of Ri T-DNA transformed roots (Wan et al. 1998). AMF spore germination in soil can be used as an early indication of the toxicity towards the ecosystem, while mycorrhizal colonisation of roots indicates toxicity at a later stage of the association between symbiotic fungi and roots. AMF can be affected by pollutants but also by soil properties such as pH and phosphorus content (Leyval et al. 1995). The lack of specificity towards heavy metals or other pollutants is often the case for other indicator organisms such as earthworms, algae, fish and plants, which are used for the assessment of ecological risk due to toxic substances. AMF could be included as additional indicator organisms in the existing battery of bioassays. There is a need for a standardized technique using arbuscular mycorrhizas as a bioassay, which should be made easier since commercial AMF inoculum is now available (Feldmann and Grotkass in this book, von Alten et al. in this book).

Conclusions

Some of the limitations for phytoremediation are the contact between roots and the pollutants, the root growth rate, and the toxicity of some of the pollutants. In that respect, AM should be beneficial because they increase the volume of soil explored by roots, improve plant growth, and contribute to alleviate toxicity of pollutants such as heavy metals.

For bioremediation of heavy metal polluted soils, metal tolerant AMF have been identified. Promising results have been obtained under laboratory conditions and in short-term experiments, showing that AMF may contribute to the phytoextraction or phytostabilisation of heavy metals and radionuclides, depending on the plant-fungus partners. Since it is often difficult to extrapolate from laboratory experiments to field situations, longer-term and field experiments should be performed to confirm the potential benefit of AMF for phytoremediation. Improving the knowledge on competitivity and survival of introduced AMF, AMF community structure and dynamics *in situ,* genetic diversity and functioning after soil disturbance by heavy metals and/or the introduction of selected endophytes, may be necessary to overcome for making the application of AMF in phytoremediation programmes successful. Methods to identify and recognise particular AMF isolates *in situ* should also be further developed. Soil parameters such as pH and P content may be as crucial for mycorrhizal establishment and efficiency as heavy metal concentrations (Weissenhorn and Leyval 1996) and should be checked before the set-up of any field experiment.

The impact of AM on PAH polluted soil is still quite uncertain, and the few results that have been obtained need verification with a wider range of soil, plants and fungi. The potential, however, is substantial as both plant establishment, survival and degradation would independently confer large advantages, and in some cases even justify the substantial cost that a field inoculation may represent. AM implication in plant tolerance to PAH does not seem improbable, as these compounds are found naturally in soil e.g. after fire. Fire is a phenomenon that occurs relatively regularly in some ecosystems, and probably frequently enough to exert a selection pressure on fungal genera that largely propagate clonally, like the Glomales.

There is still a need to improve our understanding of the mechanisms involved in the transfer and immobilisation of heavy metals by AMF, and in their contribution to organic pollutant availability and degradation. This is necessary if we are to improve the chances of successful application under practical conditions, including application in soils where both kinds of pollutants are present.

References

Barea J M, Azcón-Aguilar C, Azcón R (1997) Interactions between mycorrhizal fungi and rhizosphere microorganisms within the context of sustainable soil-plant systems. In A.C., Gange and V.K., Brown (ed.), Multitrophic interactions in terestrial systems, pp 65-77

Berreck M, Haselwandter K (2001) Effect of the arbuscular mycorrhizal symbiosis upon uptake of cesium and other cations by plants. Mycorrhiza 10:275-280

Binet P, Portal J M, Leyval C (2000a) Fate of polycyclic aromatic hydrocarbons (PAH) in the rhizosphere and mycorrhizosphere of ryegrass. Plant Soil 227:207-213

Binet P, Portal J M, Leyval C (2000b) Dissipation of 3-6-ring polycyclic aromatic hydrocarbons in the rhizosphere of ryegrass. Soil Biol Biochem 32:2011-2017

Birch, L D, Bachofen R (1990) Effects of microorganisms on the environmental mobility of radionucleides. In J.M. Bollang and G. Stozky (ed.), Soil Biochemistry, vol. 6. Marcel Dekker, New York, p.483-527

Bolan N S, Robson A D, Barrow N J (1987) Effects of phosphorus application and mycorrhizal inoculation on root characteristics of subterranean clover and ryegrass in relation to phosphorus uptake. Plant Soil 104:294-298

Brundrett M C, Ashwath N, Jasper D A (1996) Mycorrhizas in the Kakadu region of tropical Australia. I. Propaguels of mycorrhizal fungi and soil properties in natural habitats. Plant Soil 184:159-171

Criquet S, Joner E J, Léglize P, Leyval C (2000) Effects of anthracene and mycorrhiza on the activity of oxidoreductases in the roots and the rhizosphere of lucerne (*Medicago sativa* L.). Biotechnol Lett 22:1733-1737

Cunningham S D, Anderson T, Schwab A P and Hsu F C (1996) Phytoremediation of soils contaminated with organic pollutants. Adv Agron 56:55-114

Cutright T J and Lee S (1994) Microorganisms and metabolic pathways for remediation of PAH in contaminated soil. Fresenius Environ Bull 3:413-421

del Val C, Barea J M, Azcon-Aguilar C (1999) Diversity of arbuscular mycorrhizal fungus populations in heavy-metal-contaminated soils. Appl Environ Microbiol 65:718-723

Diaz G, Azcon-Aguilar C, Honrubia M (1996) Influence of arbuscular mycorrhizae on heavy metal (Zn and Pb) uptake and growth of *Lygeum spartum* and *Anthyllis cytisoides*. Plant Soil 180:241-249

Dighton J, Clint G M, Poskitt J M (1991) Uptake and accumulation of ^{137}Cs by upland grassland soil fungi: a potential pool of Cs immobilization. Mycol Res 95:1052-1056

Ebbs S, Kochian L V (1998) Phytoextraction of zinc by oat (*Avena sativa*), barley (*Hordeum vulgare*) and indian mustard (*Brassica juncea*). Environ Sci Technol 32:802-806

Ebbs S, Kochian L, Lasat M, Pence N, Jiang T (2000) An integrated investigation of the phytoremediation of heavy metal and radionuclide contaminated soils: From the laboratory to the field. In: Wise D L, Trantolo D J, Cichon E J, Inyang H I, Stottmeister U (eds) Bioremediation of contaminated soils. Dekker, New York, pp 745-769

El-Kherbawy M, Angle J S, Heggo A, Chaney R L (1989) Soil pH, rhizobia,and vesicular-arbuscular mycorrhizae inoculation effects on growth and heavy metal uptake of alfalfa (*Medicago sativa* L.). Biol Fertil Soil 8:61-65

EPA (2000) Introduction to phytoremediation, Cincinatti, Ohio p 70.

Ernst W H O (2000) Evolution of metal hyperaccumulation and phytoremediation hype. New Phytol 146:357-358

George E, Römheld V, Marschner H (1994) Contribution of mycorrhizal fungi to micronutrient uptake by plants. In: Manthey J A, Crowley D E, Luster D G (eds) Biochemistry of metal micronutrients in the rhizosphere. CRC Press, Boca Raton pp 93-109

Germida J J, Siciliano S D, de Freitas J R, Seib A M (1998) Diversity of root-associated bacteria associated with field-grown canola (*Brassica napus* L.) and wheat (*Triticum aestivum* L.). FEMS Microb Ecol 26: 43-50

Gildon A, Tinker P B (1983) Interactions of vesicular arbuscular mycorrhizal infection and heavy metals in plants. The effects of heavy metals on the development of vesicular-arbuscular mycorrhizas. New Phytol 95: 247-261

Graham J H, Leonard R T, Menge J A (1981) Membrane-mediated decrease in root exudation responsible for phosphorus inhibition of vesicular-arbuscular mycorrhiza formation. Plant Physiol 68:549-552

Griffiths B S, Ritz K, Ebblewhite N, Dobson G (1999) Soil microbial community structure: Effects of substrate loading rates. Soil Biol Biochem 31:145-153

Haselwandter K, Berreck M (1994) Accumulation of radionuclides in fungi. In: Winkelmann G, Winge D R (eds) Metal ions in fungi. Dekker, New York pp 259-277

Haselwandter K, Leyval C, Sanders F E (1994) Impact of arbuscular mycorrhizal fungi on plant uptake of heavy metals and radionuclides from soil. In: Gianinazzi S, Schüepp H (eds) Impact of arbuscular mycorrhizas on sustainable agricultural and natural ecosystems. Birkhäuser, Basel pp 179-189

Hetrick B A D, Wilson G W T, Figge D A H (1994) The Influence of mycorrhizal symbiosis and fertilizer amendments on establishment of vegetation in heavy metal mine spoil. Environ Poll

86:171-179

Hildebrandt U, Kaldorf M, Bothe H (1999) The zinc violet and its colonization by arbuscular mycorrhizal fungi. J Plant Physiol 154:709-717

Huang J, Cunningham S (1996) Lead phytoextraction: species variation in lead uptake and translocation. New Phytol 134:75-84

Jacquot E, vanTuinen D, Gianinazzi S, Gianinazzi V (1999) Monitoring species of arbuscular mycorrhizal fungi *in planta* and in soil by nested PCR: application to the study of the impact of sewage sludge. Plant Soil 226:179-188

Jasper D A, Abott L K, Robson A D (1991) The effect of soil disturbance on vesicular arbuscular mycorrhizal fungi in soils from different vegetations types. New Phytol 118:471-476

Joner E J, Johansen A, de la Cruz MAT, Szolar OJM, Loibner A, Portal JM, Leyval C Rhizosphere effects on microbial community structure, and dissipation and toxicity of PAH in spiked soil. Environ Sci Technol, submitted

Joner E J, Briones R, Leyval C (2000) Metal binding capacity of arbuscular mycorrhizal fungi. Plant Soil 226:227-234

Joner E J, Leyval C (2001a) Arbuscular mycorrhizal influence on clover and ryegrass grown together in a soil spiked with polycyclic aromatic hydrocarbons. Mycorrhiza 10:155-159

Joner E J, Leyval C (2001b) Time-course of heavy metal uptake in maize and clover as affected by different mycorrhiza inoculation regimes. Biol Fertil Soil, in press

Joner E J, Leyval C (1997) Uptake of ^{109}Cd by roots and hyphae of a *Glomus mosseae/Trifolium subterraneum* mycorrhiza from soil amended with high and low concentrations of cadmium. New Phytol 135:353-360

Kaldorf M, Kuhn A J, Schröder W H, Hildebrandt U and Bothe H (1999) Selective element deposits in maize colonized by a heavy metal tolerance conferring arbuscular mycorrhizal fungus. J Plant Physiol 154:718-728

Konoplev A V, Vitkorova N V, Virchenko E P, Popov V E, Bulgakov A A, Desmet G M (1993) Influence of agricultural countermeasures on the ratio of different chemical forms of radionuclides in soil and soil solution. Sci Total Environ 137:147-162

Koomen I, McGrath S P, Giller K (1990) Mycorrhizal infection of clover is delayed in soils contaminated with heavy metals from past sewage sludge applications. Soil Biol Biochem 22:871-873

Laheurte F, Leyval C and Berthelin J (1990) Root exudates of maize, pine and beech seedlings influenced by mycorrhizal and bacterial inoculation. Symbiosis 9:111-116

Leyval C, Binet P (1998) Effect of polyaromatic hydrocarbons (PAHs) on arbuscular mycorrhizal colonization of plants. J Environ Qual 27:402-407

Leyval C, Singh B R, Joner E J (1995) Occurrence and infectivity of arbuscular-mycorrhizal fungi in some Norwegian soils influenced by heavy metals and soil properties. Water Air Soil Poll 84:203-216

Leyval C, Turnau K, Haselwandter K (1997) Interactions between heavy metals and mycorrhizal fungi in polluted soils: physiological, ecological and applied aspects. Mycorrhiza 7:139-153

Leyval C, Weissenhorn I (1995) Root colonization of maize by a Cd-sensitive and a Cd-tolerant *Glomus mosseae* and cadmium uptake in sand culture. Plant Soil 175:233-238

Linderman R G (1988) Mycorrhizal interactions with the rhizosphere microflora: The mycorrhizosphere effect. Phytopathology 78:366-371

Linderman R G (1991) Mycorrhizal interactions in the rhizosphere. In: Keister D L, Cregan P B (eds) The rhizosphere and plant growth. Kluwer Academic Publishers, Dordrecht pp 343-348

Loth C. (1996) Abundance of arbuscular mycorrhizal fungi spores at different natives sites in dependence of sludge applications. Bodenkultur 47:89-96

McGonigle T P, Miller M H (1996) Development of fungi below ground in association with plants growing in disturbed and undisturbed soil. Soil Biol Biochem 28:263-269

McGraw A C, Gamble J F, Schenk N C (1979) Vesicular-arbuscular mycorrhizal uptake of cesium-134 in two tropical pasture grass species. Phytopathology 69:1038

Olsson P A, Bååth E, Jakobsen I and Söderström B (1996) Soil bacteria respond to presence of roots but not to mycelium of arbuscular mycorrhizal fungi. Soil Biol Biochem 28:463-470

Pawlowska T E, Chaney R L, Chin M, Charvat I (2000) Effects of metal phytoextraction practices on the indigenous community of arbuscular mycorrhizal fungi at a metal-contaminated landfill. Appl Environ Microbiol 66:2526-2530

Perry J J (1979) Microbial cooxidation involving hydrocarbons. Microbiol Rev 43:59-72

Ravnskov S, Nybroe O and Jakobsen I (1999) Influence of an arbuscular mycorrhizal fungus on

Pseudomonas fluorescens DF57 in rhizosphere and hyphosphere soil. New Phytol. 142:113-122
Reddell P, Milnes A R (1992) Mycorrhizas and other specialised nutrient-acquisition strategies - Their occurrence in woodland plants from Kakadu and their role in rehabilitation of waste rock dumps at a local uranium mine. Aust J Bot 40:223-242
Rogers R D, Williams S E (1986) Vesicular-arbuscular mycorrhiza: Influence on plant uptake of cesium and cobalt. Soil Biol Biochem 18:371-376
Salt D E, Blaylock M, Kumar N P B A, Dushenkov V, Ensley B D, Chet I, Raskin I (1995) Phytoremediation: a novel strategy for the removal of toxic metals from the environment using plants. Biotechnol 13:468-474
Salzer P, Corbiere H and Boller T (1999) Hydrogen peroxide accumulation in *Medicago truncatula* roots colonized by the arbuscular mycorrhiza-forming fungus *Glomus intraradices*. Planta 208:319-325
Schreiner R P, Bethlenfalvay G J (1995) Mycorrhizal interactions in sustainable agriculture. Crit Rev Biotechnol 15:271-285
Schwab A P, Banks M K (1994) Biologically mediated dissipation of polyaromatic hydrocarbons in the root zone. In: Anderson T A, Coats J R (eds) Bioremediation through rhizosphere technology. ACS Symposium Series 563. American Chemical Society, Washington pp 132-141
Segal M G (1993) Agricultural countermeasures following the deposition of radioactivity after a nuclear accident. Sci Total Environ 137:31-48
Shetty K G, Hetrick B A D, Figge D A H, Schwab A P (1994) Effects of mycorrhiza and other soil microbes on revegetation of heavy metal contaminated mine spoil. Environ Poll 86:181-188
Smith S E, Read D J (1997) Mycorrhizal symbiosis. Academic Press, San Diego
Thompson J P (1994) Inoculation with vesicular-arbuscular mycorrhizal fungi from cropped soil overcomes long-fallow disorder of linseed (*Linum usitatissimum* L.) by improving P and Zn uptake. Soil Biol Biochem 26:1133-1143
Tonin C., Vandenkoornhuyse P, Joner E, Straczek J, Leyval C (2001) Diversity of AM fungi in the rhizosphere of *Viola calaminaria* and effect of these fungi on heavy metal uptake by clover. Mycorrhiza 10:161-168
Turnau K, Kottke I, Oberwinkler F (1993) Element localisation in mycorrhizal roots of *Pteridium aquilinium* (L.) Kuhn collected from experimental plots treated with cadmiun dust. New Phytol 123:465-472
van Tichelen K K, Colpaert J V, Van Assche J A (1996) Development of arbuscular mycorrhizas in a heavy metal contaminated soil amended with a metal inmobilizing substance. In: Azcón-Aguilar C, Barea J M (eds) Mycorrhizae in integrated systems. from genes to plant development. European Comission, Luxemburg pp 479-482.
Vandenkoornhuyse P (1998) Effet des métaux sur la diversité des champignons mycorhiziens à arbuscules dans les sols, Ph-D Thesis, Henri Poincaré University, Nancy I
Wan M T, Rahe J E, Watts R G (1998) A new technique for determining the sublethal toxicity of pesticides to the vesicular-arbuscular mycorrhizal fungus *Glomus intraradices*. Environ Toxicol Chem 17:1421-1428
Weissenhorn I, Leyval C, Berthelin J (1995) Bioavailability of heavy metals and abundance of arbuscular mycorrhica in a soil polluted by atmospheric deposition from a smelter. Biol Fertil Soil 19:22-28
Weissenhorn I, Leyval C (1996) Spore germination of arbuscular-mycorrhizal (AM) fungi in soils differing in heavy metal content and other physicochemical properties. Eur J Soil Biol 32:165-172
White P J, Broadley M R (2000) Mechanisms of cesium uptake by plants. New Phytol 147:241-256
WHO (1983) Part I: Chemical, environmental and experimental data. In: Evaluation of the carcinogenic risk of chemicals to humans - polycyclic aromatic hydrocarbons. Volume 32. International Agency for Research on Cancer, Lyon, France
Wilson S C and Jones K C (1993) Bioremediation of soils contaminated with polynuclear aromatic hydrocarbons (PAHs): A review. Environ Pollut 88:229-249

Mycorrhizal Technology in Agriculture
ed. by S. Gianinazzi, H. Schüepp, J.M. Barea and K. Haselwandter

The contribution of arbuscular mycorrhizal fungi to the control of soil-borne plant pathogens

C. Azcón-Aguilar[1], M. C. Jaizme-Vega[2] and C. Calvet[3]

[1] *Departamento de Microbiología del Suelo y Sistemas Simbióticos, Estación Experimental del Zaidín (CSIC), Profesor Albareda 1, E-18008 Granada, Spain*
[2] *Instituto Canario de Investigaciones Agrarias, Departamento de Protección Vegetal, Apdo. 60, E-38200 La Laguna (Tenerife), Canary Islands, Spain*
[3] *Institut de Recerca i Tecnologia Agroalimentàries, Departament de Protecció Vegetal, Ctra. de Cabrils s/n, E-08348 Cabrils (Barcelona), Spain*

Keywords: Bioprotection, biocontrol system, mechanisms in bioprotection, microbial changes in the rhizosphere, activation of plant defence mechanisms

Introduction

Biological control can be defined as the directed and precise management of biological resources, to protect plants against pathogens. The control of plant diseases by applying biocontrol techniques is a multifaced scientific area of great interest because of the need to reduce chemical inputs to agriculture and significantly enhance global sustainability. Some specific groups of microorganisms are able to protect the plant against pathogens by acting through several mechanisms. Among these organisms, arbuscular mycorrhizal fungi (AMF) are promising because of their ubiquity in natural and agricultural terrestrial ecosystems (Jeffries and Barea 2001).

Arbuscular mycorrhizas, the mutualistic symbiosis between AMF and almost all plants present in terrestrial ecosystems, is by far the most universal symbiotic association involving plants. These mycorrhizas can be found in nearly all ecological situations, both in natural ecosystems, particularly in those supporting plant communities with a high species diversity, and also in normal cropping systems, especially if managed with sustainable practices (Gianinazzi and Schüepp 1994).

If AMF are stimulated to grow by the presence of a host plant, the hyphae penetrate the root system, colonize inter or intracellularly the cortex, and finally they form highly branched, short-lived hyphae (arbuscules) within the cortical cells. The complex interactions taking place between AMF and plants leading to the formation and functioning of the mycorrhizal symbiosis should be finely regulated through a series of cellular and molecular programmes in both the plant and the fungus, induced at each level of progression of the symbiosis after the adequate recognition of both partners (Gianinazzi-Pearson et al. 1996).

In a well established arbuscular mycorrhiza we can distinguish two different phases of the fungal partner: one internal, formed by the fungal mycelium biotrophically colonizing the root cortex, in intimate association with the root cells (internal mycelium), and an external phase formed by the fungal mycelium developed in the soil (external mycelium) (Harley and Smith 1983). The external mycelium helps the plant to acquire mineral nutrients from the soil. It acts as a bridge connecting the plant with the soil, exploring the soil microhabitats and interacting with the soil components and with the rhizosphere microbial populations (Barea et al. 1997).

The two phases of the symbiotic fungus show also important differences at metabolic and physiological level (Bago 2000) and their existence supports the claim by Harley and Smith (1983), who stated: "mycorrhizas are active, living components of the soil population having some properties like those of roots and some like those of microorganisms". It is important to keep this in mind because, as we will discuss later on, the role of mycorrhizas in helping the plant to control root pathogens is derived both from their role modifying roots as well as from their activity as microorganisms able to interact with other soil microbes and to alter the microbial populations developing around the mycorrhizal root.

AMF as components of a biocontrol system

In most of the published information on the topic AM associations appear to reduce damage caused by soil-borne plant pathogens. The studies have focussed mostly on fungal pathogens causing root rots and vascular damage (*Phytophthora*, *Aphanomyces*, *Fusarium*, *Verticillium*, *Sclerotium*) (Hooker et al. 1994; Azcón-Aguilar and Barea 1996) and plant parasitic nematodes causing root galls and root lesions (*Meloidogyne, Pratylenchus* and *Radophulus*) (Pinochet et al. 1996; Jaizme-Vega et al. 1997; Jaizme-Vega and Pinochet 1997). Information about diseases caused by root-pathogenic bacteria is scarce, although some studies have also shown mycorrhizal protection of tomato plants against *Erwinia carotovora* and *Pseudomonas syringae* (García-Garrido and Ocampo 1988, 1989).

However, it is difficult to generalize these bioprotective effects because they depend on many different factors and the contribution of any of them and the underlying mechanisms are not fully understood. Among the most important factors contributing to the final balance of the plant/AMF/pathogen interaction, we can include: (i) the AMF isolate, (ii) the pathogen, concerning both virulence and inoculum potential, (iii) the host plant, (iv) the growing substrate and (v) the prevailing environmental conditions. It is crucial to identify the factors contributing to the expression of the biocontrol ability of AMF if appropriate inoculation or management techniques are going to be developed.

Concerning the AMF, different studies have shown that the extent of host damage reduction depends on the isolate involved in the symbiosis, even at similar levels of inoculum and mycorrhizal colonization (Habte et al. 1999). Consequently, one of the prerequisites for the rational exploitation of the protection offered by AMF in a predictable way is the identification of the genotypes capable of protecting a particular host species against a target pathogen. The lack of relationship between the ability of AMF to protect the plant, and their ability to promote growth, underline the need for AMF screenings according to defined selection traits.

Plants have evolved to support interactions with beneficial microbes, like AMF (Smith and Goodman 1999), and host genotypes influence the response to mycorrhizal symbiosis both in terms of plant growth or protection against pathogens. This suggests that plant genes play an important role in supporting these associations, both of pathogenic or mutualistic character. Similarly, the interactions among different AMF and plant pathogens will also vary with the plant genotype (Mark and Cassells 1996).

In general, it is accepted that a mycorrhizal-induced increase in resistance or tolerance requires an extensive development of the symbiosis before the pathogen's attack. In tomato plants infected with *Phytophthora parasitica,* only a well established mycorrhizal colonization could protect plants (Cordier et al. 1996) and bioprotection by *Glomus mosseae* against *Aphanomyces euteiches* was shown to depend on a fully established symbiosis with presence of arbuscules (Slezack et al. 2000). In other studies, however, in which the effectiveness of different AMF in protecting plants against nematode damage was compared, no relationship between the extent of mycorrhizal colonization and plant protection was found (Pinochet et al. 1996; Habte et al. 1999). This finding indicates that an hyphal barrier is not the basis of resistance to nematodes in mycorrhizal plants. Both microorganisms, AMF and parasite, occupy the same sites within the root cortex, as proved by the presence of nematode eggs and adults adjacent to arbuscules and vesicles in mycorrhizal fruit tree rootstock roots (Calvet et al. 1995).

Concerning AMF-nematode interaction, the most important long-term result obtained after mycorrhizal inoculation of plant material is the suppression of nematode reproduction in the roots of the host plants, while the mycorrhizal colonization remains unaffected (Jaizme-Vega et al. 1997; Calvet et al. 2001). The direct consequence on plant development is an increased tolerance to the pathogen.

Mechanisms involved in bioprotection by AMF

The effective protection against root pathogens conferred by AMF is probably a consequence of several, and likely interacting, mechanisms. Their relative contribution is directly related with the AMF/plant genotype combination and with the environmental conditions. The mechanisms under scope can be classified as follows:

Improvement of plant nutrient status/damage compensation

It is evident that the increased capacity for nutrient acquisition by the mycorrhizal association may help the host plant to overcome a pathogen's attack. Additionally, the increase in root biomass can compensate tissue damage and decay of root sections by the pathogen and consequently reduce significantly disease symptoms. There is, however, enough evidence showing that enhanced nutrition, although obviously involved, does not explain all the control achieved (Trotta et al. 1996). Thus, other more specific mechanisms linked to the AM symbiosis are probably operating.

Competition for host photosynthates and colonization sites

The growth of both symbiotic and pathogenic organisms depends on host photosynthates, so competition for carbon compounds maybe a cause of pathogen's depression in mycorrhizal plants. However, no conclusive evidence about the relevance of carbohydrate competition in biocontrol activity of AMF has been reported.

Since both root pathogens and AMF colonize the same ecological niche, the root tissues, the possibility of some type of competition for space must be considered. Structural studies on the interactions between *Phytophthora* and AMF in tomato plants have shown that the pathogen does not penetrate arbuscule-containing cells (Cordier et al. 1996 and 1998), suggesting that AMF can exert some sort of exclusion of the pathogenic fungus from the previously or simultaneously AMF colonized cortical cells.

Changes in the anatomy and architecture of the root system

AMF exert a major impact on root system architecture, often by promoting increases in branching (Berta et al. 1995), although the opposite effect has also been described (Forbes et al. 1996). Changes in root system morphology could modify the infection dynamic pattern of a soil-borne pathogen and the progression of the disease. The significance of these changes has not been investigated in detail. Preliminary conclusions obtained from studies using strawberry as host plant indicated that changes in root architecture induced by mycorrhizal establishment were not directly involved in inducing resistance against *Phytophthora fragariae* (Norman et al. 1996).

In a recent study, roots of plants colonized with AMF and inoculated with zoospores of *P. parasitica* had fewer infection loci for the pathogen than non-mycorrhizal roots, but no changes were detected in the rate of necrosis progression within roots (Vigo et al. 2000). Moreover, no significant effects of mycorrhiza on the root system architecture were reported. The authors argue that the reduction in infection loci can be a consequence of mycorrhiza induced changes in root exudation or of any direct effects of mycorrhizal colonization on epidermal cells which would reduce their susceptibility to infection (Vigo et al. 2000).

Microbial changes in the rhizosphere

Root exudation is modified by the presence of an active mycorrhizal symbiosis (Bansal and Mukerji 1994), and soil pH is also influenced by mycorrhizal establishment and mycelium development (Bago et al. 1996; Bago and Azcón-Aguilar 1997). Consequently, certain microorganisms can be specifically stimulated or inhibited to grow in the rhizosphere of mycorrhizal plants or in the mycorrhizosphere itself (Barea 1997). There is strong evidence that mycorrhiza formation induces qualitative, quantitative and spacial shifts in the microbial populations present in the soil (Linderman and Paulitz 1990; Azcón-Aguilar and Barea 1992).

Direct or indirect competition between AMF and pathogenic microorganisms could take place near the extraradical phase of the AMF. Indirect because some of the microorganisms stimulated by the plant's mycorrhizal condition exhibit antagonistic activities against plant pathogens, and consequently, can act synergistically with AMF in protecting the plant (Filion et al. 1999). But AMF can also interact directly with the pathogenic agents. For example, sporulation of *Phytophthora fragariae* was less stimulated by root exudates from non-mycorrhizal plants than from exudates of mycorrhizal plants (Norman and Hooker 2000). Studies using compartmented monoxenic cultures of the AMF *Glomus intraradices*, which allow to grow the fungus in a compartment in the absence of the root, thus avoiding interferences by root exudates, the direct effect of *G. intraradices* altering the growth of *Fusarium oxysporum* f. sp. *chrysanthemi* was evidenced (St-Arnaud et al. 1995). Additionally, exudates obtained from the extraradical mycelium of *Glomus intraradices* influenced the growth and development of different soil-inhibiting microorganisms (Filion et al. 1999). These results show the ability of the extraradical phase of an AMF to directly influence growth of soil microbes.

Concerning microbial interactions, an interesting, frequently reported observation is that microorganisms showing antifungal activities against pathogens, do not exert any negative effect, or may even stimulate mycorrhizal colonization (Barea et al. 1998; Budi et al. 1999).

Activation of plant defence mechanisms

Bioprotection can result from the preactivation of plant defence responses by AMF. The involvement of such a mechanism is an apparent paradox because at the early stages of root colonization only a weak and transient defence response is induced in the host plant (Azcón-Aguilar and Barea 1996;

Gianinazzi-Pearson 1996) and later on this response is suppressed (Kapulnik et al. 1996). However, it seems that, in spite of the weak reactions elicited, AMF colonization sensitises the host against pathogen attack. In fact, AMF produce elicitors able to induce plant defence responses, as evidenced from the resistance events induced by AMF in non-mycotrophic plants, or in mutant lines unable to establish the symbiosis (myc$^-$ mutants) (Gollote et al. 1993). Therefore, it appears that in normal situations AMF suppress or circumvent the induction of plant defence mechanisms (Blee and Anderson 2000).

In a fully established mycorrhizal symbiosis, defence responses appear to be restricted almost exclusively to the arbuscule-containing cells, in which expression of defence-related genes has been described (Harrison and Dixon 1994). These genes probably play an important role in the intimate regulation of the symbiosis.

Accumulation of compounds involved in normal plant defence responses as a consequence of AM formation has been the subject of many studies. That information will not be analyzed here and the reader is referred to Pozo et al. in this book.

The prior activation of plant defence responses by AMF can result in a certain level of bioprotection. In this context, certain chitinolytic (Slezack et al. 2000) and β-1,3-glucanolytic (Pozo et al. 1999) isoforms, specifically induced by AMF in the mycorrhizal plant, have been involved in bioprotection. Additionally, the results of Benhamou et al. (1994) support this hypothesis, although they were obtained in artificial (monoxenic) conditions using transformed carrot roots.

Systemic effects of AMF colonization

By using a split root experimental system it has been shown recently that a decrease in the development of *Phytophthora* in mycorrhizal and non-mycorrhizal roots of mycorrhizal tomato plants was associated with accumulation of phenolics and plant cell defence responses (Cordier et al. 1998). This is the first evidence of the induction of systemic resistance by mycorrhiza formation. Arbuscule-containing cells were immune to the pathogen, and the systemically induced resistance in non-mycorrhizal root parts was characterized by elicitation of wall thickenings in reaction to the intercellular hyphae of the pathogen and by the formation of callose-rich material around hyphae penetrating root cells. None of these reactions were observed in the non-mycorrhizal pathogen-infected root systems. Reductions in root damage and in the development of the pathogen in the non-mycorrhizal root tissues of mycorrhizal systems in comparison to roots of non-mycorrhizal plants were produced as a consequence of this systemic response.

Using also a compartmented system, a similar reduction in pathogen caused symptoms and development in the non-mycorrhizal parts of mycorrhizal plants corroborated the induction of systemic responses by AMF (Pozo et al.

2002). The non-mycorrhizal parts of mycorrhizal plants showed a higher lytic activity against *Phytophthora* cell walls than the corresponding roots of non-mycorrhizal plants. The nature of the signals involved in these induced systemic responses remain unknown.

In conclusion, plants possess a variety of latent defence mechanisms conferring quantitative protection against a broad range of pathogenic microorganisms. These mechanisms can be systemically-activated following exposure to certain type of stresses, infection by pathogens or other microorganisms, including AMF (Sturz et al. 2000). Some plant defence strategies can involve endophyte-mediated *de novo* fungitoxic metabolites at sites of attempted pathogenic penetration or synthesis of structural compounds which inhibit pathogen progression. This could be the case with AMF. In consequence, the induction of resistance mechanisms by arbuscular mycorrhizas, as shown in the case of tomato plants challenged with *Phytophthora parasitica,* can result from both localized defence responses in colonized tissues and systemic responses in the non-mycorrhizal parts of mycorrhizal roots.

On the contrary, the increase in severity of diseases caused by foliar pathogens in mycorrhizal plants has been related to the suppression of the plant defence response by AMF short after the early events of root colonization (Shaul et al. 1999).

Potentialities for applying AMF in plant protection

Once demonstrated that AMF are able to contribute to the control of a specific soil-borne plant pathogen under laboratory and greenhouse conditions, it is necessary to evaluate the feasibility of including AM technology as a component of the biocontrol strategies in plant production systems. Published information on the subject supports the hypothesis that the appropriate management of AMF may have a valuable potential for disease control on an agricultural scale. However, a number of requirements must be fulfilled before exploiting AM technology in plant protection in a successful and predictable way.

A critical prerequisite for exploiting the potential of AMF in biocontrol is the identification of the most effective AMF genotypes to be used in each particular situation i. e. a given pathogen-crop-environment combination. Such a screening of AMF for their ability to protect the plant against pathogens should be done in realistic plant production and cropping conditions. The information on this side is scarce, therefore, this particular topic deserves further research before the use of particular AMF as components of a biocontrol strategy for a target disease in a given agrotechnological situation can be recommended.

On another hand, it is necessary to take into account that the successful exploitation in a predictable way of AMF in plant protection depends on the increased understanding of the diverse mechanisms governing the interactions taking place among the components of the system, i. e., AMF/pathogen/plant, and of the environmental conditions favourable for their expression.

Undoubtedly, the result of these interactions will modulate the scale and timing of the biocontrol potential and in consequence its expression and effectiveness. Cooperation at this level with plant pathologists is necessary.

An early mycorrhizal inoculation, previous to pathogen attack, has been shown to be a successful practice to increase disease tolerance/resistance in economically important crop species mainly for those involved in horticultural and fruit production systems (Jaizme-Vega et al. 1997; Pinochet et al. 1998). Such an activity has been demostrated to occur even in naturally infested replant soils (Utkhede and Smith 2000; Calvet et al. 2001).

General world awareness of the environmental impacts of chemical disease-control products is indicating the need for alternative control strategies, particularly with regard to the avoidance of plant biocides currently used in intensive agriculture. Some of the proposed alternatives are based on combining mycorrhizal inoculation and non-chemical control practices. These include either biological treatments, like the application of formulated microbial antagonists, based on rhizobacteria or saprophytic fungi (Waschkies et al. 1994; Datnoff et al. 1995), or other practices like heat disinfection by soil solarization in appropriate geographical areas (Afek et al. 1991). In this context, it has been demonstrated that the high temperatures achieved under plastic covers (55°C at soil surface) do not affect AMF inocula infectivity (Camprubí et al. 2001), nor the mycorrhizal potential in an orchard solarized to control soil-borne plant pathogens including either fungi, like for example *Sclerotinia*, *Rhizoctonia*, *Sclerotium*, *Phytophthora* and *Fusarium*, or root-knot nematodes (Afek et al. 1991).

In conclusion, it is accepted that AMF can play a key role in the protection of host plants against pathogens and are probably an important component of soil suppressiveness phenomena. Their rational exploitation in plant protection requires progress both in the understanding of their mode of action, and also in the development of appropriate technologies for the application of efficient inocula. Combination of basic and applied research strategies will be the basis for the design of management protocols for agricultural developments based on sustainable practices, finally addressed to achieve and exploit an increased capacity of AMF for pathogen suppression.

Acknowledgements
Some of the experimental work referred to in this chapter was sponsored by the projects: 1FD97-0763-C03-02, CAO99-0010-C3-03, AGF97-0549-C03 and ERB IC18 CT97-0208.

References

Afek U, Menge J A, Johnson E L V (1991) Interaction among mycorrhizae, soil solarization, metalaxyl and plants in the field. Plant Disease 75:665-671

Azcón-Aguilar C, Barea J M (1992) Interactions between mycorrhizal fungi and other rhizosphere microorganisms. In: Allen M J (ed) Mycorrhizal functioning, an integrative plant-fungal process. Routledge, Chapman and Hall Inc., New York pp 163-198

Azcón-Aguilar C, Barea J M (1996) Arbuscular mycorrhizas and biological control of soil-borne plant pathogens. An overview of the mechanisms involved. Mycorrhiza 6:457-464

Bago B (2000) Putative sites for nutrient uptake in arbuscular mycorrhizal fungi. Plant Soil 226:263-274

Bago B, Azcón-Aguilar C (1997) Changes in the rhizospheric pH induced by arbuscular mycorrhiza formation in onion (*Allium cepa* L.). Z Pflanzenernähr Bodenk 160:333-339

Bago B, Vierheilig H, Piché Y, Azcón-Aguilar C (1996) Nitrate depletion and pH changes induced by the extraradical mycelium of the arbuscular-mycorrhizal fungus *Glomus intraradices* grown in monoxenic culture. New Phytol 133:273-280

Bansal M, Mukerji K G (1994) Positive correlation between VAM-induced changes in root exudation and mycorrhizosphere mycoflora. Mycorrhiza 5:39-44

Barea J M (1997) Mycorrhiza/bacteria interactions on plant growth promotion. In: Ogoshi A, Kobayashi L, Homma Y, Kodama F, Kondon N, Akino S (eds) Plant growth-promoting rhizobacteria, present status and future prospects. OCDE, Paris pp 150-158

Barea J M, Andrade G, Bianciotto V, Dowling D, Lohrke S, Bonfante P, O'Gara F, Azcón-Aguilar C (1998) Impact on arbuscular mycorrhiza formation of *Pseudomonas* strains used as inoculants for the biocontrol of soil-borne plant fungal pathogens. Appl Environm Microbiol 64:2304-2307

Barea J M, Azcón-Aguilar C, Azcón R (1997) Interactions between mycorrhizal fungi and rhizosphere microorganisms within the context of sustainable soil-plant systems. In: Gange A C, Brown V K (eds) Multitrophic interactions in terrestrial systems. Blackwell Science, Oxford pp 65-77

Benhamou N, Fortin J A, Hamel C, St-Arnaud M, Shatilla A (1994) Resistance responses of mycorrhizal Ri T-DNA-transformed carrot roots to infection by *Fusarium oxysporum* f sp *chrysanthemi*. Phytopathol 84:958-968

Berta G, Trotta A, Fusconi A, Hooker J E, Munro M, Atkinson D, Giovannetti M, Morini S, Fortuna P, Tisserant B, Gianinazzi-Pearson V, Gianinazzi S (1995) Arbuscular mycorrhizal induced changes to plant growth and root system morphology in *Prunus cerasifera*. Tree Physiology 15: 281-293

Blee K A, Anderson A J (2000) Defense responses in plants to arbuscular mycorrhizal fungi. In: Podila G K, Douds D D (eds) Current advances in mycorrhizae research. APS Press, Minnesota pp 27-44

Budi S W, van Tuinen, D, Martinotti G, Gianinazzi S (1999) Isolation from the *Sorghum bicolor* mycorrhizosphere of a bacterium compatible with arbuscular mycorrhiza development and antagonistic towards soil-borne fungal pathogens. Appl Environ Microbiol 65:5148-5150

Calvet C, Pinochet J, Camprubí A, Fernández C (1995) Increased tolerance to the root lesion nematode *Pratylenchus vulnus* in mycorrhizal micropropagated BA-29 quince rootstock. Mycorrhiza 5:253-258

Calvet C, Pinochet J, Hernández-Dorrego A, Estaún V, Camprubí A (2001) Field microplot performance of the peach-almond hybrid GF-677 after inoculation with arbuscular mycorrhizal fungi in a replant soil infested with root-knot nematodes. Mycorrhiza 10:285-300

Camprubí A, Calvet C, Estaún V, Ferrer I, Aguado A (2001) Effect of high temperatures on the viability of the arbuscular mycorrhizal fungus *Glomus intraradices*. Journal of Plant Pathology (in press)

Cordier C, Gianinazzi S, Gianinazzi-Pearson V (1996) Colonisation patterns of root tissues by *Phytophthora nicotianae* var *parasitica* related to reduced disease in mycorrhizal tomato. Plant Soil 185:223-232

Cordier C, Pozo M J, Barea J M, Gianinazzi S, Gianinazzi-Pearson V (1998) Cell defense responses associated with localized and systemic resistance to *Phytophthora parasitica* induced in tomato by an arbuscular-mycorrhizal fungus. Mol. Plant-Microbe Interact 11:1017-1028

Datnoff L E, Nemec S, Pernezny K (1995) Biological control of *Fusarium* crown and root rot of tomato in Florida using *Trichoderma harzianum* and *Glomus intraradices*. Biological Control 5: 427-431

Filion M, St-Arnaud M, Fortin J A (1999) Direct interaction between the arbuscular mycorrhizal fungus *Glomus intraradices* and different rhizosphere microorganisms. New Phytol 141:525-533

Forbes P J, Ellison C H, Hooker J E (1996) The impact of arbuscular mycorrhizal fungi and temperature on root system development. Agronomie 16:617-620

García-Garrido J M, Ocampo J A (1988) Interaction between *Glomus mosseae* and *Erwinia carotovora* and its effects on the growth of tomato plants. New Phytol 110:551-555

García-Garrido J M, Ocampo J A (1989) Effect of VA mycorrhizal infection of tomato on damage cuased by *Pseudomonas syringae*. Soil Biol Biochem 21:165-167

Gianinazzi S, Schüepp H (eds) (1994) Impact of Arbuscular Mycorrhizas on Sustainable Agriculture and Natural Ecosystems. ALS, Birkhäuser Verlag, Basel

Gianinazzi-Pearson V (1996) Plant cell responses to arbuscular mycorrhizal fungi: Getting to the roots of the symbiosis. Plant Cell 8:1871-1883

Gianinazzi-Pearson V, Dumas-Gaudot E, Gollotte A, Tahiri-Alaoui A, Gianinazzi S (1996) Cellular and molecular defence-related root responses to invasion by arbuscular mycorrhizal fungi. New Phytol 133:45-57

Gollotte A. Gianinazzi-Pearson V., Giovannetti M., Sbrana C., Avio L., Gianinazzi S. (1993) Cellular localization and cytochemical probing of resistance reactions to arbuscular mycorrhizal fungi in a 'locus A' myc- mutant of *Pisum sativum* L. Planta 191:112-122

Habte M, Zhang Y C, Schmitt D P (1999) Effectiveness of *Glomus* species in protecting white clover against nematode damage. Can J Botany 77:135-139

Harley J L, Smith S E (eds) (1983) Mycorrhizal symbiosis. Academic Press, New York

Harrison M J, Dixon R A (1994) Spatial patterns of expression of flavonoid/isoflavonoid pathway genes during interactions between roots of *Medicago truncatula* and the mycorrhizal fungus *Glomus versiforme*. Plant J. 6:9-20

Hooker J E, Jaizme-Vega M, Atkinson D (1994) Biocontrol of plant pathogens using arbuscular mycorrhizal fungi. In Gianinazzi S, Schüepp H (eds) Impact of Arbuscular Mycorrhizas on Sustainable Agriculture and Natural Ecosystems. ALS, Birkhäuser Verlag, Basel, pp 191-200

Jaizme-Vega M C, Pinochet J (1997) Growth response of banana to three mycorrhizal fungi in *Pratylenchus goodeyi* infested soil. Nematropica 27:69-76

Jaizme-Vega M C, Tenoury P, Pinochet J, Jaumot M (1997) Interactions between the root-knot nematode *Meloidogyne incognita* and *Glomus mosseae* in banana. Plant Soil 196:27-35

Jeffries P, Barea J M (2001) Arbuscular Mycorrhiza – a key component of sustainable plant-soil ecosystems. In: Hock B. (ed) The mycota. volm. IX fungal associations. Springer-Verlag, Berlin, Heidelberg pp 95-113

Kapulnik Y, Volpin H, Itzhaki H, Ganon D, Galili S, David R, Shaul O, Elad Y, Chet I, Okon Y (1996) Suppression of defence responses in mycorrhizal alfalfa and tobacco roots. New Phytologist 133: 59-64

Linderman R G, Paulitz T C (1990) Mycorrhizal-rhizobacterial interactions In: Hornby D, Cook R J, Henis Y, Ko W H, Rovira A D, Schippers B, Scott P R (eds) Biological control of soil-borne plant pathogens. Wallingford, UK, CAB International pp 261-283

Mark G L, Cassells A C (1996) Genotype-dependence in the interaction between *Glomus fistulosum, Phytophthora fragariae* and the wild strawberry (*Fragaria vesca*). Plant Soil 185:233-239

Norman J R, Atkinson D, Hooker J E (1996) Arbuscular mycorrhizal fungal-induced alteration to root architecture in strawberry and induced resistance to the root pathogen *Phytophthora fragariae.* Plant Soil 185:191-198

Norman J R, Hooker J E (2000) Sporulation of *Phytophthora fragariae* shows greater stimulation by exudates of non-mycorrhizal than by mycorrhizal strawberry roots. Mycol Res 104:1069-1073

Pinochet J, Calvet C, Camprubi A, Fernandez C (1996) Interactions between migratory endoparasitic nematodes and arbuscular mycorrhizal fungi in perennial crops: A review. Plant Soil 185:183-190

Pinochet J, Camprubí A, Calvet C, Fernández C, Rodríguez-Kábana R (1998) Inducing tolerance to the root-lesion nematode *Pratylenchus vulnus* by early mycorrhizal inoculation of micropropagated Myrobalan 29 C plum rootstock. J Am Soc Hort Sci 123: 342-347

Pozo M J, Azcon-Aguilar C, Dumas-Gaudot E, Barea J M (1999) _-1,3-glucanase activities in tomato roots inoculated with arbuscular mycorrhizal fungi and/or *Phytophthora parasitica* and their possible involvement in bioprotection. Plant Sci 141:149-157

Pozo M J, Cordier C, Dumas E, Gianinazzi S, Barea J M, Azcón-Aguilar C (2001) Localized *vs* systemic effect of arbuscular mycorrhizal fungi on defence responses to *Phytophthora* infection in tomato plants. J Exp Bot in press

Shaul O, Galili S, Volpin H, Ginzberg I, Elad Y, Chet I, Kapulnik Y (1999) Mycorrhiza-induced changes in disease severity and PR protein expression in tobacco leaves. Mol Plant Microb Interac 12:1000-1007

Slezack S, Dumas-Gaudot E, Paynot M, Gianinazzi S (2000) Is a fully established arbuscular mycorrhizal symbiosis required for bioprotection of *Pisum sativum* roots against *Aphanomyces euteiches*? Mol Plant Microb Interac 13:238-241

Smith K P, Goodman R M (1999) Host variation for interactions with beneficial plant-associated microbes. Ann Rev Phytopathol 37:473-491

St-Arnaud M, Hamel C, Vimard B, Caron M, Fortin J A (1995) Altered growth of *Fusarium oxysporum* f. sp. *chrysanthemi* in an *in vitro* dual culture system with the vesicular arbuscular mycorrhizal fungus *Glomus intraradices* growing on *Daucus carota* transformed roots. Mycorrhiza 5:431-438

Sturz A V, Christie B R, Nowak J (2000) Bacterial endophytes: potential role in developing sustainable systems of crop production. Crit Rev Plant Sci 19: 1-3

Trotta A, Varese G C, Gnavi E, Fusconi A, Sampo S, Berta G (1996) Interactions between the soil-borne root pathogen *Phytophthora nicotianae* var *parasitica* and the arbuscular mycorrhizal fungus *Glomus mosseae* in tomato plants. Plant 185:199-209

Utkhede R S, Smith E M (2000) Impact of chemical, biological and cultural treatments on the growth and yield of apple in replant-disease soil. Australasian Plant Pathology 29:129-136

Vigo C, Norman J R, Hooker J E (2000) Biocontrol of the pathogen *Phytophthora parasitica* by arbuscular mycorrhizal fungi is a consequence of effects on infection loci. Plant Pathol 49:509-514

Waschkies C, Schropp A, Marschner H (1994) Relations between grapevine replant disease and root colonization of grapevine (*Vitis* sp.) by fluorescent pseudomonads and endomycorrhizal fungi. Plant Soil 162:219-227

Mycorrhizal Technology in Agriculture
ed. by S. Gianinazzi, H. Schüepp, J.M. Barea and K. Haselwandter

Mycorrhization as a stress adaptation procedure

M. Tsimilli-Michael[1,2] and R.J. Strasser[1]

[1] *Bioenergetics Laboratory, University of Geneva, CH-1254 Jussy-Geneva, Switzerland*
[2] *Cyprus Ministry of Education and Culture, CY-1434 Nicosia, Cyprus*

Keywords: Mycorrhization, optimisation strategy, photosystem II behaviour, polyphasic fluorescence transient, stress adaptation

Introduction

Mycorrhizal symbiosis is attracting a continuously increasing interest from the scientific world, due to the beneficial role it plays in plants' growth and development in natural, semi-natural and agricultural plant communities. Mycorrhizal associations are very essential in helping plants to survive in nutrient-deficient, degraded habitats or during stress periods and can be very advantageous in sustainable agriculture. Hence, it has become a major subject of both basic and applied research (see Gianinazzi and Schüepp 1994; Varma and Hock 1998).

The aim of this paper is to integrate the mycorrhization of a plant in our general conceptual frame for stress and stress adaptation of plants. Any living system is an energy convertor with a complex structure in which flow of energy and matter never ceases and irreversible processes occur. It can thus be considered as an open thermodynamic system, i.e. a dissipative structure and, hence, we assume that its behaviour can be described on the basis of inferences from non-equilibrium thermodynamics.

One of the most important insights of non-equilibrium thermodynamics is that for dissipative structures the *steady state* is the state at which the entropy production assumes the minimal value consistent with the restraints imposed (Prigogine 1947). This is equivalent with minimal free energy dissipation. Thus, steady states are *optimal states* and the system tends to achieve and maintain them. In this sense they are states of *stability* and they are, therefore, considered to play a role in the thermodynamics of irreversible processes similar to that played by the equilibrium states in classical thermodynamics.

Based on this theoretical prediction we approach the adaptation of plants to the varying environmental conditions, and more generally the adaptive strategies and evolutionary behaviour of organisms, as expressions of optimisation strategies dictated by the thermodynamic demand for minimal entropy production (Strasser 1988). Our analysis utilises the *JKB Trilogy Concept* (Strasser 1985, 1988) which, without losing from its generality, was formulated in a way that creates a logic bridge between the thermodynamic directives and the bioenergetics of the photosynthetic machinery (for an extended article see Tsimilli-Michael et al. 1996).

Mycorrhization has multiple effects on the physiology of the plant at different levels. We focus our interest on the responses of the photosynthetic apparatus. The experimental access to the bioenergetics of the photosynthetic apparatus is achieved by the analysis of the fast fluorescence kinetics O-J-I-P exhibited by all photosynthetic organisms upon illumination (Strasser et al. 1995). The analysis, based on a simple model and the *Theory of Energy Fluxes in Biomembranes* (Strasser 1978, 1981), provides a quantification of the behaviour and performance of the photosynthetic apparatus (Strasser et al. 2000).

The conceptual approach of stress and stress adaptation

The stress concept: description and definitions

Though stress is exactly defined in physics, it appears with differing meanings in biology, which mainly converge in attributing stress to any environmental factor 'unfavourable' for the living organism (Levitt 1980). Accordingly, the ability of the organism to survive the unfavourable factor has been called *stress resistance*.

Our approach is different in principle (Strasser 1988; Tsimilli-Michael et al. 1996). We consider that *stress* has a relative meaning, with *non-stress* as the reference condition, i.e. we consider *stress* as a deviation from the *non-stress* situation. The latter is dynamically defined as the situation at which the plant is *adapted* to its environment or, equivalently, the plant is at its thermodynamically *optimal state*, i.e. the state of minimal entropy production, under a given constellation of environmental conditions. The system is then characterised as being in *harmony* with its environment (Strasser 1988). This characterisation expresses the dialectics of optimality, pointing out that it refers to the relation – interaction of the system with its environment. Any change in the environmental input which disturbs the achieved harmony and leads to *disharmony* with the environment is a stressful event and is hence defined as *stressor*. Such disturbances lead the system to *suboptimality*.

Because of the thermodynamic demand for optimality, suboptimality creates a *driving force*, under which the system undergoes *state changes* towards a new constellation of conformational parameters, denoted as the *attraction point* for this *state change walk* (Strasser 1988). The force is hence called *state change force.* If the system succeeds to reach the attraction point the state change force vanishes and a new optimality is established, i.e. a state adapted to the new environmental conditions. *Stress adaptation* is hence defined as the sequence of processes that lead to a new optimality. Different adaptive strategies are employed to regulate different functional and structural parameters of the system. Any change of these parameters caused by the conformational changes that realise adaptation is denoted as *strain.*

However, the environmental conditions never cease to manifest alterations and, thus, the system is perpetually undergoing *stress – stress adaptation* processes, searching and approaching harmony with its environment.

In this concept, based on the dynamic relation between organism and environment and keeping from physics the concept of *action-reaction*, no environmental factor is considered *a priori* as unfavourable and the plant has not to 'resist', but it simply reacts. As far as the system manages to adapt, which means that the attraction point is within realistic limits (Strasser 1988), stress is not only harmless but, even more, *constructive* because it results in improved resistance and adaptive evolution; if the adaptability of the system is overtaxed, then stress is *destructive*, leading to permanent damages or even to death, as stated by Larcher (1987).

The stress concept has also been quantitatively elaborated on the basis of the J-K-B trilogy (Strasser 1988), as shown below. This offers the possibility for analytical description and establishment of links with the bioenergetics of the photosynthetic apparatus, both on the phenomenological and the biophysical level.

The J-K-B trilogy

The efficiency (η) of a biological system at any given time, i.e. the ratio of its biological activity (υ) or output to the energetic input, has been described by Strasser (1985) as defined by three terms, $\eta = f(J,K,B)$.

The terms of this trilogy are:

J all energetic inputs,

K the constellation of structural – conformational parameters that determine the kinetic pathways for energy conservation and dissipation; K is thus an extensive parameter corresponding to the *hardware* of the system.

B the relative level of the energy flow through the system; hence, B is an intensive parameter and can be regarded as an expression of the *behaviour* of the system.

Since B is determined by J and K through a function defined by the model that describes the system, the efficiency can be expressed as a function of only two terms, i.e. $\eta = f(J,K)$ or $\eta = f(B,K)$. Hence, if the conformation (the K-term) remains constant, the efficiency of the system can be expressed either as a function of the external factor J or, equivalently, of the internal response B, i.e. $\eta = f_K(J)$ or $\eta = f_K(B)$. These are the *state functions* of the system for the given K, each K defining another set of them. Thus, the conversion of the system to a different *state* can be equivalently ascribed as the shift to a different set of *state functions*, or as a *K-change*.

Stress and stress-adaptation in terms of the J-K-B trilogy

The thermodynamic prediction for the steady state can be rephrased in terms of the J-K-B trilogy as follows: For a living system, the *optimal* state corresponds to a unique pair of J and B for which, compared to all pairs of J and B expressed by the same state function – i.e. having the same K-, dissipation is minimal. Hence, at the optimal state the energy conservation, expressed by the biological efficiency (η) is maximal.

This can be visualised by the multiparametric presentation of $\eta = f(J,K,B)$ demonstrated for a theoretical example in Figure 1. Circles and squares indicate the constellations corresponding to the chosen energetic inputs J_1 and J_2 respectively. Two different conformational states of the biological system are

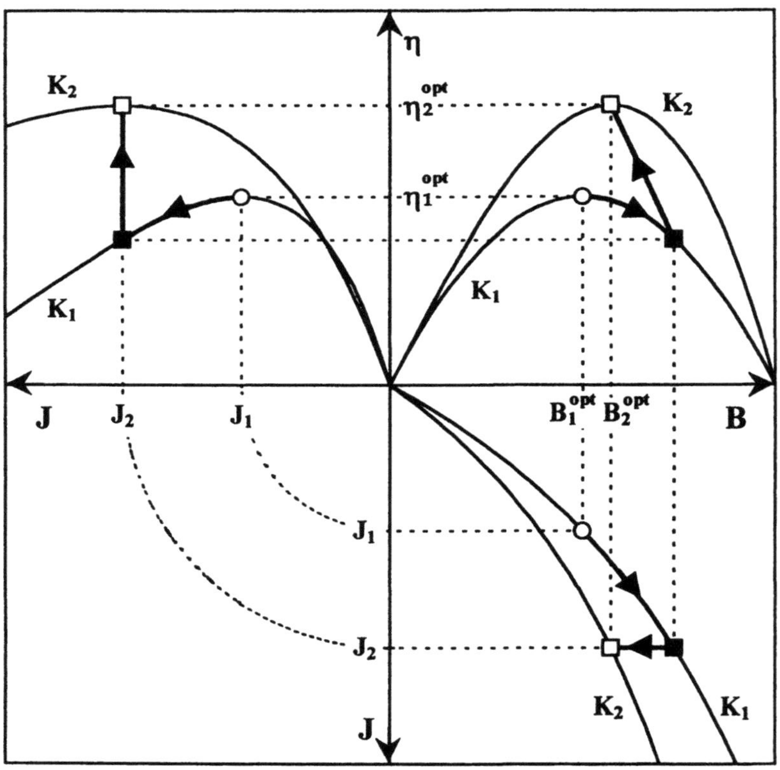

Figure 1. The multiparametric presentation of η-B-J for two different states K_1 and K_2. J is the energetic input, B the dynamic behaviour and η a parameter expressing the biological efficiency of the system. The values K_1 and K_2 of the conformational term K denote the conformational states adapted to the energetic inputs J_1 and J_2 respectively. Circles and squares indicate the constellations corresponding to J_1 and J_2 respectively. Open symbols correspond to optimality and closed symbols to suboptimality. The heavy lines with the arrows present the state change walk provoked by the environmental change $J_1 \rightarrow J_2$.

considered, one adapted to J_1 and one to J_2. Hence, their conformation term K assumes different values, denoted accordingly as K_1 and K_2. For each conformational state, the functions $\eta = f_K(J)$, $B = f_K(J)$ and $\eta = f_K(B)$ are plotted. The equivalency between optimal and adapted state is demonstrated by the position of the maxima of $\eta = f_K(J)$, being at $J = J_1$ for K_1 and $J = J_2$ for K_2. It is worth pointing out that, at the optimal states, B is not maximised but is simply optimised. The optimal constellations η_2^{opt}-J_2-K_2-B_2^{opt} and η_2^{opt}-J_2-K_2-B_2^{opt} are marked in Figure 1 with open symbols.

Let us now focus on the plot η-J and start watching the system when it is at the optimal state adapted to J_1 (open circle). If a change from J_1 to J_2 is provoked, the biological efficiency η, being a function of J, will consequently change, as shown by the heavy line with the arrow on the curve $\eta = f_K(J)$ corresponding to K_1. This means that the system deviates from its optimum and is led to *suboptimality* (closed square), a state of thermodynamic instability that it has to overcome. The system can neither change the external factor J, nor the internal response B which defines the activity υ and, concomitantly, the efficiency η; it has, however the capability to change its conformation, i.e. the K term. This is an advantage of living systems or, generally, of self organising systems over ordinary conservative systems which are obliged to perform with a fixed state function – a constant K – and, hence, show an activity strictly defined by the value of the input J. The conformational changes are, however, not abrupt and the search for the attraction point η_2^{opt}-J_2-K_2-B_2^{opt}, which expresses the adapted to J_2 state, is a "walk" via changes of K, consequent changes of B and resulting changes of the efficiency. The state change walk is presented schematically in the plot η-J by the vertical heavy line with the arrow leading from the K_1 to the K_2 curve. The two heavy lines with arrows, presenting the whole of the stress – stress adaptation procedure, are also projected in the plots η-B and B-J of Figure 1.

The generalised stress concept: A schematic presentation

In Figure 1 optimality was assumed as equivalent to maximality of the efficiency η. However, this is not always the case. In general, any performance parameters can be maximised at the optima while the others are simply optimised, like B in Figure 1. Hence, in a generalised approach we can follow, instead of η, any performance parameter Y maximised at the optima. Moreover, instead of J, we can refer to any environmental factor E. The Y-values are denoted as $^{n}Y_m$ to indicate that the system is at a state adapted to an environmental factor E_n and exposed to an environmental factor E_m. The function $^{n}Y = f(E)$ expresses the dependence of Y on E within the conformational state adapted to E_n or, equivalently, is the state function for $K = K_n$. Hence, optimal states are those for which $n = m$ (see Tsimilli-Michael et al. 1996). Figure 2 summarises in a schematic way the stress concept and all the relevant definitions, using the generalised notations.

Fluorescence kinetics: an insight into structure and function

Chlorophyll (Chl) *a* fluorescence, though corresponding to a very small fraction of the dissipated energy from the photosynthetic apparatus, has been proven to be a very useful, non-invasive tool for the investigation of its structure and function. At ambient temperature Chl *a* fluorescence is basically emitted by photosystem II and the transient it exhibits upon illumination, usually called the Kautsky transient, consists of a rise completed in less than 1 s and a subsequent slower decline towards a steady state. A lot of information derived during the last sixty years from the fluorescence transient (see e.g. Krause and Weis 1991; Govindjee 1995).

Transients recorded with high time-resolution fluorimeters, e.g. with the PEA-instrument (data acquisition every 10 μs for the first 2 ms and 1 ms thereafter), have provided additional and/or more accurate information. It was shown that the fluorescence rise kinetics is polyphasic exhibiting clearly, when plotted on a logarithmic time scale, the steps J (at 2 ms) and I (30 ms) between the initial O (F_0) and maximum P level (F_P); moreover, detection of F_0, the fluorescence value at the onset of illumination when all reaction centres are open, and of the initial slope, which offers a link to the maximum rate of photochemical reaction, has become more precise (Strasser and Govindjee 1991; Strasser et al. 1995).

An analysis of the O-J-I-P transient has been developed, named as the 'JIP-test' after the steps of the transient (Strasser and Strasser 1995), which leads to the calculation of conformational and functional parameters quantifying the PSII behaviour (see Strasser et al. 2000). The derivation of the equations relating the parameters with the experimental signals is based on the *Theory of Energy Fluxes in Biomembranes* (Strasser 1978, 1981). The shape changes of the O-J-I-P transient, due to the establishment of different physiological states provoked by different stressors (see e.g. Srivastava and Strasser 1996, 1997; Tsimilli-Michael et al. 1996, 1999, 2000; Krüger et al. 1997; Clark et al. 1998, 2000), are thus translated to quantitative changes of the parameters via the JIP-test.

Hence, states adapted to any value E_n of an environmental factor E can be compared by comparing the respective values of any parameter. Their difference or ratio can also serve as additional performance parameter, like e.g. the relative stress-tolerance (Romano et al. 1996). Depending on the case, any experimentally accessible parameter Y can be used to probe the state changes triggered by stress. It is worth clarifying that what we experimentally follow is the optimal values ${}^{n}Y_n$ at the different environmental values E_n of the environmental factor.

Mycorrhization in the frame of the stress concept

The analysis of the fluorescence transient O-J-I-P by the JIP-test has also been used to screen the effects of mycorrhization on the behaviour and performance of the photosynthetic apparatus (Romano et al. 1996; Köves-Péchy et al. 1998;

Calantzis 1999, 2000a, b; Tsimilli-Michael et al. 2000). All these studies demonstrated the beneficial role of mycorrhizal symbiosis on the photosystem II activity, revealed by increased values of various activity and/or performance parameters.

In terms of the stress concept, it can be stated that mycorrhizal inoculation acts as a stressor, which triggers conformational changes leading to a state adapted to the new conditions, i.e. to the internal environment modified by the establishment of symbiosis.

In Figure 2 the non-mycorrhiza condition refers to E_1 and the mycorrhizal symbiosis to E_2. At the corresponding adapted states the parameter Y accomplishes the values ${}^1Y_1$ and ${}^2Y_2$ respectively, where ${}^2Y_2 > {}^1Y_1$.

However, there have been reported cases where mycorrhizal inoculation provoked first a decrease of performance parameters, e.g. of the relative stress-

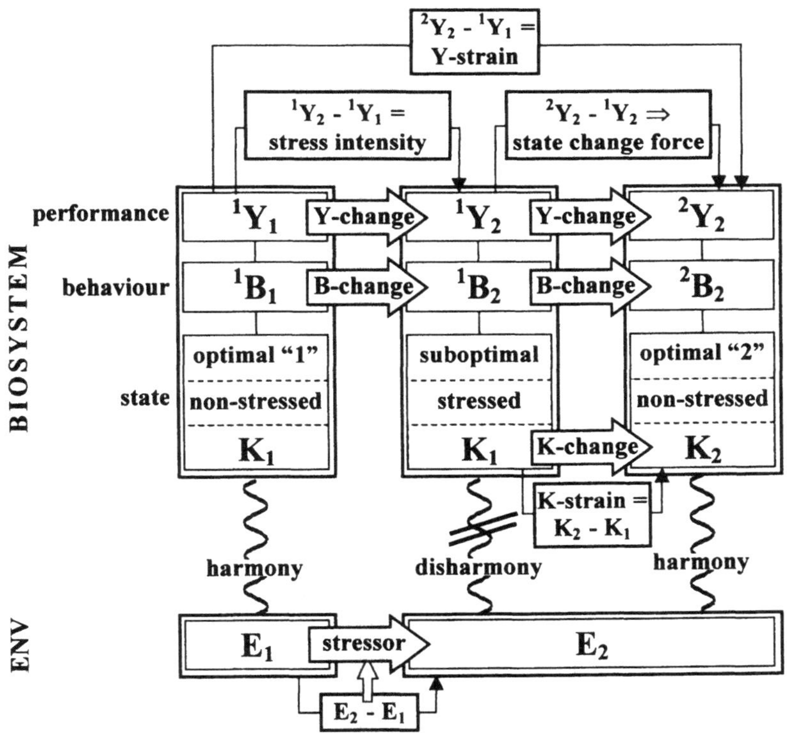

Figure 2. The schematic presentation of the stress concept refers to a biosystem adapted and exposed to environment (ENV) E_1, hence being at the optimal "1" state (= non-stressed = adapted to E_1 = in harmony with E_1) described by $K_1 - {}^1B_1 - {}^1Y_1$. The system is subjected to a stressor = $\Delta E = E_2 - E_1$, and is led to a suboptimal state (= stressed = in disharmony with E_2), described by $K_1 - {}^1B_2 - {}^1Y_2$. The difference ${}^1Y_2 - {}^1Y_1$ is a measure of the stress intensity. The attraction point is the state adapted to E_2, i.e. the optimal "2" state described by $K_2 - {}^2B_2 - {}^2Y_2$. Under the state change force created by the difference ${}^2Y_2 - {}^1Y_2$, the system undergoes K-changes and consequent Y- and B-changes, until it reaches the attraction point $K_2 - {}^2B_2 - {}^2Y_2$. The strain is expressed either as Y-strain = ${}^2Y_2 - {}^1Y_1$, or as K-strain = $K_2 - K_1$.

tolerance (Romano et al. 1996) or the performance index PI (Calantzis et al. 2000a, b; for the PI definition and meaning see Strasser et al. 2000; Tsimilli-Michael et al. 2000); the initial decrease was followed by an increase that not only counter-balanced the negative effect, but finally led to a wide net increase of the parameters. It thus appears that the sequence of events provoked by the mycorrhizal inoculation comprised two basic phases, which can be considered as the infection phase and the symbiosis' establishment phase.

Based on the stress concept, we can assume that each of the two phases represents a stress adaptation procedure induced by an imposed stress and leading to the corresponding optimal state. This is demonstrated in Figure 3, where the schematic presentation, based on Figure 2, integrates the two successive phases triggered by the plant's inoculation with arbuscular mycorrhizal fungi (AMF).

The environment is distinguished in external (EXT ENV) – soil, and internal (INT ENV) – plant. The initial condition, i.e. of no MF, the phase of infection and the phase of symbiosis' establishment are indicated by 'n', 'i' and 's', respectively.

The BIOSYSTEM refers to the plant component under investigation, i.e. to the photosynthetic apparatus, monitored by the O-J-I-P fluorescence transient. The biosystem is described by (a) its state, i.e. optimal = non-stressed or sub-optimal = stressed, (b) its relation with the environment, i.e. harmony or disharmony and (c) its performance quantified by a parameter Y, The values of the conformational K-term are included in the state description, and the values of the B-term are also mentioned.

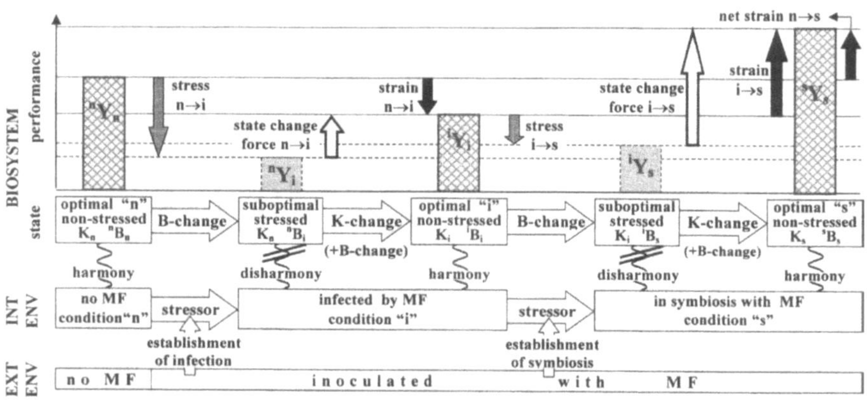

Figure 3. A schematic presentation of the stress concept applied for cases where the inoculation of a plant with mycorrhiza fungi (MF) provokes two successive stress effects, i.e. infection by MF (indicated by "i") and establishment of symbiosis with MF ("s"). The initial condition, i.e. of no MF, is indicated by "n". The environment is distinguished in external (EXT ENV) – soil, and internal (INT ENV) – plant. The BIOSYSTEM, referring to the experimentally monitored plant component (here the photosynthetic apparatus), is monitored by a performance parameter Y (see Fig. 2). For each stress effect, the stress intensity (grey arrow), state change force (white arrow with heavy black line) and Y-strain (black arrow) are also demonstrated. For more details see text.

All the values of K, B and Y can be deduced via the JIP-test. The Y values at the three optima are: the initial ${}^{n}Y_{n}$ (for no mycorrhiza), the subsequent ${}^{i}Y_{i}$ for mycorrhizal infection and the final ${}^{s}Y_{s}$ for mycorrhizal symbiosis, where ${}^{s}Y_{s} > {}^{n}Y_{n} > {}^{i}Y_{i}$. For each stress effect, the stress intensity (grey arrow), state change force (white arrow with heavy black line) and Y-strain (black arrow) are also demonstrated.

Combination of mycorrhization with other stressors

The JIP-test was also applied to study the combination of mycorrhiza with other stressors, e.g. heavy metals (Calantzis et al. 1999) or nitrogen-fixing bacteria (Köves-Péchy et al. 1998; Tsimilli-Michael et al. 2000). Figure 4 presents, as an example, the changes in the electron transport per leaf area of alfal-

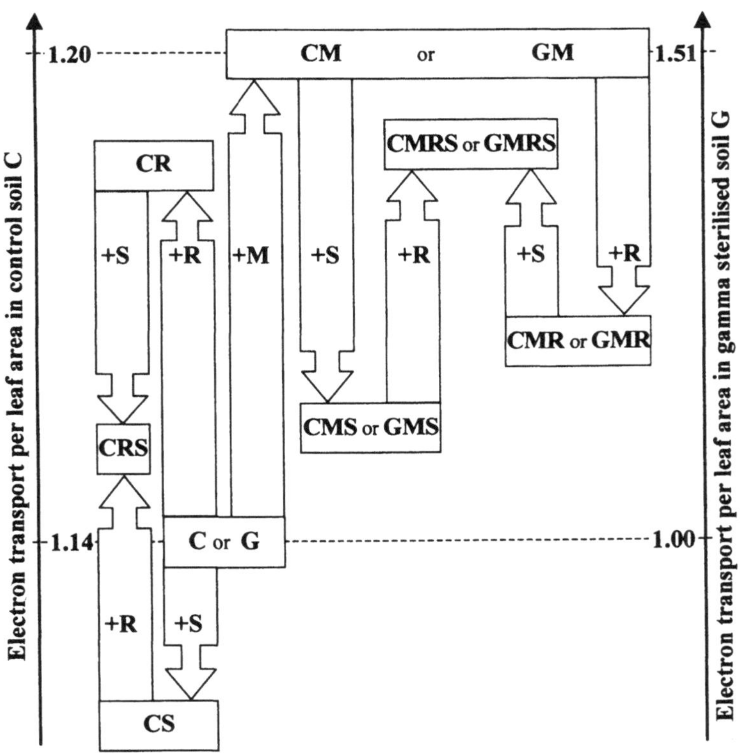

Figure 4. Changes in the electron transport per leaf area of alfalfa, induced by inoculation with arbuscular mycorrhizal fungi (M) without and with *Azospirillum* (S) and/or *Rhizobium* (R) as N_2 fixing bacteria, in control soil carrying all the native microflora (C) or gamma-sterilised soil (G), as well as by inoculation only with S and/or R in control soil. The electron transport values are normalised on the value exhibited with the G soil and no inoculation. Two scales of different sensitivity are used, the left for the C and the right for the G soil.

fa (*Medicago sativa*), inducedby inoculation with the AMF *Glomus fasciculatum* (M) without and with the associative *Azospirillum brasilense* (S) and/or the symbiotic *Rhizobium meliloti* (R) nitrogen-fixing bacteria (Tsimilli-Michael et al. 2000). The results refer to plants grown in control soil carrying all the native microflora (C) or gamma-sterilised soil (G). The effects of inoculation only with the diazotrophs S and/or R in control soil are also presented for comparison. The electron transport values are normalised on the value exhibited without any inoculation in the G soil.

The beneficial effect of mycorrhiza is clearly revealed as well as the antagonism by both bacteria. In the presence of each other, the diazotrophs are no more antagonistic to mycorrhiza. It is interesting to note that the trend of the changes is the same in the two soils, while their amplitude is about 9 times smaller in the C soil, as demonstrated by the different sensitivity of the two scales used; this means that the control soil has a buffering capacity, due to the fact that it already carries all the microbe microflora. The observed antagonistic and synergistic effects of AMF and the diazotrophs *Azospirillum* and *Rhizobium* are in accordance with characteristic interactions between stressors, mainly recognised so far for cases of abiotic stressors.

Acknowledgements
M. Tsimilli-Michael thanks the Minister of Education and Culture of Cyprus, Mr. Ouranios Ioannides, for giving her the opportunity to carry out this work in Geneva. R. J. Strasser acknowledges financial support from the Swiss National Foundation (3100-052541.97 and 3100-057046.99) and the Swiss Office for COST-Action 838 (C98.0048 and C99-0075).

References

Calantzis C, Rivera-Becerril F, Gianinazzi-Pearson V, Gianinazzi S, Strasser R J (1999) Mycorrhiza buffer cadmium-induced stress in pea plants: vitality probing by the fast fluorescence transient O-J-I-P. In: COST-Action 838 Meeting: Arbuscular mycorrhizas and plant health under abiotic stress. Nancy, France. http://www.inra.fr/Internet/Centres/Dijon/cost/cost838/scientific_meetings/abstractsNancy.pdf p3

Calantzis C, Trouvelot S, van Tuinen D, Gianinazzi-Pearson V, Gianinazzi S, Strasser R J (2000a) A non destructive method for evaluating the mycorrhizal status of micropropagated vine root-stock. In: COST-ACTION 838 Meeting: Genetic and Cell Programme and Mycorrhizal Technology. Santiago de Compostela, Galicia, Spain. http://www.inra.fr/Internet/Centres/Dijon/cost/cost838/scientific_meetings/AbstractsSantiago.pdf p 24

Calantzis C, Trouvelot S, van Tuinen D, Gianinazzi-Pearson V, Gianinazzi S, Strasser R J (2000b) A non destructive method for evaluating the mycorrhizal status of micropropagated vine root-stock. In: Mycorrhiza and root research in Switzerland, Bulletin SGPW/SSA No 14, p 8

Clark A J, Landolt W, Bucher J, Strasser R J (1998) The response of *Fagus sylvatica* to elevated CO_2 and ozone probed by the JIP-test based on the chlorophyll fluorescence rise OJIP. In: De Kok L J, Stulen I (eds) responses of plant metabolism to air pollution and global change. Backhuys Publishers, Leiden, The Netherlands pp 283-286

Clark A J, Landolt W, Bucher J, Strasser R J (2000) Beech (*Fagus sylvatica*) response to ozone exposure assessed with a chlorophyll *a* fluorescence performance index. Environmental Pollution 109:501-507

Gianinazzi S, Schüepp H (eds) (1994) Impact of arbuscular mycorrhizas on sustainable agriculture and natural ecosystems. Birkhäuser Verlag AG, Basel, Switzerland

Govindjee (1995) Sixty-three years since Kautsky: Chlorophyll *a* fluorescence. Aust J Plant Physiol 22:131-160

Köves-Péchy K, Biró B, Vörös I, Takács T, Osztoics E. Strasser R J (1998) Enhanced activity of microsymbiont-alfalfa system probed by the fast fluorescence rise OJIP. In: Garab G (ed) Photosynthesis: mechanisms and effects. Kluwer Academic Publishers, The Netherlands, Vol IV pp 2765-2768

Krause G H, Weiss E (1991) Chlorophyll fluorescence and photosynthesis: the basics. Annu Rev Plant Physiol Plant Mol Biol 42:313-349

Krüger G H J, Tsimilli-Michael M, Strasser R J (1997) Light stress provokes plastic and elastic modifications in structure and function of photosystem II in camellia leaves. Physiol Plant 101:265-277

Larcher W (1987) Stress bei Pflantzen. Naturwissenschaften 74:58-167

Levitt J (1980) Responses of plants to environmental stresses. Academic Press, N York

Prigogine I (1947) Etude thermodynamique des phénomènes irréversibles. Dunod - Paris et Desoer-Liège.

Romano A, Strasser R J, Eggenberg P, Martins-Loução M A (1996) Mycorrhization of cork oak micropropagated plantlets: its vitality during acclimatization measured with fast fluorescence techniques. In: Azcon-Aquilar C, Barea J M (eds) European Commission - EUR 16728 - Mycorrhizas in integrated systems from genes to plant development. Luxemburg pp 373-376

Srivastava A, Strasser R J (1996) Stress and stress management of land plants during a regular day. J Plant Physiol 148:445-455

Srivastava A, Strasser R J (1997) Constructive and destructive actions of light on the photosynthetic apparatus. J Sci Ind Res 56:133-148

Strasser B J, Strasser R J (1995) Measuring fast fluorescence transients to address environmental questions: The JIP - Test. In: Mathis P (ed) Photosynthesis: from light to biosphere. Kluwer Academic Publishers, The Netherlands, Vol V pp 977-980.

Strasser R J (1978) The Grouping Model of Plant Photosynthesis. In: Akoyunoglou G, Argyroudi-Akoyunoglou J (eds) Chloroplast Development. Elsevier/North Holland, Biomedical Press, Amsterdam pp 513-524

Strasser R J (1981) The grouping model of plant photosynthesis: heterogeneity of photosynthetic units in thylakoids. In: Akoyunoglou G (ed) Photosynthesis III. Structure and molecular organisation of the photosynthetic apparatus. Balaban International Science Services, Philadelphia pp 727-737

Strasser R J (1985) Dissipative Strukturen als Thermodynamischer Regelkreis des Photosyntheseapparates. Ber Deutsche Bot Ges Bd 98:53-72

Strasser R J (1988) A concept for stress and its application in remote sensing. In: Lichtenthaler H K (ed) Applications of chlorophyll fluorescence. Kluwer Academic Publishers pp 333-337

Strasser R J, Govindjee (1991) The F_0 and the O-J-I-P fluorescence rise in higher plants and algae. In: Argyroudi-Akoyunoglou J H (ed) Regulation of chloroplast biogenesis. Plenum Press, New York pp 423-426

Strasser R J, Srivastava A, Govindjee (1995) Polyphasic chlorophyll a fluorescence transient in plants and cyanobacteria. Photochem Photobiol 61:32-42

Strasser R J, Srivastava A, Tsimilli-Michael M (2000) The fluorescence transient as a tool to characterize and screen photosynthetic samples. In: Yunus M, Pathre U, Mohanty P (eds) Probing photosynthesis: mechanism, regulation and adaptation. Taylor and Francis, London, UK pp 443-480

Tsimilli-Michael M, Krüger G H J , Strasser R J (1996) About the perpetual state changes in plants approaching harmony with their environment. Archs Sci Genève 49:173-203

Tsimilli-Michael M, Pêcheux M, Strasser R J (1999) Light and heat stress adaptation of the symbionts of temperate and coral reef foraminifers probed *in hospite* by the chlorophyll *a* fluorescence kinetics O-J-I-P. Z Naturforsch 54C:671-680

Tsimilli-Michael M, Eggenberg P, Biro B, Köves-Pechy K, Vörös I, Strasser R J (2000) Synergistic and antagonistic effects of arbuscular mycorrhizal fungi and *Azospirillum* and *Rhizobium* nitrogen-fixers on the photosynthetic activity of alfalfa, probed by the chlorophyll *a* polyphasic fluorescence transient O-J-I-P. Applied Soil Ecology 15:169-182

Varma A, Hock B (eds) (1998) Mycorrhiza. Structure, function, molecular biology and biotechnology, 2nd edition, Springer-Verlag, Berlin

Mycorrhizal Technology in Agriculture
ed. by S. Gianinazzi, H. Schüepp, J.M. Barea and K. Haselwandter

Arbuscular mycorrhizal fungi in low input agriculture

D. Atkinson[1], J.A. Baddeley[2], N. Goicoechea[3], J. Green[1], M. Sánchez-Díaz[3] and C. A. Watson[2]

[1] *SAC, West Mains Road, Edinburgh, EH9 3JG, UK*
[2] *SAC, Craibstone Estate, Aberdeen, AB21 9YA, UK*
[3] *Departamento de Fisiologia Vegetal, Universidad de Navarra, Irunlarrea s/n E-31008, Pamplona, Spain*

Keywords: organic farming, integrated crop management, drought, nutrient supply, intensification, conventional farming

Introduction

The features of various types of European farming systems have recently been reviewed by Atkinson and Watson (2000) and Tinker (2000). Atkinson and Watson (2000) suggested that systems varying from intensive arable through IPM/ICM (integrated pest management/integrated crop management) and mixed farming to organic could be regarded as a gradation in which the importance of minimal environmental impact and of the management of crop and soil ecological processes increased (Tab. 1). Conventional high input agricultural systems tend to be dominated by simple, usually linear, management models. Here the identification of a limitation in nutrient supply is followed by the application of that nutrient as a fertiliser. This philosophy gives rise to a system usually involving high levels of externally derived inputs. "It permits the use of any chemical found to be beneficial" (Tinker 2000). Arbuscular mycorrhizal fungi (AMF) are usually of limited value in such systems. 'Low input' systems, such as organic, are harder to define in simple terms.

Table 1. A gradation of farming systems based on external resource use and their development needs

Development type			
Intensive arable →	Environmentally aware cropping (IPM/ICM) →	Mixed farming →	Organic
Development needs			
• Cost reduction and management simplification • AMF unimportant	• Optimisation of element use and biological control • AMF role	• Additionally between enterprises AMF role	• Management of biological and ecological cycles AMF critical

Organic systems are often defined in terms of what they are not (Woodward and Lampkin 1990). In practice they function through the management of nature and by balancing resource demand with availability through the integration of processes. Organic systems follow a defined set of rules. Other low input systems aim to maximise the use of renewable resources, but without some defined rules.

These include integrated cropping systems and many of the systems found in arid situations. In all low input situations effective production requires the crop to have access to nutrients and water and to be maintained relatively free of the adverse impact of pests and diseases, but with minimal recourse to added inputs. The provision of all of these can be influenced or provided by AMF. This Chapter discusses the role of AMF in the productivity of such farming systems. This role is illustrated by assessing the roles of AMF in temperate crops, grassland systems, and in Mediterranean systems.

Temperate mixed cropping systems

The core of the lay-arable system, and of most other low input systems is the crop rotation. A typical rotation is shown in Figure 1. One of the aims of crop rotation is to modify the soil micro-flora, including AMF, and through this and plant activity to change the availability of soil nutrients and to beneficially reduce deleterious organisms to the benefit of crop health. Agricultural man-

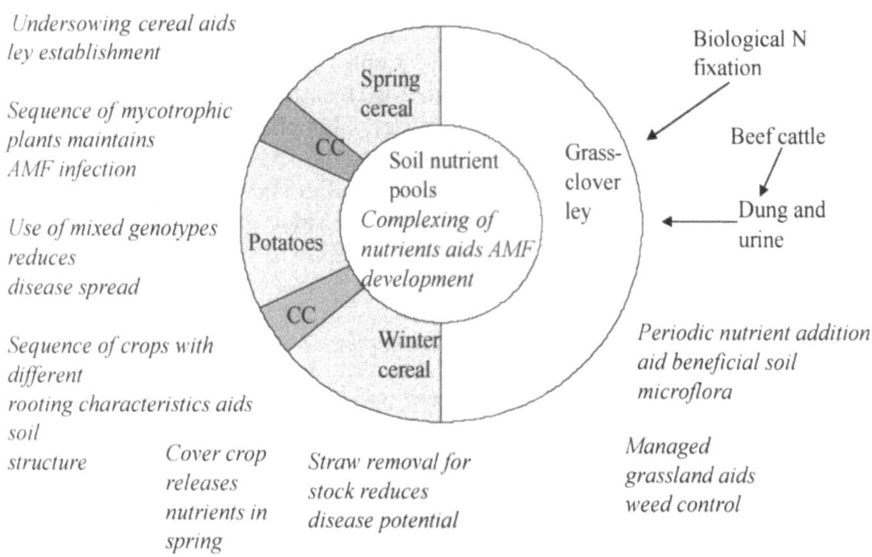

Figure 1. A typical mixed stock/crop rotation and some of its influences including those mediated by AMF.

agement thus influences both the presence of AMF and their activity. Organic agriculture is used, here, as a model for other low input systems.

Around 60 years ago Sir Albert Howard, a founder of the organic movement suggested that "the presence of an effective mycorrhizal symbiosis is essential to plant health" (Howard 1943). The potential benefits of AMF to crop nutrition, health, stress resistance and the development of soil structure have been discussed by Bethlenfalvay and Lindermann (1992), Miller and Jastrow (1990), Azcón-Aguilar and Barea (1996). Although AMF low-input systems have previously been reviewed (e.g. Bethlenfalvay and Schüepp 1994; Johnson and Pfleger 1992) current increases in organic agriculture in Europe require their reconsideration. Table 2 summarises some of the potential effects of agricultural management practices on AMF in the field. A major difficulty in translating the results of research into practical recommendations is the interaction between factors affecting the AMF symbiosis and the separation of cause and effect. The future success of low input agriculture depends on an improved understanding of the dynamic relationships between agricultural practice and soil biology.

Several observational studies have assessed differences between colonisation levels (Ryan et al. 1994; Satelmacher et al. 1991), spore populations (Douds et al. 1993; Kahiluoto and Vestberg 1998; Kurle and Pfleger 1994) and species abundance and diversity (Douds et al. 1993; Kurle and Pfleger 1996) in differently managed agricultural systems. A number of these suggest higher colonisation rates and greater spore populations in lower input systems. Interpretation is, however, complicated by the multiple factors involved. A Swiss trial (DOC) compared conventional, organic and biodynamic systems using identical crop rotations and tillage but different fertilisation and plant protection strategies (Mader et al. 2000). Root colonisation was highest in the low input treatments, which was in part attributable to soil chemical properties in the different systems. Ryan et al. (1994) highlighted the importance of changes such as modified fungicide use and of lowering the levels of available P. The nature of the comparisons made can further increase the difficulty in interpreting results, for example, Kahiluoto and Vestberg (1998) compared conventionally managed continuous cereals with an organically managed lay-arable rotation. In a recent study, Franke-Synder et al. (2001) compared the structure of AMF populations associated with maize or soybeans in conventional or low input systems. The results of the study showed that a 15 year long application of the different system had little effect on the fungal communities. Most of the 15 fungal species found occurred in all treatments. Although there were some differences in sporulation the species structure of most communities was not affected by treatment.

An underlying principle of biodynamic agriculture is the use of preparations based on material of plant and animal origin to stimulate soil biological activity. Critical studies of this form of agriculture are rare although Penfold et al. (1995) examined the effect of preparation BD500 on soil microbial biomass. No effects were detected during a 2-year period, although it is suggest-

Table 2. Summary of some of the potential effects of agricultural management practices on Arbuscular Mycorrhizal Fungi in the field

Factors	Potential effects on:				Relevant references
	Effectivity of symbiosis	Host presence	Spore populations and viability	Extra radical hyphae	
Amendments					
Inoculants	√		√		Kahiluoto and Vestberg 1998
Lime (pH changes)	√		√	√	Frey and Ellis 1997
Fungicides and fumigants	√		√		Fitter 1986; Jawson et al. 1993; Johnson and Pfleger 1992
Herbicides		√	√		Kurle and Pfleger 1998; Schwab et al. 1993; Johnson and Pfleger 1992
Seed dressings	√		√		Ryan et al. 1994
Nitrogen	√	√	√		
Phosphorus	√				Abbott and Robson 1998; Hayman 1986
Manures and crop residues	√				Douds et al. 1997
Biodynamic preparations	√				No known studies
Management					
Crop choice		√			Black and Tinker 1977; Douds et al. 1997; Harinikumar and Bagyaraj 1988
Variety choice	√			√	
Sequence (rotation)	√	√	√	√	Baltruschat and Dehne 1988
Use of winter cover crop		√	√		Boswell et al. 1998; Galvez et al. 1995; Kabir and Koide 2000
Tillage		√		√	Kabir et al. 1997; McGonigle and Miller 2000
Drainage	√			√	
Fallow period		√	√		Thompson 1987; Kabir and Koide 2000
Farming systems					
Organic, low input, conventional	√	√	√	√	Douds et al. 1993; Kahiluoto and Vestberg 1998; Kurle and Pfleger 1994; Johnson and Pfleger 1992; Ryan and Ash 1999; Ryan et al. 1994; Sattelmachar et al. 1991.

ed that these preparations stimulate the biomass and/or activity of specific organisms.

The movement towards more sustainable forms of agriculture will result in an increased use of organic sources of nutrients including crop residues, farm and urban wastes. The ability of AMF to utilise organic forms of N (e.g. Hawkins et al. 2000) and P (e.g. Kahiluoto and Vestberg 1998) is likely to be a key-determining factor in the impact of these wastes on crop productivity. St John et al. (1983) suggested that organic compounds in manure may stimulate AMF and Kabir et al. (1997) showed that liquid dairy manure increased the densities of both total and viable hyphae in clay soils. Douds et al. (1997) cite studies of the effects of different composts on spore populations. The results depend on both fungal species and the nature of the organic material and are likely to be related to the concentration and availability of nutrients. Within crop rotations, colonised roots and hyphae are an important source of inoculum for the cropping sequence (Boswell et al. 1998; Sylvia 1992). Thus fallow periods (Thompson 1987), the inclusion of non-mycorrhizal species (Douds et al. 1997) and the use of herbicides (Schwab et al. 1998) can have similar effects on subsequent cropping, mediated through a reduction in the availability of AMF-host plants. Additional recorded detrimental effects of herbicide use include lower spore viability (Schwab et al. 1983) and potentially selective multiplication of different fungal species. Weeds within the crop can thus have positive advantages where there are non-mycorrhizal crops within the rotation and cover crops can provide a reservoir of mycorrhizal inoculum. Kabir and Koide (2000) suggested that mycotrophic weeds, e.g. dandelion, have potential as a cover crop.

There is widespread evidence for the so-called 'rotation effect' in addition to the straight nutritional beneficial effects of N-fixation by legumes in rotations (Altieri 1995). Cook (1986) attributes the 'rotation effect' to improved root health, in which AMF may have an important role. Azcón-Aguilar and Barea (1996) have addressed the beneficial effects of AMF in reducing disease susceptibility. As a result of finding large changes in mycorrhizal communities associated with different rotations under the same soil and climatic conditions, Hendrix et al. (1995) suggest a role for AMF in the rotation effect. One of the compromises of combining rotations with low input systems is the need for cultivation to control weeds. There is evidence that tillage can restrict P nutrition early in the season by disrupting the extraradical hyphae (McGonigle et al. 1999; McGonigle and Miller 2000). Tillage may cause a shift in AMF and host plant communities (Douds et al. 1995).

Temperate grassland systems

Grassland systems share many of the features of lay/arable systems although the duration of the period under grass is usually longer than in a rotational system. AMF form associations with over 85% of the *Gramineae* as well as with many of the dicots found in such systems. Although the importance of AMF is documented for a number of ecosystems their influence in agricultural grasslands is less well known. As with crop systems, the challenge within lower input or organic grassland systems is to achieve acceptable production with reduced inputs. In practice this means increased reliance on legumes.

AMF have been shown in some circumstances to increase plant N uptake (eg Azcón and Barea 1992) although this has not always been found. This property is not universal: N capture by *Plantago* was unaffected by AMF despite increased root growth (Hodge et al. 2000) and the uptake of clover-derived N by *Lolium* was independent of AMF (Rogers et al. 2001). Nitrogen is not, however, the only relevant element. That AMF associations are more abundant in P limited situations and can increase P uptake is well established. The efficiency with which AMF acquire P can be so great that it has been suggested as a factor in the N limitation seen in many natural ecosystems (Grogan and Chapin 2000).

Intensive grassland management has led to a marked decline in grassland biodiversity. There is interest in reversing this trend. Intensively managed hay meadows in northern Britain have a lower soil fungal biomass than traditionally managed meadows (Donnison et al. 2000). Van der Heijden et al. (1998) showed that plant species diversity in two different grassland ecosystems could be manipulated by changing the AMF taxa present in the soil. A strong association between plant diversity and population structures and AMF was demonstrated in alpine meadows (Barni and Siniscalco 2000), tallgrass prairie (Wilson and Hartnett 1997; Eom et al. 2000) and calcareous grasslands (Streitwolf-Engel et al. 1997).

Mediterranean systems

In contrast to the systems described above the primary challenge to crop production in many Mediterranean areas is the imbalance between the supply of and demand for water. While in these systems there are also problems with nutrient supply, in this section, the focus is placed on the role of AMF in helping plants cope with drought.

Aschmann (1973) defined the Mediterranean regions of the world as areas where at least 65% of annual precipitation takes place in winter, the annual precipitation varies between 275 and 900 mm, and the median temperature in winter is equal to or lower than 15 °C, although periods with temperatures <0 °C must be less than 3% of the year. Mediterranean ecosystems are thus subjected to extreme environmental conditions, eg dry summers followed by heavy rainfall (Blanca and Morales 1991). Erratic precipitation, combined with high temperatures, may disturb the vegetation cover resulting in soil erosion, the loss of mineral nutrients and high evapotranspiration rates; direct evaporation from the soil during the cropping season can represent 30 to 50% of annual rainfall (Sadler and Turner 1994).

Several *Glomus* species have been found in the semi-arid soils of the Mediterranean regions: the genus *Glomus* is best represented in terms of species variability (Sánchez-Díaz and Honrubia 1994). The maintenance of mycorrhizal fungi diversity is critical for successful production (Jeffries and Barea 2001). Unfortunately, crop monoculture, the use of soil disinfectants such as fungicides (Johnson et al. 1992; Jawson et al. 1993) and soil erosion and degradation (Herrera et al. 1993; Jeffries and Barea 1994) have all been shown to cause a reduction in diversity (Johnson et al. 1992). In addition, colonization of host-plant roots and spore production vary seasonally as a function of climate and host plant (Giovanetti 1985).

In general terms, the transpiration of a given crop is positively correlated with dry matter production (Tanner and Sinclair 1983), and so reduced transpiration is undesirable (Sadler and Turner 1994). In addition, water stress usually causes a decrease of nutrient uptake while a decline in soil moisture produces a reduction in the diffusion rate of nutrients, particularly P, from soil to root (Pinkerton and Simpson 1986). The ability of mycorrhizal hyphae to transport nutrients into host plants (Koide 1991) can improve plant water relations under drought (Fitter 1988) and root hydraulic conductivity (Andersen et al. 1988).

The majority of studies of the influence of AMF on drought avoidance and tolerance have shown that AMF increase plant resistance to water stress (Sánchez-Díaz and Honrubia 1994). There is, however, no consensus as to the mechanisms involved (Barea et al. 1993; Bryla and Duniway 1997b). A number of different mechanisms each important in different situations seem to operate (Sánchez-Díaz and Honrubia 1994; Subramanian et al. 1997). The AMF association seems to benefit from both the avoidance and tolerance of drought. It has been suggested that improved water relations are a consequence of greater phosphorus uptake (Nelsen and Safir 1982). Enhanced leaf phosphorus in AMF plants has been implicated in increased stomatal conductance and transpiration (Fitter 1988 et al. 1986), while higher phosphorus levels can lead to improved osmotic adjustment and stomatal control in plants, allowing more efficient water use, or drought tolerance (Augé et al. 1986a, 1986b; Augé and Stodola 1990).

Other than through the indirect efforts detailed above AMF colonisation may affect plant water uptake from the soil by altering root system architecture, by direct hyphal flow, and by changing root resistance to water flow. Root systems with AMF seem in general to have a greater surface area in a given soil volume and so are able to utilize soil localised water resources more rapidly. The role of the extraradical hyphae of AMF for direct water uptake remains equivocal (Ruiz-Lozano and Azcón 1995). Direct evidence for water uptake by hyphae comes from studies where plants were grown in containers in which water was applied to compartments inaccessible to roots, but into which AMF could penetrate (Faber et al. 1991; Ruiz-Lozano and Azcón 1995). AMF plants were found to have greater water uptake and so it was concluded that the hyphae played a significant role. The increased efficiency of water uptake in AMF plants may also be due to increases in the hydraulic conductance of the root system (Koide 1993).

Apart from their ability to translocate nutrients from soil to host plant, mycorrhizal hyphae are crucial for the formation of stable soil aggregates (Bethlenfalvay and Schüepp 1994). Hyphae produce important amounts of a glycoprotein positively correlated with soil aggregate stability (Wright and Upadhyaha 1998).

Salinity is another frequent stressful condition in Mediterranean environments (Sánchez-Díaz and Aguirreolea 1991). The direct effects of salts on plant growth may be divided into three categories: reduction in the soil osmotic potential with a subsequent reduction in plant available water, deterioration in the physical structure of the soil and specific ion toxicity (Dudley 1994). Several studies concerning AMF effects on plant growth under saline conditions indicate an improvement of growth in some species, including sour orange (Ezz and Nawar 1994), onion (Ojala et al. 1983) and cucumber (Rosendahl and Rosendahl 1991). Unfortunately, high salt concentrations in soil can negatively affect the germination of AMF spores and subsequent hyphal growth (Juniper and Abbott 1993).

To extrapolate from pot experiments to field situations represents a considerable challenge. A number of mechanisms of drought amelioration may be acting in concert and may vary in the different systems. Low-input agricultural systems which maintain a high inoculum potential for crops (Thompson 1994) can result in increased levels of AMF colonisation (Boswell et al. 1998), leading to improved water relations. The drought ameliorating effects of AMF remain to be thoroughly validated under field conditions. However, low-input systems which maximise the AMF colonisation of crops seem to have the potential to improve yields not just in arid areas, but in any system where water availability limits growth (Dodd and Jeffries 1986). In degraded mediterranean soils a decrease in the AMF inoculum-potential is often combined with severe summer drought conditions. This provides, therefore, a model system in which to investigate AMF-inoculation and drought interactions. The challenge now is to demonstrate the link between AMF and drought amelioration in the field (Green et al. 1999).

Conclusions

In order to increase our ability to optimise management of AMF in field situations there is a need for more basic information on the seasonal variation in the development of AMF in different crop species, and how this is influenced by agricultural practices. The development of a diverse AM fungal population which can adapt to management and environmental changes is likely to be a key factor in improving the sustainability of low input and organic cropping systems. One of the difficulties in interpreting data from comparative studies of farming systems has been the reliance on the use of spore morphology to characterise AMF which cannot give a complete picture of the systems (Douds et al. 1999). Molecular techniques, used together with a knowledge of mycorrhizal ecology and crop physiology, hold the key to improved understanding of the functioning of AMF in sustainable crop production systems.

References

Abbott L K, Robson A D (1984) The effect of mycorrhizae on plant growth. In: Powell CL and Bagyaraj D J (eds) VA Mycorrhiza, CRC Press, Boca Raton pp 113-130

Altieri M A (1995) Agroecology: the science of sustainable agriculture. Intermediate Technology Publications, London

Andersen C P, Markhart A H III, Dixon R K, Sucoff E I (1988) Root hydraulic conductivity of vesicular-arbuscular mycorrhizal green ash seedlings. New Phytol 109: 465-471

Aschmann H (1973) Distribution and peculiarity of mediterranean ecosystems. In: de Castri F, Mooney HA (eds) Mediterranean type ecosystems. Springer-Verlag, Berlin, pp 11-19

Atkinson D, Watson C A (2000) The research needs of organic farming: distinct or just the same as other agricultural research. Proceeding BCPC Conference – Pest and Diseases pp 151-158

Augé R M, Stodola J W (1990) An apparent increase in symplastic water contributes to greater turgor in mycorrhizal roots of droughted rosa plants. New Phytol 115:285-295.

Augé R M, Schekel K A, Wample R L (1986a) Greater leaf conductance of well-watered va mycorrhizal rose plants is not related to phosphorus-nutrition. New Phytol 103:107-116.

Augé R M, Schekel K A, Wample R L (1986b) Osmotic adjustment in leaves of VA mycorrhizal and nonmycorrhizal rose plants in response to drought stress. Plant Physiol 82:765-770.

Azcón R, Barea J M (1992) Nodulation, N-2 fixation (N-15) and N-nutrition relationships in mycorrhizal or phosphate-amended alfalfa plants. Symbiosis 12:33-41.

Azcón-Aguilar C, Barea J M (1996) Arbuscular mycorrhizas and biological control of soil-borne plant pathogens - An overview of the mechanisms involved Mycorrhiza 6:457-464

Baltruschat H, Dehne H W (1988) The occurrence of vesicular arbuscular mycorrhiza in agro-ecosystems. 1. Influence of nitrogen fertilisation and green manure in continuous monoculture and in crop rotation on the inoculum potential of winter wheat. Pl Soil 107:279-284

Barea J M, Azcón R, Azcón-Aguilar C (1993) Mycorrhiza and crops. Advances in plant pathology 9:167-189.

Barni E, Siniscalco C (2000) Vegetation dynamics and arbuscular mycorrhiza in old-field successions of the Western Italian Alps. Mycorrhiza 10:63-72.

Bethlenfalvay G J, Lindermann R G (1992) Mycorrhizae in sustainable agriculture. ASA Special Publication No 54, Madison, Wisconsin

Bethlenfalvay G J, Schüepp H (1994) Arbuscular mycorrhizas and agrosystem stability. In: Gianinazzi S and Schüepp H (eds) Impact of arbuscular mycorrhizas on sustainable agriculture and natural ecosystems. Birkhauser Verlag, Basel, pp 117-132

Black R L B, Tinker P B (1977) Interactions between effects of vesicular arbuscular mycorrhiza and fertilizer phosphorus on field grown potatoes. Nature 267:510-511

Blanca G, Morales C (1991) Impactos sobre la flora y medidas de protección. In: Servicio de Publicaciones de la Universidad de Granada (ed) Flora del parque natural de la Sierra de Baza, 1st ed Universidad de Granada, Granada, Spain, pp 349-362

Boswell E P, Koide R T, Shumway DL, Addy H D (1998) Winter wheat cover cropping, VA mycorrhizal fungi and maize growth and yield. Agr Ecosyst and Environ 67:55-65

Bryla D R, Duniway J M (1997) Growth, phosphorus uptake, and water relations of safflower and wheat infected with an arbuscular mycorrhizal fungus. New Phytol 136:581-590.

Cook R J (1986) Interrelationships between plant health and the sustainability of agriculture, with special reference to plant diseases. American Journal of Alternative Agriculture 1:19-25

Dodd J C, Jeffries P (1986) Early development of vesicular arbuscular mycorrhizas in autumn-sown cereals. Soil Biology and Biochemistry 18:149-154.

Donnison L M, Griffith G S, Hedger J, Hobbs P J, Bardgett R D (2000) Management influences on soil microbial communities and their function on botanically diverse haymeadows of Northern England and Wales. Soil Biology and Biochemistry 32:253-263.

Douds D D, Millner P (1999) Biodiversity of arbuscular mycorrhizal fungi in agroecosystems. Agriculture, Ecosystems and Environment 74:77-93

Douds D D, Galvez L, FrankeSnyder M, Reider C, Drinkwater L E (1997) Effect of compost addition and crop rotation point upon VAM fungi. Agr, Ecosyst and Environ 65:257-266

Douds D D, Galvez L, Janke R R, Wagoner P (1995) Effect of tillage and farming system upon populations and distribution of vesicular-arbuscular mycorrhizal fungi. Agr Ecosyst and Environ 52:111-118

Douds D D, Janke R R, Peters S E (1993) VAM fungus spore populations and colonisation of roots of maize and soybean under conventional and low-input sustainable agriculture. Agr, Ecosyst and Environ 43:325-336

Dudley L M (1994) Salinity in the soil environment. In: Pessarakli M (ed) Handbook of plant and crop stress. Marcel Dekker, New York

Eom A H, Hartnett D C, Wilson G W T (2000) Host plant species effects on arbuscular mycorrhizal fungal communities in tallgrass prairie. Oecologia 122:435-444.

Ezz T, Nawar A (1994) Salinity and mycorrhizal association in relation to carbohydrate status, leaf chlorophyll and activity of peroxidase and polyphenol oxidase enzymes in sour orange seedlings. Alexandria J Agric Res 39:263-280

Fitter A H (1986) Effect of benomyl on leaf phosphorus concentration in alpine grasslands - A test of mycorrhizal benefit. New Phytol 103:767-776

Fitter A H (1988) Water relations of red clover *Trifolium pratense* L. as affected by VA mycorrhizal infection and phosphorus supply before and during drought. J Exp Bot 39:595-603

Franke-Snyder M, Douds D D, Glalvez L, Phillips J G, Wagoner P, Drinkwater L, Morton J B (2001) Diversity of Communities of Arbuscular Mycorrhizal (AM) fungi present in conventional versus low input agricultural sites in Eastern Pennsylvania, USA. Appl Soil Ecol 16:35-48.

Frey J E, Ellis J R (1997) Relationship of soil properties and soil amendments to response of *Glomus intraradices* and soybeans. Can J Bot 75:483-491

Galvez L, Douds D D, Wagoner P, Longnecker L R, Drinkwater L E, Janke R R (1995) An overwintering cover crop increases inoculum of VAM fungi in agricultural soil. American Journal of Alternative Agriculture 10:152-156

Giovanetti M (1985) Seasonal variation of vesicular-arbuscular mycorrhizas and endogonaceous spores in a maritime sand dune Trans Br Mycol Soc 84:679-684

Green J J, Vallejo R, Serrasolses I, Martins-Loucao M A, Hatzistathis A, Barea J M, Cortina J, Watson C A (1999) Restoration of degraded ecosystems in Mediterranean regions (REDMED) In: Papanastasis V P, Frame J and Nastis A S (eds) Grasslands and woody plants in Europe. HERPAS, Thessaloniki, pp 281-286

Grogan P, Chapin F S (2000) Nitrogen limitation of production in a Californian annual grassland: The contribution of arbuscular mycorrhizae. Biogeochemistry 49:37-51.

Harinikumar K M, Bagyaraj D J (1988) Effect of crop-rotation on native vesicular arbuscular mycorrhizal propagules in soil. Pl Soil 110:77-80

Hawkins H J, Johansen A, George E (2000) Uptake and transport of organic and inorganic nitrogen by arbuscular mycorrhizal fungi. Pl Soil 226:275-285

Hayman D S (1986) VA mycorrhizas in field crop systems. In: Safir G R (ed) Ecophysiology of VA Mycorrhizal plants,. XCRC Press, Florida pp 172-193

Hendrix J W, Guo B Z, An ZQ (1995) Divergence of mycorrhizal fungal communities in crop production systems Pl Soil 170:131-140

Herrera M A, Salamanca C P, Barea J M (1993) Inoculation of woody legumes with selected arbuscular mycorrhizal fungi and rhizobia to recover desertified mediterranean ecosystems. Appl Environ Microbiol 59:129-133

Hodge A, Robinson D, Fitter A H (2000) An arbuscular mycorrhizal inoculum enhances root proliferation in, but not nitrogen capture from, nutrient-rich patches in soil. New Phytol 145:153-159.

Howard A 1943. An agricultural testament. Oxford University Press, London.

Jawson M D, Franzlubbers A J, Galusha D K, Aiken R M (1993) Soil fumigation within monoculture and rotations - response of corn and mycorrhizae. Agron J 85:1174-1180

Jeffries P, Barea J M (1994) Biogeochemical cycling and arbuscular mycorrhizas in sustainability of plant-soil systems. In: Gianinazzi S, Schüepp H (eds) Impact of arbuscular mycorrhizas on sustainable agriculture and natural ecosystems. Birkhäuser, Basel, pp 101-115

Jeffries P, Barea J M (2001) Arbuscular mycorrhiza -a key component of sustainable plant-soil ecosystems. In: Esser K (ed) The mycota IX. Fungal Associations. Springer-Verlag, Berlin, Heidelberg, pp 95-113

Johnson C R, Copeland P J, Crosskston R K, Pfleger F L (1992) Mycorrhizae: a possible explanation for yield decline associated with continuous cropping of corn and soybean. Agron J 84:387-390

Johnson N C, Pfleger F L (1992) Vesicular-arbuscular mycorrhizae and cultural stress. In: Bethlenfalvay G J and Lindermann R G (eds) Mycorrhizae in sustainable agriculture ASA Special Publication No 54, Madison, Wisconsin pp 71-100

Juniper S, Abbott L (1993) Vesicular-arbuscular mycorrhizas and soil salinity. Mycorrhiza 4:45-57

Kabir Z, Koide R T (2000) The effect of dandelion or a cover crop on mycorrhiza inoculum potential, soil aggregation and yield of maize. Agr, Ecosyst and Environ 78:167-174

Kabir Z, Ohalloran I P, Fyles J W, Hamel C (1997) Seasonal changes of arbuscular mycorrhizal fungi as affected by tillage practices and fertilization: hyphal density and mycorrhizal root colonization. Pl Soil 192:285-293

Kahiluoto H, Vestberg M (1998) The effect of arbuscular mycorrhiza on biomass production and phosphorus uptake from sparingly soluble sources by leek *(Allium porrum l.)* in finnish field soils. Biol Agr and Hortic 16:65-85

Koide R T (1991) Nutrient supply, nutrient demand and plant response to mycorrhizal infection. New Phytol 117:365-386

Koide, R T (1993) Physiology of the mycorrhizal plant. Advances in plant pathology 9:33-54.

Kurle J E, Pfleger F L (1994) Arbuscular mycorrhizal fungus spore populations respond to conversations between low-input and conventional management practices in a corn-soybean rotation. Agron J 86:467-475

Kurle J E, Pfleger F L (1996) Management influences on arbuscular mycorrhizal fungal species composition in a corn-soybean rotation. Agron J 88:155-161

Mader P, Edenhofer S, Boller T, Wiemken A, Niggli U (2000) Arbuscular mycorrhizae in a long-term field trial comparing low-input (organic, biological) and high-input (conventional) farming systems in a crop rotation. Biol Fert Soils 31:150-156

McGonigle T P, Miller M H (1999) Winter survival of extraradical hyphae and spores of arbuscular mycorrhizal fungi in the field. Appl Soil Ecol 12:41-50

McGonigle T P, Miller M H (2000) The inconsistent effect of soil disturbance on colonization of roots by arbuscular mycorrhizal fungi: a test of the inoculum density hypothesis. Applied Soil Ecology 14:147-155

Miller R M, Jastrow J D(1990) Hierarchy of root and mycorrhizal fungal interactions with soil aggregation. Soil Biol Biochem 22:579-584

Nelsen, C E, Safir G R (1982) Increased drought tolerance of mycorrhizal onion plants caused by improved phosphorus-nutrition. Planta 154:407-413

Ojala J C, Jarrell W H, Menge J A, Johnson E L V (1983) Influence of mycorrhizal fungi on the mineral nutrition and yield of onion in saline soil. Agron J 75: 255-259

Penfold C M, Miyan M S, Reeves T G, Grierson I T (1995) Biological farming for sustainable agricultural production. Aust.J.Exp.Agric. 35:849-856

Pinkerton A, Simpson J R (1986) Interactions of surface drying subsurface nutrients affecting plant growth on acidic soil profile from an old pasture. Aust J Exp Agr 26:681-686

Rogers JB, Laidlaw A S, Christie P (2001) The role of arbuscular mycorrhizal fungi in the transfer of nutrients between white clover and perennial ryegrass. Chemosphere 42:153-159.

Rosendahl C N, Rosendahl S (1991) Influence of vesicular-arbuscular mycorrhizal fungi (*Glomus* spp.) on the response of cucumber *(Cucumis sativus L.)* to salt stress. Environ Exp Bot 31:313-318

Ruiz-Lozano, J M, R Azcón (1995) Hyphal contribution to water uptake in mycorrhizal plants as affected by the fungal species and water status. Physiol Plant 95:472-478

Ryan M, Ash J (1999) Effects of phosphorus and nitrogen on growth of pasture plants and vam fungi in se australian soils with contrasting fertiliser histories (conventional and biodynamic) Agr, Ecosyst Environ 73:51-62

Ryan M H, Chilvers G A, Dumaresq DC (1994) Colonisation of wheat by VA-mycorrhizal fungi was found to be higher on a farm managed in an organic manner than on a conventional neighbour. Pl Soil 160:33-40

Sadler E J, Turner N C (1994) Water relationships in a sustainable agricultural system. In: Hatfield J L, Karlen D L (eds) Sustainable agriculture systems. CRC Press, Boca Raton, Florida, pp 21-46

Sánchez-Díaz M, Aguirreolea J (1991) Fijación de nitrógeno en ambientes áridos y semiáridos. In: Megías M, Palomares A, Ruiz Berraquero F (eds) Aportaciones a la biología de la fijación del nitrógeno atmosférico. Publicaciones Universidad de Sevilla, Spain, pp 405-418

Sánchez-Díaz M, Honrubia M (1994) Water relations and alleviation of drought stress in mycorrhizal plants. In: Gianinazzi S, Schüepp H (eds) Impact of arbuscular mycorrhizas on sustainable agriculture and natural ecosystems. Birkhäuser, Basel, pp 167-178

Sattelmacher B, Reinhard S, Pomikalko A (1991) Differences in mycorrhizal colonisation of rye *(Secale cereale L)* grown in conventional or organic (biological-dynamic) farming systems. Journal of Agronomy and Crop Science 167:350-355

Schwab S M, Menge J A, Leonard RT (1983) Quantitative and qualitative effects of phosphorus on extracts of sudangrass roots in relation to vesicular-arbuscular mycorrhizae formation. Plant Physiol 73:761-765

St John T V, Coleman D C, Reid C P P. (1983) Association of vesicular-arbuscular mycorrhizal hyphae with soil organic particles. Ecology 64:957-959

Streitwolf-Engel R, Boller T, Wiemken A, Sanders I R (1997) Clonal growth traits of two *Prunella* species are determined by co-occurring arbuscular mycorrhizal fungi from a calcareous grassland. J Ecol 85:181-191

Subramanian K S, Charest C, Dwyer L M, Hamilton R I (1997) Effects of arbuscular Mycorrhizae on leaf water potential, sugar content, and P content during drought and recovery of maize. Can J Bot 75:1582-1591.

Sylvia D M (1992) Quantification of external hyphae of vesicular-arbuscular mycorrhizal fungi. In: Norris J R, Read D J, Varma A K (eds) Methods in microbiology 24: Techniques for the study of mycorrhiza, Academic Press, London, pp 53-65

Tanner C B, Sinclair T R (1983) Efficient water use in crop. In: Taylor H M, Jordan W R, Sinclair T R (eds) Limitations to efficient water use in crop production. Agron Monogr American Society of Agronomy, Madison, Wis, pp 1-27

Thompson J P (1994) What is the potential for management of mycorrhizas in agriculture? In: Robson A D, Abbott L K and Malajczuk N (eds) Management of mycorrhizas in agriculture, horticulture and forestry. Kluwer academic publishers, Netherlands, pp 191-200

Thompson J P (1987) Decline of vesicular-arbuscular mycorrhizae in long fallow disorder of field crops and its expression in phosphorus deficiency of sunflower. Aust J Agr Res 38:847-867

Tinker P B (2000) Introduction. In: Tinker P B (ed) Shades of Green: A review of UK farming system RASE, Stoneleigh, UK

Van der Heijden M G A, Klironomos J N, Ursic M, Moutoglis P, Streitwolf-Engel R, Boller T, Wiemken A, Sanders I R (1998) Mycorrhizal fungal diversity determines plant biodiversity, ecosystem variability and productivity. Nature 396:69-72.

Wilson G W T, Hartnett D C (1997) Effects of mycorrhizae on plant growth and dynamics in experimental tallgrass prairie microcosms. Am J Bot 84:478-482.

Woodward L, Lampkin N (1990) Organic Agriculture in the United Kingdom. In: Crop Protection and Low Input Agriculture: Options for reducing agrochemical usage, BCPC Monograph 45 pp 19-29

Wright S F, Upadhyaya A (1998) A survey of soils for aggregate stability and glomalin, a glycoprotein produced by hyphae of arbuscular mycorrhizal fungi. Plant Soil 198:97-107

Mycorrhizal Technology in Agriculture
ed. by S. Gianinazzi, H. Schüepp, J.M. Barea and K. Haselwandter

Arbuscular mycorrhizal fungi and micropropagation of high value crops

M. Vestberg[1], A.C. Cassells[2], A. Schubert[3], C. Cordier[4] and S. Gianinazzi[4]

[1] *Agricultural Research Centre of Finland, MTT, Laukaa Research and Elite Plant Station, Antinniementie 1, FIN-41330 Vihtavuori, Finland*
[2] *Department of Plant Science, National University of Ireland, Cork, Ireland*
[3] *Università di Torino, Dipartimento Colture Arboree, Via Leonardo de Vinci 44, I-10095 Grugliasco, Italia*
[4] *Laboratoire de Phytoparasitologie INRA-CNRS, INRA, BV 1540, F-21034 Dijon Cedex, France*

Keywords: Biotization, optimatization of micropropagation, multimicrobial inoculum, mycorrhization, weaning stress

Introduction

Micropropagation has established its position as a way of propagating large numbers of uniform plants. For some plant species that are difficult to propagate by seeds or by conventional cuttings, this technique provides the only possible way of producing high quality plants. Micropropagation is widely used for propagation of high value crops like ornamentals, fruits, vegetables, plantation crops and spices (Vestberg and Estaún 1994). The micropropagation industry was growing fast in Europe up to 1992 (O'Riordain 1992) but after that the micropropagation industry seems to have stabilized its position although a slight increase in production of microplants was still recorded for the period 1993–1996 (O'Riordain 2000). *Prunus* is the most important plant in both commercial and official laboratories (O'Riordain 2000).

Microplants or microcuttings are traditionally produced *in vitro* in a 'biological vacuum' and have to be acclimatized in high moisture conditions before they can be transplanted to the field. Acclimatization, i.e. weaning is therefore the most critical phase in production of plants using micropropagation. Microplants often suffer from weaning stress that can decrease survival considerably.

The arbuscular mycorrhizal fungi (AMF) are a natural microbial component of most soils. They colonize the roots of plants and form a mutual symbiosis with the majority of plant species in the plant kingdom including almost all the plants currently micropropagated. Mycorrhizae have been shown to positively affect plant growth by improving nutrient uptake, especially phosphorus uptake, to increase the rhizosphere volume and to alleviate biotic and abiotic stress. Thus, the AMF have three main ways of function as biofertilizers, as biocontrol agents and as bioregulators (Lovato et al. 1996).

Micropropagation eliminates not only pathogens and pests from the plants but unfortunately also beneficial micro-organisms like the AMF. The peat-based substrates commonly used for weaning of microplants are also usually devoid of AMF propagules. Micropropagated plants are therefore often lacking a functioning mycorrhiza by the time of transplanting into the field or into pots in the greenhouse. On the other hand, because of their lack of naturally occurring mycorrhizae, micropropagated plants can benefit highly positively from inoculation with AMF. This has been demonstrated in many plant species including for example avocado (Azcón-Aguilar et al. 1992), banana (Jaizme-Vega et al. 1998), cassava (Azcón-Aguilar et al. 1997), *Gerbera* (Wang et al. 1993), grapevine (Schubert et al. 1990), *Hortensia* (Varma and Schüepp 1994), *Prunus* (Estaún et al. 1999) and strawberry (Vestberg et al. 2000).

A range of factors affect the success of AMF inoculation in micropropagated high value plants. Such factors are for example timing and method of inoculation, efficiency of AMF strain, mycorrhizal receptiveness of plant species and cultivar, physical and chemical composition of growth substrate and environmental factors like light and moisture. In order to achieve an optimal benefit from AMF inoculation to micropropagated plants, all these factors must be considered. Several review papers deal with many of these aspects (Budi et al. 1998; Lovato et al. 1996; Vestberg and Estaún 1994), so they will not be reviewed in detail here. Instead, we will discuss strategies for decreasing weaning stress on micropropagated plants. Such strategies include also optimisation of micropropagation itself. Possibilities for combined use of AMF with other beneficial micro-organisms will be also discussed.

Strategies to improve the physiological quality of microplants, including mycorrhization *in vitro* and *in vivo*

It is generally recognised that poor microplant physiology underlies many of the quality problems affecting the micropropagation industry (Grunewaldt-Stoecker 1997). Poor physiological quality is expressed in the need for specialised microplant weaning facilities (George 1996) and results in a high incidence of seedling-type ('damping-off') diseases, plant growth checks and weak post establishment performance with expression of prolonged juvenility (Preece and Sutter 1991). Disease control at weaning is complicated as the plants may show high sensitivity to the normal pesticides used in horticulture to control pests and diseases (Werbrouck et al. 1999). At the most extreme, the poor quality of microplants has been associated with nursery failures and has been the subject of official enquiry (Grunewaldt-Stoecker 1997).

Poor microplant quality is linked with a number of *in vitro* factors and is family, genus, species and cultivar specific (Preece and Sutter 1991; George 1993, 1996). *In vitro* morphological aberration is expressed as hyperhydration of the tissues, apical necrosis, basal callusing of the explant and basal branching. This syndrome is termed 'hyperhydricitiy' or 'vitrification' (Ziv 1991).

There may also be loss of control of stomatal function and low photosynthetic competence which is reflected in microplant failure at weaning due to dessication (Cassells and Walsh 1994) and/or lack of photoautotrophic competence (Davies and Santamaria 2000). Less is known about the quality of *in vitro* roots, especially those produced in agar (Davies and Santamaria 2000). Frequently when microplants establish successfully, they undergo a growth check, this may be associated with the loss of the '*in vitro*' leaves and/or roots and their replacement by new *in vivo* former tissues (Preece and Sutter 1991). It should also be emphasised that hyperhydrated microplant cells may have thin walls and thus be especially sensitive to faculatative pathogens in establishment substrates. As mentioned in the introduction, if aseptic, they represent a biological vacuum. If introduced into a non-sterilised growth substrate they may be very susceptible to facultative pathogens such as *Pythium* spp. and *Rhizoctonia* (Williamson et al. 1997). If introduced into a sterilised substrate, particularly under high humidity condition and with a growth check, they may be equally vulnerable to pathogen attack following colonization of the substrate by environmental organisms.

Various factors have been implicated in the physiological aberrations in microplants. These include media factors e.g. high cytokinin concentrations, high nitrogen and poor vessel design e.g. the lack of gaseous exchange resulting in ethylene build-up in the vessel, high vessel humidity and lack of development of a transpiration stream in the microplants *in vitro* which results in stomatal malfunction *ex vitrum* (Ziv 1991). It has been argued that the underlying cause of hyperhydricity and epigenetic changes in *in vitro* culture is oxidative stress (Cassells and Curry 2000).

Tissue culture strategies to improve the quality of microplants have focussed mainly on optimising the culture medium and improvements in container design (Cassells 2000). In the former, the benefits of supplementation of the medium with high sucrose (3%) prior to establishment has been shown for some genotypes to increase the survival potential (Van Huylenbroeck et al. 1998). The principle is to treat the microplants like seedlings and to increase their energy reserves by sucrose priming to support the development of new and, or, replacement tissues at establishment. In the latter case, the strategy involves either the use of gas permeable vessels, lids or bottom cooling to induce a transpiration in the microshoots *in vitro* and replacement of agar by polyurethane foam or alternative tissue support (Cassells 2000) or the avoidance of any requirement for a tissue support, by the use of temporary immersion (Teisson and Alvard 1995).

Little research has been done on the *in vitro* inoculation of microplants with beneficial micro-organisms (Cassells et al. 1996; Duffy et al. 1999). In the case of inoculation with AMF, the problems are two fold; firstly, optimised micropropagation media generally contain high levels of phosphate which may inhibit mycorrhization; and secondly, there is a requirement for aseptic inoculum to avoid bacterial over-growth of the sugar-containing culture media. In contrast to the conventional aseptic heterotropic, or more correctly mixotrophic, culture system used universally in micrporopagation, recent years have seen an expansion of interest in autotrophic culture systems ('aseptic microhydroponics') (Cassells 2000). Autotrophic culture has the advantage that it is sugar-free and is carried out in simple mineral nutrient solutions, including low phosphate formulations. Aseptic microhydroponic culture systems offer the potential to establish a microbially balanced rhizosphere before return of microplants to the environment. Strategies to improve microplant weaning based on physiological improvement of microplant quality by *in vitro* procedures and incorporating mycorrhization *in vitro* and *in vivo* are summarised in Figure 1.

Microplants are generally stressed during the weaning stage. Some plants, however, suffer only for a short time then they quickly form new roots, new leaves and start photoautotrophic growth. In such microplants, survival rate after weaning can be close to 100%. Strawberry is an example of an 'easy' plant to acclimatize. Some plants however suffer more from weaning. These plants often have a slow development of new roots and photoheterotrophic (mixotrophic) growth for some time from the start of weaning. Survival rate after weaning can be decreased even to 50% in such plants. Inoculation with AMF has increased low weaning survival in many studies (Naqvi and Mukerji 1998; Subhan et al. 1998; Vidal et al. 1992).

In some cases, AMF have not increased survival rate but even decreased it (Azcón-Aguilar et al. 1994; Uosukainen and Vestberg 1994). Working with

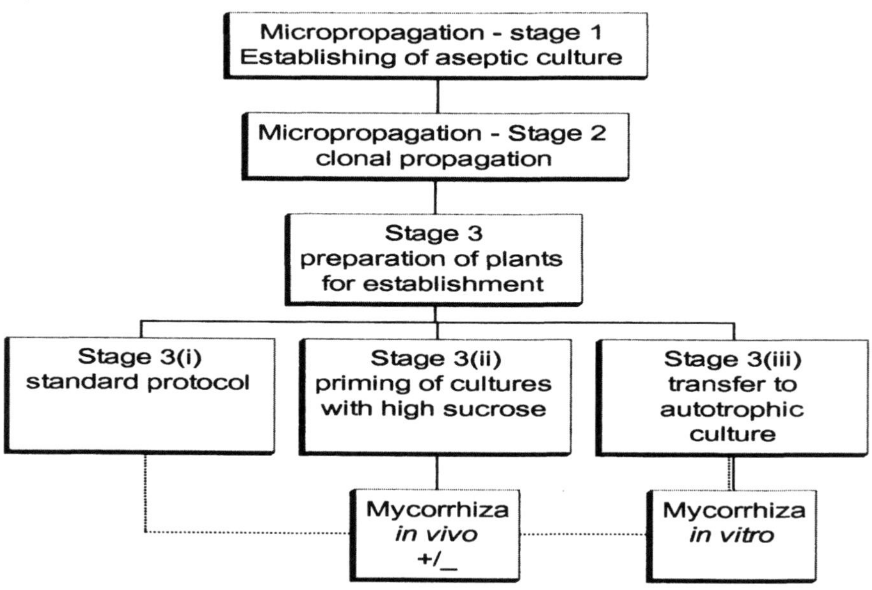

Figure 1. Flow diagramme showing alternative strategies for mycorrhization of microplants *in vitro* and *in vivo*.

Malus that has low microplant survival rates, Uosukainen and Vestberg (1994) found that AMF inoculation further decreased the plant survival. Therefore they started to develop the micropropagation procedure itself in order to get basically stronger microcuttings that would better stand the carbon drain caused by inoculation with AMF.

Sucrose tanking (sucrose priming) and use of rooted *vs* unrooted micro-cuttings were found to considerably improve weaning survival (Fig. 2). However, AMF inoculation still decreased weaning survival in rooted microcuttings. This problem was solved by inoculating the microplants not at start of weaning but at about two weeks later (results not shown here). By optimisation of both micropropagation and mycorrhization techniques it was thus possible to produce high quality microplants of *Malus*.

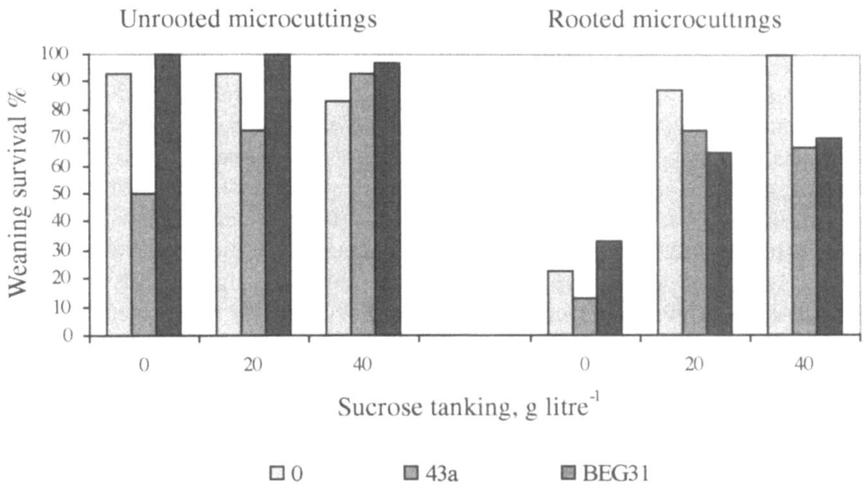

Figure 2. Effect of sucrose tanking on weaning survival of unrooted and rooted microcuttings of *Malus* inoculated with *Glomus lamellosum* V43a, *G. claroideum* BEG31 or uninoculated (unpublished data).

Co-inoculation of AMF with other beneficial soil micro-organisms

Following field transplantation, the rhizosphere of micropropagated plants is usually colonized with a variety of micro-organisms not always beneficial for plant growth. Techniques used for plant micropropagation, either by *in vitro* or green cuttings, and subsequently for plant growth under greenhouse conditions have led to the development of roots weakly colonized by beneficial micro-organisms. There is considerable interest in studying, not only the impact of AMF on plant growth but also that obtained by coupling AMF with other beneficial rhizospheric micro-organisms in binary or multiple combinations

(Cordier et al. 2000a). The objective is to develop microbial mixtures able to fully colonize the rhizosphere at the first stage of microplant development and to improve plant growth and health better than that already obtained with single microbial inoculation. Different strategies for mixing AMF with other beneficial soil micro-organisms such as fungi (*Trichoderma* spp., *Gliocladium* spp.) or bacteria (*Pseudomonas fluorescens*, *Bacillus* spp.) are actually tested from binary to quaternary microbial combinations in several European laboratories. The different micro-organisms are introduced during the acclimatization phase, either at the same time as AMF inoculum, or a few days after mycorrhizal inoculation.

Work on green cuttings of *Hibiscus* showed that inoculation of *Glomus* spp. did not affect the growth of three different varieties in terms of height, branching and fresh biomass (Cordier et al. 2000b). In contrast, binary microbial inoculation containing the AMF together with other beneficial micro-organisms such as *Trichoderma harzianum* DB11, *Gliocladium catenulatum* (commercial inoculum Gliomix) and *Pseudomonas fluorescens* C7r12 significantly increases the development of *Hibiscus* plants. For example, a positive synergistic effect with *Glomus* sp. and *T. harzianum* was observed in *Hibiscus* fresh biomass, either with shoot weight (*Hibiscus* var. Jaune Tahiti), branching (*Hibiscus* var. Tivoli) or root weight (*Hibiscus* var. Sunshine). Moreover, the three *Hibiscus* varieties responded differently to co-inoculation with mycorrhizal fungus and *G. catenulatum*; the percentage of branching (*Hibiscus* var. Tivoli) and plant height (*Hibiscus* var. Jaune Tahiti) was significantly increased whilst these effects were not observed with the variety Sunshine. Only the variety Sunshine responded positively to inoculation with a binary combination of AMF and *Pseudomonas fluorescens*, with an improvement in fresh shoot biomass. Moreover, it should be indicated that when *T. harzianum*, *G. catenulatum* or *P. fluorescens* were introduced alone, only *G. catenulatum* significantly increased the height and fresh shoot biomass of the variety Jaune Tahiti. These results, obtained with *Hibiscus*, show that the binary microbial mixture containing an AMF and another beneficial micro-organism more often results in better plant growth than with the single inoculation of each micro-organism. Furthermore, such positive effects also depend on the plant variety, indicating that an appropriate multimicrobial biotization of microplants requires screening of the most appropriate micro-organisms.

The interest in using more than one micro-organism was also evaluated on microplants with the aim of improving plant survival rate and rooting during the acclimatization phase as well as plant health. Experiments with micropropagated peach rootstock GF677 showed that a better development could be achieved with microbial mixtures than with single micro-organism inoculation (Girard 1999).

Inoculation of *in vitro* raspberry var. Meeker with AMF, *P. fluorescens* and *A. radiobacter* also results in better growth and in protection against *Phytophthora fragariae* var. *rubi* as 40% of the biotized plants survived eleven weeks after being challenged with the pathogen compared to only 20% of the control plants. Moreover, the mix including four micro-organisms (AMF, *P. fluorescens,* non-pathogenic *F. oxysporum* strain and *A. radiobacter)* protects raspberry even better against the same pathogen (80% of plant survival). In spite of these promising results, when raspberry microplants were primed with AMF or *A. radiobacter* alone a full bioprotection was obtained, indicating that multimicrobial inoculation is not always the best solution (Lemoine et al. 2000). Interestingly, quantification of the development of the four micro-organisms into the growing substrate of raspberry microplants three weeks after acclimatization showed that these two fungi and two bacteria, introduced at the beginning of the weaning stage, did not interfere negatively with each other in their reciprocal development: satisfactory densities of *P. fluorescens* (10^4–10^5 cfu/g of dry soil), *A. radiobacter* (10^4–10^5 cfu/g of dry soil) and non-pathogenic *F. oxysporum* strain (10^5–10^6 cfu/g of dry soil) were found and mycorrhiza developed between 20% to 30% infection intensity. The ability of these micro-organisms to develop together without negative interactions opens the possibility of creating a given diversified microbial-rich environment around roots, beneficial for microplant development and therefore increasing the level of tolerance of microplants for a large spectrum of biotic and abiotic stresses, linked to their transplantation into the field.

In conclusion, multimicrobial biotization could be a promising biotechnology to improve the quality of microplants and to ensure an appropriate development within the context of sustainable agriculture. However, better knowledge of the compatibility between the different micro-organisms interacting in the rhizosphere is required for a successful application of this biotechnology.

Future perspectives

The application of AMF inoculum to micropropagated plants can by now be considered an established technique, supported by a considerable mass of experimental data and by the commercial production of AM inoculum by several companies in Europe and abroad. Nevertheless this technique can be expanded and improved in several ways, some of which at present are under investigation.

In the first place, the number of plants which are at the moment micropropagated is very high: a report by the European COST Action 822 (O'Riordain 1999) lists 1966 genotypes (species and cultivars) under experimental tissue culture in Europe, and the aim of most of these experiments is to set up rapid micropropagation protocols. Arbuscular mycorrhizal inoculation has been tested on many commercially important micropropagated crops, but in a survey of available literature since 1993 we could count not more than 54 plant genotypes involved in published experiments. It is clear that many other species remain to be subjected to experimental evaluation of the effects of mycorrhizal inoculation. Among these are horticultural plants, pot-grown ornamentals for outdoor and indoor use and medicinal plants. A particularly interesting group of plants are those from endangered habitats, which are particularly sensitive to pedoclimatic conditions and which absolutely need that the natural root microflora be reconstituted before definitive transplanting (Cassells et al. 1999).

The fungal endophytes used for inoculation should also be subjected to further exploration. Again only a small fraction of the known *Glomales* is employed for inoculation of micropropagated plants. This could be a limitation to the efficacy of inoculation in the light of the fact that several nurseries more and more outsource micropropagation and weaning, in particular of ornamentals, to facilities located in developing countries, in order to cut production costs. If these plants are shipped after weaning, the conditions of the acclimatization phase could be quite different from those of further growth, and inocula should be optimized for being efficient in both conditions.

Although the large body of experimental literature available shows that AM inoculation is in general beneficial to micropropagated plants, it cannot in all cases solve all the problems which can arise in the rhizosphere of such artificially obtained plants. As discussed above and elsewhere (Cordier et al. 2000a), coinoculation of AM fungi with micro-organisms active in biological control of pathogens is a technique which can improve fitness of the roots after transplanting in unsterile soil. However, another component of the inoculum which can be improved is the substrate. Commonly little attention is paid to the inoculation substrate, which is usually similar to the pot substrate used for micropropagation or is chosen simply on the purpose of facilitating inoculum production. It is today possible to design substrates of different chemical composition, of different ion exchange capacity and of different ability to release adsorbed nutrients and finally of different textural characteristics. Examples of such substrates are artificially expanded clays and naturally expanded clays like zeolites. An inoculum containing a 'cocktail' of micro-organisms in a substrate which can release the amount of nutrients required by plant and fungus and which can improve soil texture and water retention in the pot is nearly a 'complete treatment' to prevent the problems which can arise in the roots during acclimatization and nursery growth of micropropagated plants.

The ultimate frontier in AMF inoculation of micropropagated plants is the development of sterile inocula. This aim has been pursued in the last 30 years, but unfortunately no commercially viable techniques have been published or patented up to now. A technique of sterile culture of mycorrhizal clover seedlings was first published in 1962 (Mosse 1962). A considerable breakthrough was the development of self-sustaining mycorrhizal root cultures transformed with *Agrobacterium rhizogenes* (Mugnier and Mosse 1987; Bécard and Fortin 1988), but even these systems show slow root and fungal development and are difficult to be used as a source of sterile inoculum. Clearly the problem of understanding the molecular processes, which lead to mutual recognition and growth of the AM partners, have to be elucidated to improve this technique. The isolation of a root exudate fraction able to enhance the growth of germinating arbuscular fungi was recently reported (Buee et al.

2000). These compounds could be lacking *in vitro* and their addition to the medium could improve fungal growth in sterile binary cultures.

Another way to circumvent the problem of slow growth of root/fungus binary cultures is to reverse the strategy again to mycorrhizal plant cultures. A new technical tool which can open new possibilities in this field is the availability of sterile growth vessels with controlled air exchange (see above). Careful management of environmental conditions like sugar concentration in the nutrient solution and light intensity could lead to acceptable rates of mycorrhizal root development (Debergh et al. 2000).

Mycorrhizal inoculation of micropropagated plants can be improved on the technical and the commercial point of view. The development of specialized inoculum producers will be needed to develop much more complex inocula than those which have been used in the early experiments. Plant producers in this way will have at their disposition a biological tool to improve and protect growth of their plantlets in diverse artificial and natural ecosystems.

References

Azcón-Aguilar C, Barceló A, Vidal M T, De la Viña G (1992) Further studies on growth and development of micropropagated avocado plants. Agronomie 12:837-840

Azcón-Aguilar C, Cantos M, Troncoso A, Barea J M (1997) Beneficial effect of arbuscular mycorrhizas on acclimatization of micropropagated cassava plantlets. Scient Hort 72:63-71

Azcón-Aguilar C, Encina C L, Azcón R, Barea J M (1994) Effect of arbuscular mycorrhiza on the growth and development of micropropagated *Annona cherimola* plants. Agric Sci Finl 3:281-288

Bécard G, Fortin J A (1988) Early events of vesicular-arbuscular mycorrhiza formation on Ri T-DNA transformed roots. New Phytol 108:211-218

Buee M, Rossignol M, Jauneau A, Ranjeva R, Bécard G (2000) The pre-symbiotic growth of arbuscular mycorrhizal fungi is induced by a branching factor partially purified from plant root exudates. Mol Plant Micr Inter 13:693-698

Budi S W, Cordier C, Trouvelot A, Gianinazzi-Pearson C, Gianinazzi S. (1998) Arbuscular mycorrhiza as a way of promoting sustainable growth of micropropagated plants. Acta Hortic 457:71-77

Cassells A C (2000) Aseptic microhydroponics. Acta Hortic 530:187-194

Cassells A C, Curry R F (2000) Oxidative stress and physiological, epigenetic and genetic variability in plant tissue culture. Plant Cell Tiss Org Cult (in press).

Cassells A C, Walsh C (1994) The influence of the gas permeability of the culture lid on calcium uptake and stomatal function in *Dianthus* microplants. Plant Cell Tiss Org Cult 37:171-178

Cassells A C, Mark G L, Periappuram C (1996) Establishment of arbuscular mycorrhizal fungi in autotrophic strawberry culture in vitro: comparison with inoculation of microplants *in vivo*. Agronomie 16:233-239

Cassells A C, Walsh C, Belin M, Cambornac M, Robin J R, Lubrano C (1999) Establishment of a plantation from micropropagated *Arnica chamissonis*, a pharmaceutical substitute for the endangered *A. montana*. Plant Cell Tiss Org Cult 56:139-144

Cordier C, Lemoine M C, Lemanceau P, Gianinazzi-Pearson V, Gianinazzi S (2000a) The beneficial rhizosphere: a necessary strategy for microplant production. Acta Hortic 530:259-268

Cordier C, Lemoine M C, Jaubertie J P, Gianinazzi S (2000b) Application of beneficial rhizosphere micro-organisms in horticulture. Third International Congress on Symbiosis, Philipps-Universität, G, 13-19 August, p. 57 (Abstr.).

Davies W J, Santamaria J M (2000) Physiological markers for microplant shoot and root quality. Acta Hortic 530:363-376

Debergh P C, Topoonyanont N, Van Huylenbroeck J, Moreira Da Silva H, Oyaert E (2000) Preparation of microplants for *ex-vitro* establishment. Acta Hortic 530:269-273

Duffy E M, Hurley E M, Cassells A C (1999) Weaning performance of potato microplants following bacterization and mycorrhization. Potato Res 42:521-527

Estaún V, Calvet C, Camprubi A, Pinochet J (1999) long-term effects of nursery starter substrate and AM inoculation of micropropagated peach X almond hybrid rootstock GF677. Agronomie 19:483-489

George E F (1993) Plant Propagation by Tissue Culture. Part 1 – The Technology. Exegetics, Basingstoke

George E F (1996) Plant Propagation by Tissue Culture. Part 2 – In Practice. Exegetics, Basingstoke

Girard C (1999) Effet de la biotisation sur la croissance de pêcher-amandier et isolement d'agents pathogènes d'hybrides de pêchers. Mémoire BTS Biotechnologie (Lycée Lumière, Luxeuil les Bains (F), Unité de prédéveloppement *in vitro*, INRA/Agri Obtentions, Dijon, France p 25

Grunewaldt-Stoecker G (1997) Problems with plant health of *in vitro* propagated *Anthurium* spp. and *Phalaenopsis* hybrids. In: Cassells A C (ed) Pathogens and microbial contamination management in micropropagation. Kluwer, Dordrecht pp 363-370

Jaizme-Vega, M C, Sosa Hernández B, Hernández J M (1998) Interaction of arbuscular mycorrhizal fungi and the soil pathogen *Fusarium oxysporum* f.sp. *cubense* on the first stages of micropropagated Grande Naine banana. Acta Hortic. 490:285-295

Lemoine M C, Cordier C, Gianinazzi S, Jaubertie J P, Alabouvette C (2000) Lutte biologique chez le framboisier : Résultats prometteurs et intérêt potentiel de lutte contre *Phytophthora fragariae* var. *rubi*. L'arboric Fruit 538 :19-23

Lemoine M C, Jaubertie J P, 1999 Mise au point d'une technique de multiplication végétative herbacée chez l'Hibiscus. Contrat région 1999-2000 p10

Lovato P E, Gianinazzi-Pearson V, Trouvelot A, Gianinazzi S (1996) The state of art of mycorrhizas and micropropagation. Adv Hort Sci 10:46-52

Mosse B (1962) The establishment of vesicular-arbuscular mycorrhiza under aseptic conditions. J Gen Microbiol 27:509-520

Mugnier J, Mosse B (1987) Vesicular-arbuscular mycorrhizal infections in transformed Ri T-DNA roots grown axenically. Phytopathol 77:1045-1050

Naqvi N S, Mukerji K G (1998) Mycorrhization of micropropagated *Leucaena leucocephala* (Lam.) de Wit. Symbiosis 24:103-114

O'Riordain F (1992) The European plant tissue culture industry. Agronomie 12:743-746

O'Riordain F (1999) Directory of European plant tissue culture laboratories 1996-1997. COST 822. Office for official Publications of the European Communities, Luxembourg pp. 1-176

O'Riordain F (2000) The directory of plant tissue culture. Acta Hortic. 530:33-38

Preece J E, Sutter E G (1991) Acclimatization of micropropagated plants to the greenhouse. In: Debergh P C, Zimmerman R H (eds) Micropropagation: technology and application. Kluwer, Dordrecht pp. 71-94

Schubert A, Mazzitelli M, Ariusso O, Eynard I (1990) Effects of vesicular-arbuscular mycorrhizal fungi on micropropagated grapevines: influence of endophyte strain, P fertilization and growth medium. Vitis 29:5-13

Subhan S, Sharmila P, Pardha Saradhi P (1998) *Glomus fasciculatum* alleviates transplantation shock of micropropagated *Sesbania sesban*. Pl Cell Rep 17:268-272

Teisson C, Alvard D (1995) A new concept of plant *in vitro* cultivation liquid medium: temporary immersion. In: Terzi M, Cella R, Falavigna A (eds) Current issues in plant molecular and cellular biology. Kluwer, Dordrecht pp 105-109

Uosukainen M, Vestberg M (1994) Effect of inoculation with arbuscular mycorrhizas on rooting, weaning and subsequent growth of micropropagated *Malus* (L.) moench. Agric Sci Finl 3:269-279

Van Huylenbroeck J M, Piqueras A, Debergh P C (1998) Photosynthesis and carbon metabolism in leaves formed prior and during *ex vitro* acclimatization of micropropagated plants. Plant Sci 134:21-30

Varma A, Schüepp H (1994) Positive influence of arbuscular mycorrhizal fungus on *in vitro* raised hortensia plantlets. Angew Bot 68:108-115

Vestberg M, Kukkonen S, Neuvonen E-L , Uosukainen M (2000) Mycorrhizal inoculation of micropropagated strawberry – Case studies on mineral soil and on a mined peat bog. Acta Hortic 530: 297-304

Vestberg M, Estaún V (1994) Micropropagated plants, an opportunity to positively manage mycorrhizal activities. In: Gianinazzi S, Schüepp H (eds) Impact of arbuscular mycorrhizas on sustainable agriculture and natural ecosystems. Birkhäuser Verlag Basel, Switzerland pp 217-225

Vidal M T, Azcón-Aguilar C, Barea J M (1992) Mycorrhizal inoculation enhances growth and development of micropropagated plants of avocado. HortSci 27:785-787
Wang H, Parent S, Gosselin A, Desjardins Y (1993) Vesicular-arbuscular mycorrhizal peat-based substrates enhance symbiosis establishment and growth of three micropropagated species. J Amer Soc Hort Sci 118:896-901
Werbrouck S, Goethals K, Van Montague M, Debergh P (1999) Surprising micropropagation tools. In: Altman A, Ziv M, Izhar S (eds) Plant biotechnology and *in vitro* biology in the 21st Century. Kluwer, Dordrecht pp 667-672
Williamson B, Cooke D E L, Duncan J M, Leifert C, Breese W A, Shattock R C (1997) Fungal infections of micropropagated plants at weaning: a problem exemplified by downy mildew in *Rubus* and *Rosa*. In: Cassells A C (ed.) Pathogens and microbial contamination management in micropropagation. Kluwer, Dordrecht pp 309-320
Ziv M (1991) Vitrification: morphological and physiological disorders of in vitro plants. In: Debergh P C, Zimmerman R H (eds) Micropropagation: technology and application. Kluwer, Dordrecht, pp 47-70

Mycorrhizal Technology in Agriculture
ed. by S. Gianinazzi, H. Schüepp, J.M. Barea and K. Haselwandter

Ecological considerations for successful application of arbuscular mycorrhizal fungi inoculum

M. Vosatka[1] and J.C. Dodd[2]

[1] *Institute of Botany, Academy of Sciences of the Czech Republic, 252 43 Pruhonice, Czech Republic*
[2] *The International Institute of Biotechnology, 1-13 Innovation Building 1000, Sittingbourne Research Centre, Sittingbourne, Kent ME9 8HL, UK*

Keywords: Ecology of AMF, edaphic stress, indigenous mycorrhiza, inoculum application, tuning of AMF inocula

Introduction

Research on arbuscular mycorrhizal fungi (AMF) in the 1970s and 1980s was dominated by the search for 'superstrains' capable of increasing plant biomass under any environmental and soil conditions. The desire to exploit AMF as natural biofertilizers for the agricultural biotechnology industry was understandable, but it became clear that more knowledge was needed of the fungal biology to allow commercial exploitation. Many inoculant companies have tried to commercialise the use of AMF in the past but with an increasingly environmentally-aware market developing for mycorrhizal products, greater care is needed in producing inocula which can lead to mycorrhization of plants in most circumstances (a minimum requirement from a commercial mycorrhiza product). Many mycorrhizal fungi inoculants, used in research or sold, comprise the same fungal consortia and aim to mycorrhize plants in all target environments. The problem being, that changing environmental conditions can sometimes reveal the limited adaptation of the fungi used. There is also the problem of where inocula are to be used, in monoculture production (agriculture or horticulture) or in land restoration, where complex communities of plants need to be encouraged to establish in degraded soil conditions. Van der Heijden et al. (1998a, b) showed that belowground diversity of AMF is a major factor in the maintenance of plant biodiversity and to ecosystem stability and function. AMF can enter the roots of many plant species in the same community resulting in simultaneous colonisation by several species of AMF and the importance, therefore, of functional compatibility between plant and fungi is relevant to both low diversity and high diversity plant systems. What is the best approach for the selection of AMF for their successful application in research or in the commercial markets? An approach to this has been summarised before (Dodd and Thompson 1994) and its practice implemented in one exceptional traunche of work with tropical crops (Sieverding 1991) but few others have implemented such screenings to produce effective consortia of AMF. There also remains the question of what do we mean by 'effective' given all the possible markets? Simply answered it is that it is a positive effect we are looking for using AMF inoculum and that can be any of a long list of potential benefits:

I. Increased plant nutrient uptake via the AMF e.g. P
II. Increased tolerance of root pathogens by the plant system
III. Increased tolerance of water stress and adverse environmental conditions (e.g. heavy metal pollutants)
IV. Increased efficacy of N-fixation by *Rhizobium* (Plant nodulating bacteria which fix atmospheric N_2 for legumes)
V. Increased plant biodiversity in restored ecosystems
VI. Increased stability of soils (erosion control)

Some of these points have been discussed in more detail in other chapters in this book (von Alten et al., Feldmann and Grotkass in this book). However, AMF are not simply an alternative to chemical fertilizers in intensive agricultural or horticultural production where the varieties of plants used have been bred for high yields from high inputs and we would refer readers to other information sources for discussions of the role of AMF in agriculture (Dodd 2000; Hamel 1996). The choice of employing AMF as a plant management strategy will necessitate the appropriate choice of P input (source or levels applied). Ecological approaches will require a thorough overview of the compatibility of plant and fungus (and soil) and more evidence is emerging that AMF should be selected from environments similar to those of their intended use. An example of the success of this approach has been seen in recent work on the reclaimed platform produced from the Eurotunnel link between the UK and France (Dodd et al. 2001). In these studies the grass *Elymus pycnanthus* was inoculated with AMF isolated from plants on the same site and 2 years later those plants which had been inoculated produced nearly seven times the number of flower spikes compared with non-inoculated or commercially produced non-mycorrhizal plants (Dodd et al. 2001).

There is now an increasing body of evidence (Koide 2000) highlighting an effect of AMF on plant reproduction, growing in natural systems, in P-limiting soils. Natural populations of AMF had no significant effect on P-uptake of field-grown *Vulpia ciliata* (Carey et al. 1982) but did increase seed production. Furthermore, seeds produced by mycorrhizal *Avena fatua* were lighter than those of non-mycorrhizal plants but contained more P (Lu and Koide 1991; Koide and Lu 1992). The seeds from mycorrhizal plants also produced greater offspring vigour than those from non-mycorrhizal plants (Lu and Koide 1991). The effect of early colonisation by AMF may, therefore, be not only to increase seed production but also to improve seed quality in some plant species. The interesting point here is that continued monitoring of perennial plants is necessary to establish mycorrhiza benefits. Monitoring of outplanted inoculated seedlings over one growth season may miss the true importance of the symbiosis to plant populations and communities.

There are many effects which can be expected using inoculation of AMF, however, a crucial fact is that successful root colonization should occur and mycorrhiza formation follows. We will discuss several issues which we believe are relevant to an ecological approach to AMF selection for commercial and research use but it is not intended as an all encompassing review of the subject.

AMF species and isolates

There are more than 150 AMF species described now, some are erroneous but there is agreement that at least 70 or more morphospecies do exist (Morton, Rosendahl, Dodd unpublished). There are differences between geographical isolates of the same species of AMF in the way they function under experimentation, probably caused by adaptation of indigenous isolates to local edaphic conditions. The use of species names only, therefore, in publications is not correct and an isolate code should always be supplied. Recent work has revealed the large variation in the D2 region of the large sub-unit rRNA gene of morphologically similar species of *Glomus* and even between isolates of the same species (Clapp et al. 2001). There was also evidence for the loss of sequences of an isolate of the same AMF when maintained under different growth conditions in individual laboratories (Clapp et al. 2001). This may indicate that there is a gradual selection for sub-populations of nuclei as the AMF adapts to the prevailing biotic conditions.

If spores are used as propagules it suggests that genetic diversity can change rapidly under different culturing conditions and supports the genetic drift between isolates of *Glomus mosseae* BEG12 as noted by Wyss and Bonfante (1993), using RAPD analysis, which has been maintained long-term in different European laboratories. These studies provide evidence of a mechanism by which AMF can quickly gain tolerance to particular environmental stresses, and conversely how quickly an AMF can potentially lose gained-tolerance when maintained under non-stress conditions.

In recent years the potential role of AMF in helping to increase a plant's tolerance of heavy metals and for the restoration of degraded natural ecosystems has been investigated. Work on heavy metal tolerant strains of AMF, isolated from polluted sites, has provided evidence for their rapid adaptation to contaminated soils. Joner and Leyval (1997) found, using a Cadmium (Cd)-tolerant *G. mosseae* isolate, that the extraradical mycelium (ERM) was capable of uptake and transport of Cd. In contrast, some isolates of AMF have shown very high plasticity and an ability to tolerate soil stresses as well as the indigenous ones (Batkhugyin et al. 2000; Gonzalez-Chavez and Dodd unpublished). The adaptation can differ also according to the developmental stage of the fungus. There seems to be little effect of polluted or degraded soils on AMF spore germination as was found in studies of the germination of 6 different isolates when germinated on membranes inserted into 4 polluted soils and sand (Figure. 1).

However, the effect of polluted soils was more pronounced on root colonization. The isolates of *Glomus intraradices* (BEG98) and *G. mosseae* (BEG99) when cultivated on maize in 7 degraded soils and sand, had greater

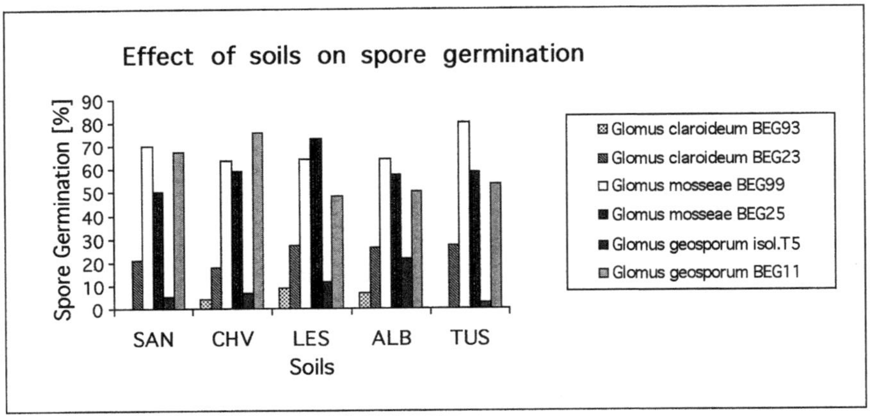

Figure 1. The effect of polluted soil on the percentage germination of spores of AMF isolates from polluted soils (BEG93, BEG99, T5) or isolates from non-polluted soils (BEG23, BEG25, BEG11). SAN -sand, CHV -pyrite waste site, LES -acid rain polluted site, ALB -clay spoil bank, TUS -fly ash deposit). No significant difference found within each isolate germinated in different soils. Columns represent the means of 6 replicates, 20 spores in each treatment.

colonisation in the soil they originated from, than in soils from other degraded sites (Fig. 2).

The diversity in function between AMF has also been seen recently working with tropical legumes where species of AMF from different genera had unique effects on plant growth (Boddington and Dodd 1998; Boddington and Dodd 1999). The explanation for such effects may be clearly related to the

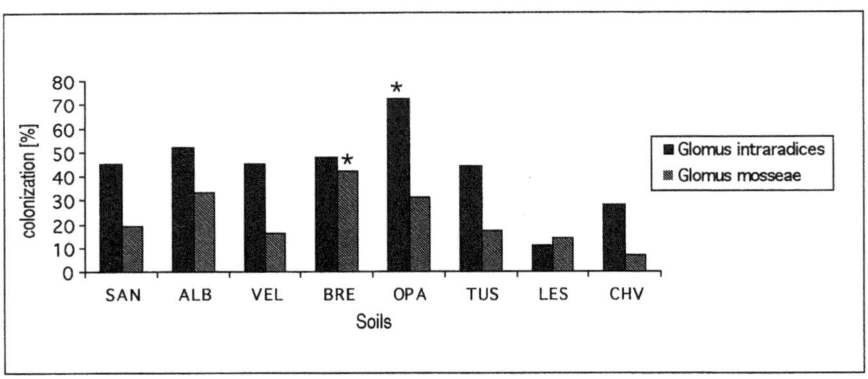

Figure 2. Two isolates of AMF, *G. mosseae* BEG99 from BRE soil and *G. intraradices* BEG98 from OPA soil, cultivated for 12 weeks on maize in sand or in 7 soils from disturbed ecosystems had the highest percentage colonisation of roots in soils from which they were isolated (indicated by star) compared with colonization by a particular isolate in other soils. SAN – sand; ALB, VEL, BRE – spoil banks; OPA, TUS – fly ash deposits, LES – acid rain polluted site; CHV – pyrite waste site (Columns represent the means of 6 replicates).

unique life-cycle strategies (Boddington and Dodd 1999) of these AMF, what is reflected in the differences observed in the phosphate metabolism and architecture of their mycelium *in planta* and in the soil (Dodd et al. 2000).

Edaphic and climatic factors influencing AMF development and mycorrhiza effects

There are many factors influencing the development of AMF and growth response of plants to mycorrhizal fungal inoculation. Even if there is no obvious specificity between plants and AMF there is clear evidence for preferential sporulation (selection) of AMF under certain crops (Dodd et al. 1990a, b). There is also a great difference in the development of different AMF species in the same cultivation system, as shown for many crops in the tropics (Sieverding 1991) as well as for onion inoculated with 10 different AMF (Vosatka 1995) and potato inoculated with 9 different combinations of AMF and bacteria (Vosatka and Gryndler 2000). This response can significantly depend upon the AMF isolate and its origin as shown for alder cultivated under flooded or dry conditions and inoculated with either a non-indigenous isolate or an isolate from a wet or dry habitat (Fig. 3). Trees were inoculated with either a non-indigenous *Glomus mosseae* BEG25 or with two *G. mosseae* isolates from flooded or dry habitat and grown in greenhouse pot experiment for 6 months. Plants had a higher growth response to inoculation with the indigenous AMF isolate originating from soil with conditions similar to simulated treatment compared with the non-indigenous fungus.

Over the years there have been misinterpretations of growth room experimentation due to plants being grown under light-limiting conditions where P

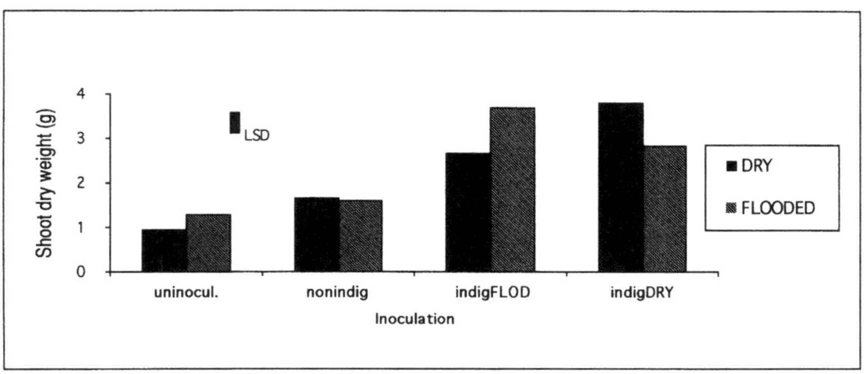

Figure 3. Growth responses to inoculation with two indigenous isolates of *Glomus mosseae* from a flooded or dry site (indigFLOD, indigDRY) or a non-indigenous isolate *G. mosseae* BEG25 on alder growth in two water regimes simulating two types of site where the indigenous AMF were isolated. Columns represent means of 12 replicates, LSD level at $p < 0.05$.

and C supply combine to produce growth depressions (e.g. Olsen et al. 1999) and hence the conclusion that certain isolates of AMF were inefficient. Whereas prolonged experimentation has shown how initial transient growth depressions in greenhouses can lead to subsequent mycorrhizal benefits when planted out (Dodd et al. 2001) as long as the AMF introduced are adapted to their outplanting environment. The stage at which plants become mycorrhizal or rely on mycorrhizal fungi can also differ as shown in an experiment where mature nurse plants of *Calamagrostis epigejos* were cultivated in the centre of a rhizobox system and seedlings were planted on either side of this to become inoculated via the ERM produced from the mother plant. Plants in two developmental stages (mature nurse plants in central compartment, 24 weeks old *versus* young seedlings 12 weeks old infected in the side compartments of rhizobox by ERM radiating from nurse plants) showed significantly different responses to inoculation in sand or spoil bank substrate Brezno (Fig. 4).

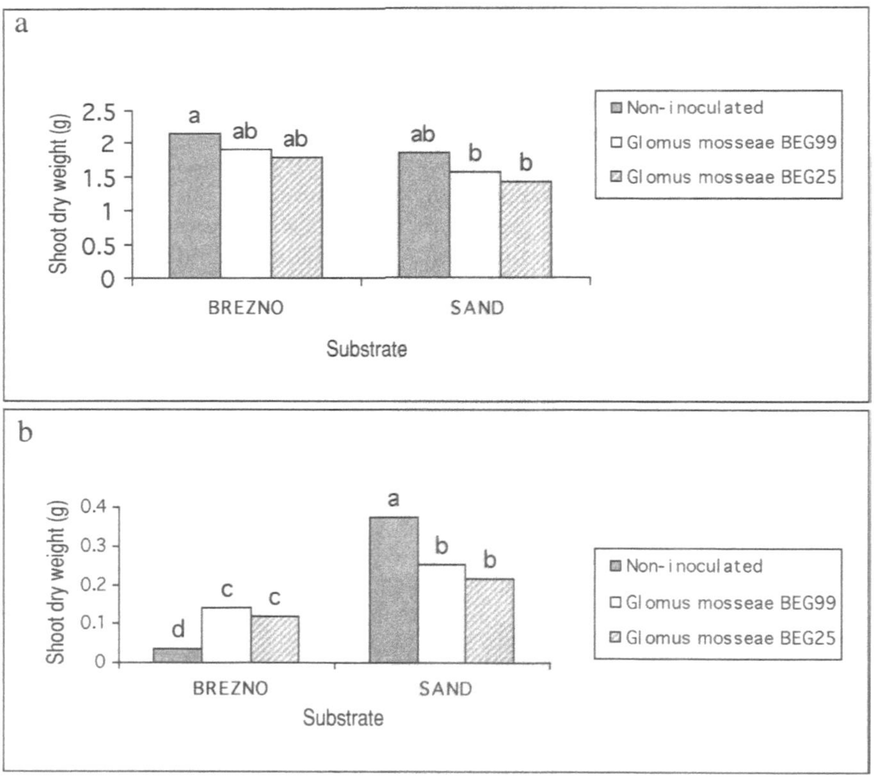

Figure 4. Growth response of mature nurse plants (a) or young seedlings (b) of grass *Calamagrostis epigejos* to AMF inoculation with indigenous (*G. mosseae* BEG99) or non-indigenous (*G. mosseae* BEG25) isolates in soil from spoil bank Brezno or in sand. Columns represent means of 10 replicates, those indicated by the same letter are not significantly different according to Duncan's test, $p < 0.05$ (Malcová et al. 2001 In press).

Maintaining and tuning the inoculum for specific applications

It appears that environmental stress causes selection pressure on exposed AMF populations. Some fungal species apparently are more stress tolerant than others. During the survey of AMF from degraded and polluted ecosystems in Central Europe about 180 trap cultures were established. Soils from mine spoil banks, metallurgical smelter deposits, sedimentation ponds of fly ash, acid rain polluted sites, landfills etc. yielded very few species of AMF. Those frequently identified included isolates of *Glomus claroideum*, *G. etunicatum*, *G. geosporum*, *G. intraradices* and *G. mosseae* (Vosatka and Dodd unpublished data, isolates BEG91-100). These same species have been frequently isolated from early successional plant communities on a spoil bank at Samphire Hoe, UK comprising material recovered from Eurotunnel excavations (Dodd et al. 2001) and on tin mines in UK (Gonzalez-Chavez et al. 2000). This indicates that these consortia of AMF can be used to inoculate plants when restoring such disturbed or polluted sites. The aim of tuning of inoculant mixtures for certain types of environment is the result of data showing that non-tolerant AMF isolates failed to help plant establishment in extremely stressed soils (Vosatka et al. 1999). Inoculation with non-indigenous mixtures of AMF in lignite spoil banks with extremely low pH in Germany failed to produce any significant improvement of plant performance (Vosatka et al. 1999).

When attempting to increase the establishment and diversity of plants on extreme sites a tuning of inocula, to incorporate AMF with proven adaptation to the stresses, would seem to be the only approach which can enable successful reintroduction of mycorrhizas. If an appropriate mixture of AMF is introduced it would be expected that a number of benefits resulting would include better plant establishment, improved plant diversity and maybe fecundity (Koide 2000). The long-term benefit, however, often overlooked will be the development of a widespread mycelial network which will allow easier ingress by more mycorrhizal-responsive plants, which in turn will help stabilise the soil structure (Miller and Jastrow 2000). This approach for inoculum producers obviously leads to the need for large libraries of cultures, as undertaken by PlantWorks Ltd., UK to meet the needs of inoculum applications in environments highly diverse edaphically and climatically. This 'tuning' approach was established to select specific consortia of AMF isolates capable of surviving under stress or otherwise specific conditions of the target ecosystem. This approach is of course difficult as it requires not only numerous isolations from different types of target ecosystems or cultivation systems but also it encompasses the need to check that maintenance in culture under well-watered conditions with favourable temperatures and no perceived stresses keeps the tolerance traits required. This is an ongoing concern of inoculum producers that the stability of tolerance to a given stress, which was screened for, is maintained following subculturing. The inoculum cannot be routinely produced in substrates with target pollutants due to the regulations on the product entering the market. However, it seems to be appropriate that mother cultures of isolates from extremely degraded soils e.g. polluted with heavy metals, should be kept and used regularly as the source inoculum for the bulking-up process. We do not know how quickly certain types of tolerance are gained or lost when an isolate of AMF is subcultured away from its soil and environment of origin and further research aimed at monitoring related gene activity (Lapopin and Franke 2000) could prove invaluable for producers in the future.

Last, but not least, the production processes should be modified for some specific isolates adapted, for example, to certain soil pHs e.g. isolates from within the Gigasporaceae isolated from acid tropical soils etc. or common in sand dune ecosystems worldwide (Koske 1987). There are particular production problems associated with using species of *Acaulospora* and *Scutellospora* which is the result of difficulties in routinely obtaining good cultures, since spores are their primary propagules. This is complicated by the frequent observations that dormancy factors are prevalent in the spores of some of these species and where AMF are not adapted to the environmental conditions. Much more research is needed on these AMF as they appear to be bioindicators for stable climax ecosystems.

Influence of host plant

This could generate a chapter on its own but we would prefer to point readers to other tomes for more detailed discussions on this issue (Dodd 2000; Hamel 1996; Parke and Kaeppler 2000). It suffices to say that the range of plant diversity within species (ecotypes) alone can complicate predictions of the extent to which AMF colonise roots and the resulting effects on the plant growth and development. Crop species and cultivars of plant species can differ dramatically in their ability to respond to AMF (Sieverding 1991). In agriculture and horticulture this is the result of traditional plant breeding practices overlooking the symbiosis and selecting for efficiency of nutrient uptake and use under high input conditions (Hetrick et al. 1993). When combined with screening for disease tolerance, particularly of root pathogens (Toth et al. 1990), then looking to develop reduced input systems appears ridiculous, if based on this same germplasm. It is also worthy of note that plants often reported as being non- or weakly mycorrhizal plants e.g. semi-aquatics, which can often invade wet degraded sites, can become mycorrhizal when soil conditions are dry and in this respect can aid the build-up of the mycelial network in the soil if the drying process continues (Oliveira et al. 2001). These data point to the need to avoid sweeping generalisations and for increased applied research to support the exploitation of AMF in all plant production systems. A good example of functional significance of AMF was shown by Van der Heijden (2000) who found that even a very low percentage of *Glomus mosseae* colonisation (less than 5% root length) of *Salix repens* grown from cuttings in a sand-perlite substrate resulted in huge plant increases in shoot and root growth and P uptake as the low levels of colonisation occurred irrespective of the quantities of AMF inoculum used.

Expected effects of inoculum application

Inoculum users often have great expectations of that they will obtain a large plant growth response as an outcome of inoculation. The best output of inoculation is indeed considerable increase in yields i.e. for an agricultural crop. In field or grower trials, however, it is usually necessary to keep the same technologies commonly used in agriculture or horticulture, as any changes of these practices are difficult to impose on the growers. A significant yield increase may result from inoculation for some highly dependent crops like leek (Fig. 5). Two-months-old leek seedlings transplanted from plastic boxes with a peat-based substrate pre-inoculated with two AMF showed significant yield increases after transplantation to the field and grown under normal agriculture practice for 3 months (Vosatka et al., unpublished). In horticultural systems plants of cyclamen, a species which usually shows a medium response to inoculation, showed greater tolerance to pathogen attack by *Gleosporium cyclaminis* leading to a reduction in plant losses (Vosatka et al., unpublished). There were 40 plants in each treatment and the ones inoculated with a mixture of four AMF showed 16% of mortality while non-inoculated plants had 33% mortality. The plants inoculated with AMF together with *Trichoderma harzianzum*, a biocontrol agent, showed zero mortality. Therefore it seems that

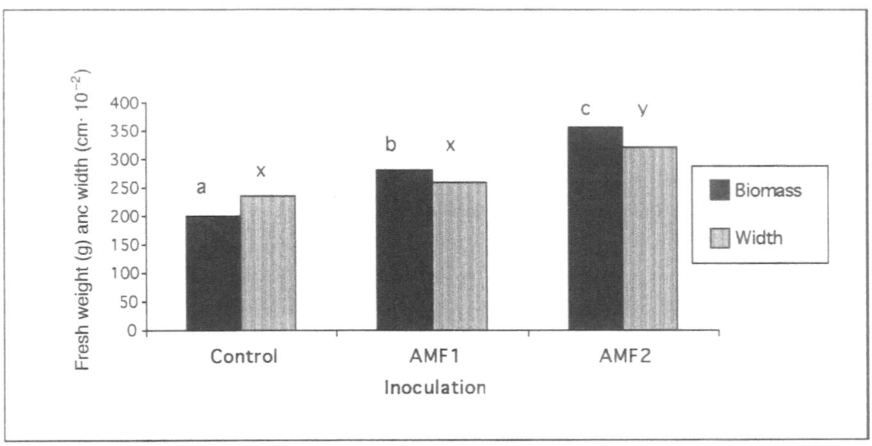

Figure 5. Effect of pre-inoculation of leek plants with two AMF (AMF1: *Glomus etunicatum* BEG92, AMF2: *G. mosseae* BEG95) on fresh weight and width of seedlings transplanted to the field. Columns represent a mean of 500 plants as the replicates were grown in 100 m^2 blocks. Columns indicated by the same letter are not significantly different within each growth parameter according to Duncan's test, $p < 0.05$.

the inoculation acted in this particular experiment rather as a plant health insurance under unfavourable and stress conditions than as a fertiliser (Vosatka et al. unpublished). There are certainly some general types of environment, and even cultivation systems, where plant growth response to inoculation is likely, for example sandy, low fertility soils, particularly in dry climates, sand dune environments, some tropical soils (oxisols) with extremely low availability of phosphates. At the other end of the spectrum are high input intensive cultivation systems, such as greenhouse production of vegetables in rock wool hydroponics systems, where the introduction of mycorrhizal fungi represents a difficult task irrespective of a belief that growth increases might result. It is necessary, however, always to consider all aspects and functions of mycorrhizas apart from nutrition and growth effects. There are certain types of environment where increased aggregation of soil particles and increased erosion control can be far more important than any effect of AMF on the plant biomass. In cold arid or wet deserts, where plants grow extremely slowly and erosion can be a major limiting factor in promoting vegetation cover, the establishment of a functional network of ERM is most important. The ERM is able not only to increase amounts of water stable aggregates (Miller and Jastrow 2000; Perez-Solis and Vosatka, unpublished) but also to interconnect roots or rhizosphere of plants in a plant community and influence the whole vegetation structure (Malcová et al. in press; Malcová et al. 1999; Newman 1990).

Conclusion

There are many other areas of research and development where better targeted screening could yield 'beneficial' AMF. Should we be selecting for AMF to utilize their ability to lock up carbon (Staddon and Fitter 1998)? Species of AMF from different genera have very different carbon contents (Vosatka and Dodd unpublished) and combinations of these fungi could be vital in sequestering carbon in the future elevated levels of carbon dioxide in our atmosphere. This could increase the rates at which C is fixed if new forests or stable plant communities were produced on degraded lands. There are commercial as well as political considerations attached to such an approach. The theory behind an ecological approach to selecting adapted mycorrhizal fungi for plant production systems is one that we believe should be an imperative of all inoculum producers if the market for these products is to be maintained in the future. These fungi play a pivotal role in ecosystems and maybe, in the future, reduced input plant production. An approach for choosing either inoculation or management of the mycorrhizal fungi can be used as a guideline for planners intent on using AMF (Fig. 6) and this will require scientists and growers to understand more about manipulation of the organisms. We are generating many new young scientists interested in AMF and molecular biology whereas what we also need an equal number interested in the biology and ecology of these organisms.

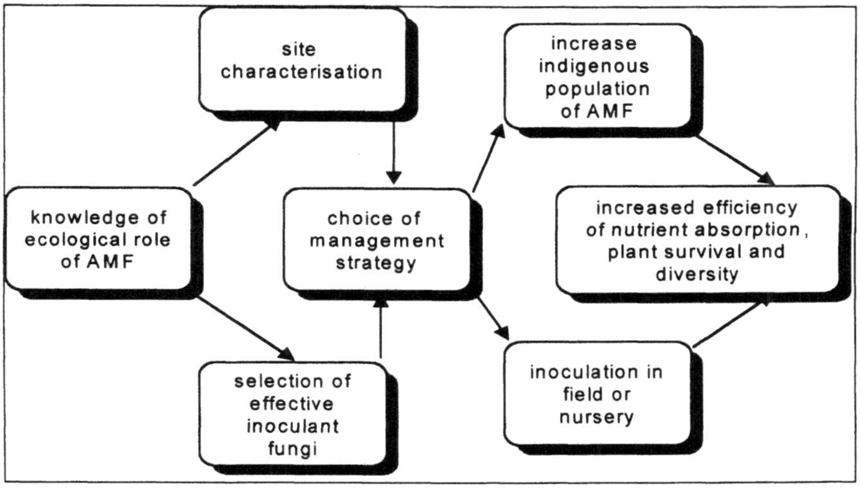

Figure 6. Summary of ecological considerations needed for the implementation of mycorrhizal fungi inoculation in plant production systems.

Acknowledgments
Some data are from experimental studies supported by the Grant Agency of the Czech Republic (grant 526/99/0895) and grant COST 838 of Ministry of Education, Youth and Sports of the Czech Republic and M. Vosatka acknowledges the financial support.

References

Batkhugyin E, Rydlová J, Vosátka M (2000) Effectiveness of indigenous and non-indigenous isolates of arbuscular mycorrhizal fungi in soils from degraded ecosystems and man-made habitats. Applied Soil Ecology 14:201-211

Boddington C L, Dodd J C (1998) A comparison of the development and metabolic activity of mycorrhizas formed by arbuscular mycorrhizal fungi from different genera on two tropical forage legumes. Mycorrhiza 8:149-157

Boddington C L, Dodd J C (1999) Evidence that differences in phosphate metabolism in mycorrhizas formed by species of *Glomus* and *Gigaspora* may be related to their life-cycle strategies. New Phytologist 142:531-538

Carey P D, Fitter A H, Watkinson A R (1982) A field study using the fungicide benomyl to investigate the effect of mycorrhizal fungi on plant fitness. Oecologia 90:550-555

Clapp J P, Rodriguez A, Dodd J C (2001) Inter- and intra-isolate rRNA large sub-unit (LSU) variation in spores of *Glomus coronatum* compared with morphologically similar species of *Glomus*. New Phytologist in press).

Dodd J C (2000) The role of arbuscular mycorrhizal fungi in agro- and natural ecosystems. Outlook in Agriculture 29:55-62

Dodd J C, Arias I, Koomen I, Hayman D S (1990a) The management of vesicular-arbuscular mycorrhizal populations in acid-infertile soils of a savanna ecosystem. I. The effect of pre-cropping and VAMF inoculation on plant growth and nutrition in the field. Plant and Soil 122:229-240

Dodd J C, Arias I, Koomen I, Hayman D S (1990b) The management of vesicular-arbuscular mycorrhizal populations in acid-infertile soils of a savanna ecosystem. II. The effects of inoculation and pre-crops on the native VAMF spore populations. Plant and Soil 122:241-247

Dodd J C, Boddington C L, Rodriguez A, Gonzales-Chavez C, Mansur I (2000) Mycelium of arbuscular mycorrhizal fungi (AMF) from different genera: form, function and detection. Plant and Soil 226:131-151

Dodd J C, Dougall T A G, Clapp J P, Jeffries P (2001) The role of arbuscular mycorrhizal fungi in plant community establishment at Samphire Hoe, Kent, UK –The reclamation platform created during the building of the Channel Tunnel. Biodiversity and Conservation (in Press).

Dodd J C, Thompson B (1994) The screening and selection of inoculant arbuscular-mycorrhizal and ectomycorrhizal fungi. Plant and Soil 59:149-158

Gonzalez-Chavez C, D'Haen J, Vangronsveld J, Dodd J C (1999) Copper adsorption and accumulation by the external mycelium of three arbuscular mycorrhizal fungi from polluted soils. In: Abstracts COST 8.38 Meeting, Nancy, France, 1999 p 4

Hamel C (1996) Prospects and problems pertaining to the management of arbuscular mycorrhizae in agriculture. Agriculture, Ecosystems and Environment 60:197-210

Hetrick B A D, Wilson G W T, Cox T S (1993) Mycorrhizal dependence of modern wheat cultivars and ancestors: a synthesis. Canadian Journal of Botany 70:512-518

Joner E J, Leyval C (1997) Uptake of ^{109}Cd by roots and hyphae of a *Glomus mosseae/Trifolium subterraneum* mycorrhiza from soil amended with high and low concentrations of cadmium. New Phytologist 135:353-360

Koide R T, Lu X (1992) Mycorrhizal infection of wild oats: maternal effects on offspring growth and reproduction. Oecologia 90:218-226

Koide R T (2000) Mycorrhizal symbiosis and plant reproduction. In: Kapulnik Y, Douds D (eds) Arbuscular mycorrhizas: physiology and function. Kluwer Academic Publishers, The Netherlands pp 19-46

Koske R E (1987) Distribution of VA mycorrhizal fungi along a latitudinal temperature gradient. Mycologia 79:55-68

Lapopin L, Franken P (2000) Modification of Plant Gene Expression. In: Kapulnik Y, Douds D (eds) Arbuscular mycorrhizas: physiology and function. Kluwer Academic Publishers, The Netherlands pp 69-84

Lu X, Koide R T (1991) *Avena fatua* L. seed and seedling nutrient dynamics as influenced by mycorrhizal infection of the maternal generation. Plant Cell Environment 14, 931-939.

Malcová R, Vosátka M, Albrechtová J (1999) Influence of arbuscular mycorrhizal fungi and simulated acid rain on the growth and coexistence of the grasses *Calamagrostis villosa* and *Deschampsia flexuosa*. Plant and Soil 207:45-57

Malcová R, Albrechtová J, Vosátka M (2001) The role of extraradical mycelium of arbuscular mycorrhizal fungi on the establishment and growth of *Calamagrostis epigejos* in industrial waste substrates. Applied Soil Ecology (in press)

Miller R M, Jastrow J D (2000) Mycorrhizal fungi influence soil structure. In: Kapulnik Y, Douds D (eds) Arbuscular mycorrhizas: Physiology and function. Kluwer Academic Publishers, The Netherlands pp 3-18

Oliveira R S, Dodd J C, Castro P M L (2001) The mycorrhizal status of *Phragmites australis* in several polluted soils and sediments of an industrialised region of Northern Portugal. Mycorrhiza 10:241-247

Olsen J K, Schaefer J T, Edwards D G, Hunter M N, Galea V J, Muller L M (1999) Effects of mycorrhizae, established from an existing intact hyphal network, on the growth response of capsicum (*Capsicum annuum* L.) and tomato (*Lycopersicon esculentum* Mill.) to five rates of applied phosphorus. Australian Journal of Agriculture Research 50:223-237

Parke J L, Kaeppler S W (2000) Effects of genetic differences among crop species and cultivars upon the arbuscular mycorrhizal symbiosis. In: Kapulnik Y, Douds D (eds) Arbuscular mycorrhizas: physiology and function. Kluwer Academic Publishers, The Netherlands pp 131-146

Sieverding E (1991) Vesicular-Arbuscular Mycorrhizal Management in Tropical Agrosystems, GTZ Publishers, Germany 371 pp

Staddon P L, Fitter A H (1998) Does elevated atmospheric carbon dioxide affect arbuscular mycorrhizas? Tree 13:455-458

Toth R, Toth D, Stark D, Smith D R (1990) Vesicular-arbuscular mycorrhizal colonisation in *Zea mays* affected by breeding for resistance to fungal pathogens. Canadian Journal of Botany 68:1039-1044

Van der Heijden L (2000) Mycorrhizal symbioses of Salix repens: Diversity and functional significance.Thesis University of Wageningen, The Netherlands pp 89-109

Van der Heijden M G A, Boller T, Wiemken A, Sanders I R (1998a) Different arbuscular mycorrhizal fungal species are potential determinants of plant community structure. Ecology 79:2082-2091

Van der Heijden M G A, Klironomos J N, Ursic M, Moutoglis P, Streitwolf-Engel R, Boller T, Wiemken A, Sanders I R (1998b) Mycorrhizal fungal diversity determines plant biodiversity, ecosystem variability and productivity. Nature 396:69-72

Vosatka M (1995) Influence of inoculation with arbuscular mycorrhizal fungi on the growth and mycorrhizal infection of transplanted onion. Agriculture, Ecosystem and Environment 53:151-159

Vosátka M, Gryndler M (2000) Response of micropropagated potatoes transplanted to peat media to post-vitro inoculation with arbuscular mycorrhizal fungi and soil bacteria. Applied Soil Ecology 15:145-152

Vosátka M, Rydlová J, Malcová R (1999) Microbial inoculations of plants for revegetation of disturbed soils in degraded ecosystems. In: Kovár P (ed) Nature and culture in landscape ecology. The Karolinum Press, Prague pp 303-317

Waaland M E, Allen E B (1987) Relationships between VA mycorrhizal fungi and plant cover following surface mining in Wyoming. Journal of Range Management 40:271-276

Weissenhorn I, Leyval C, Berthelin J (1993) Cd-tolerant arbuscular-mycorrhizal (AM) fungi from heavy-metal polluted soil. Plant and Soil 157:247-256

Wyss P, Bonfante P (1993) Amplification of genomic DNA of arbuscular mycorrhizal (AM) fungi by PCR using short arbitrary primers. Mycological Research 97:1351-1357

Mycorrhizal Technology in Agriculture
ed. by S. Gianinazzi, H. Schüepp, J.M. Barea and K. Haselwandter

Selecting arbuscular mycorrhizal fungi for field application

V. Estaún[1], A. Camprubí[1] and E.J. Joner[2]

[1] *Dpt. Protecció Vegetal, Centre de Cabrils IRTA, E-08348 Cabrils (Barcelona) España*
[2] *Centre de Pédologie Biologique – CNRS, 17, rue N.D. des Pauvres, BP 5, F-54501 Vandoeuvre-les-Nancy Cédex, France*

Keywords: Sustainable land use, soil management practices link between plants and soil, horticulture, land-reclamation inoculum production and application indigenous population, monoculture, chemical inputs, soil characteristics, infective propagules

Introduction

Sustainable land use requires that soil management practices be adopted to conserve and augment soil resources. Arbuscular mycorrhizal fungi (AMF) are important in the soil ecosystem as a key link between plants and soil, mediating many interactions. The effects of mycorrhizas in agriculture, horticulture and land-reclamation are almost exclusively beneficial. There is increasing evidence that AMF can enhance crop productivity in the field, and they are also important in the way crops respond to fertilisation regimes. Direct inoculation of AMF is currently limited to small-scale, high value crops, and there is scarce information on large-scale field inoculations, partially due to the fact that these fungi are obligate symbionts which hinders production of inoculum. This picture is, however, rapidly changing with new approaches and methods to improve both, inoculum production and application. In this chapter we discuss the factors which need to be considered for successful field inoculation with AMF. There are mainly two areas where the introduction of selected fungi could be considered: agriculture-horticulture and land reclamation. Though these represent quite different scenarios, the same key factors need to be evaluated before deciding on the appropriateness of inoculation. These key factors can be classified as parameters associated with soil characteristics and as parameters associated with plant characteristics. A selection process must then be implemented to select the most suitable fungi with respect to the prevailing soil conditions and the vegetation objectives set for each situation.

Virtually all soils contain AMF, but inoculum density and fungal species differ. The distribution of many endophyte species is worldwide, but localised distributions patterns can be very variable. A relationship has been established between plant community and soil AMF indigenous population (Wilson and Hartnett 1997; van der Heijden et al. 1998). The AMF community can thus change rapidly when undisturbed soil is brought into cultivation. The change from a natural environment to an agro-ecosystem implies, in many cases, a monoculture along with chemical inputs which will modify two factors known to have a profound impact on AMF populations: plant community and soil characteristics.

The abiotic soil characteristics that can affect native populations of AMF have not been examined systematically, however studies show that soil mineral content and structure can influence the species richness and diversity of fungi. When considering field inoculation the conditions have to be identified where the indigenous AMF are insufficiently abundant or effective.

Population of native AMF

Colonisation of roots by AMF can arise from three sources of inoculum: spores, infected root fragments and hyphae, which are collectively termed propagules. Several methods may be used to obtain indications of the indigenous AMF population: spore enumeration, root infection ratings from native plants and the most probable number (MPN) method. Molecular methods are being developed (Sanders et al. 1995; Redecker et al. 1997; Chelius and Triplett 1999), but are not yet available for routine tests.

Spore enumeration

Spores are a well defined source of inoculum, and, for the time being, the only propagules that can be identified to species level. The procedures used to extract and enumerate spores are based on the wet-sieving method first developed by Gerdeman and Nicolson (1963). This method has the advantage of allowing the study of the species richness, and the relative frequency of each fungal species recovered. These parameters are of importance when assessing field data from an ecological standpoint. AMF spore densities show seasonal variations due to formation of new spores, hibernation and die-off of preformed spores (Clapp et al. 1995) and thus spores are not always the dominating propagules (Giovanetti 1985; Louis and Lim 1987). The quality of the spores for inoculum purposes may vary greatly between fungal species and is further affected by storage due to species- specific (or isolate-specific) differences in dormancy and tolerance to the imposed storage conditions (temperature, humidity, time etc.). Finally, important symbionts may be found that inherently do not sporulate, or in which sporulation is repressed due to the conditions of the soil or the climate (Merryweather and Fitter 1998). Spores of these fungi are thus absent in field-collected samples. In this way, spore populations only provide a partial account of the propagules involved in the symbiosis that are functioning at a given site, and other methods need to be used to obtain a complete image of the mycorrhizal infectivity potential of a given soil.

Root infection ratings

The root colonisation of native plants in a soil targeted for AMF inoculation may provide useful information on propagule density, but such colonisation depends on plant characteristics and soil conditions on the site. Thus infection ratings mainly give an indication as to whether or not AMF are present, unless bait plants are used under standardised conditions, or intracellular AMF biomass is estimated and compared to reference sites.

Most probable number measurements

The most probable number method (MPN) establishes an approximate measure of the inoculum density, and was first adapted from plant pathology to studies of AMF by Porter (1979). This procedure involves soil dilutions and mixing of samples. Though these treatments destroy the hyphal network, it is relatively easy to conduct and quantifies all the infective propagules to give an estimate including the non-sporulating AMF population. To assess the infectivity of undisturbed soil, including the contribution of hyphal networks is more difficult, but can be done using whole soil cores (Gianinazzi–Pearson et al. 1985; Requena et al. 1996). Both systems are bioassays that determine the presence of viable propagules. The relative importance of the different type of propagules cannot be distinguished, and the diversity of the fungi present in the population may only be evaluated from root colonisation patterns in special cases (see e.g. Brundrett and Abbott 1994, 1995).

Effectivity of indigenous AMF and selected isolates

The concept of effectivity can be defined in a simple manner as the ability of a given AMF to increase plant growth and nutrient uptake. This is a host-specific factor that needs to be considered in conjunction with AMF inoculum potential. These measurements should preferably be made under the edaphoclimatic conditions of the site and assayed with the range of plants that are to be used.

To determine the effectivity of different fungi, Sieverding (1991) measured the critical inoculum level needed for a response to inoculation with AMF in cassava. The critical level was defined as the AMF propagule concentration (calculated as MPN) below which application of AMF inoculum increased growth and above which there was no further growth response. He found that this value changed with the soil and the fungal species considered. Sanders et al. (1977), comparing four different endophytes, related fungal effectivity to root colonisation, showing that a slow root colonising fungus, *Glomus microcarpum*, had no effect on plant P-uptake and growth, while the other fungi studied, which were rapid colonisers, mediated an increase in P-uptake and plant growth. Abbott and Robson (1982) argued that differences between fungi in their effectivity at increasing plant growth were directly correlated with their abilities to infect roots and to form an extensive network of extraradical hyphae. This was confirmed by Schweiger and Jakobsen (1999) who showed a strong correlation between extraradical hyphal length of 15 isolates of *G. fistulosum* and symbiotic P-uptake. McGee et al. (1999) determined the critical density of *G. mosseae* propagules needed to reach the colonisation plateau characteristic for the symbiosis between this fungus and cotton. They found that 100 propagules per 95 g of soil placed underneath the cotton seedling gave 85% colonisation after 15 days, while 10 propagules per 95 g of soil achieved the same colonisation only after 42 days. A rapid spread of root colonisation is important for the production of an extraradical mycelium which will allow exploration of a larger volume of soil. Root colonisation development exhibits a curvilinear response to the inoculum density (Smith and Walker 1981; Wilson 1982; McGee et al. 1999) but the maximum level achieved can vary between fungi, and with different environmental conditions. High propagule densities reduce the lag phase in the curve of colonisation *versus* time, and therefore it is essential that the number of propagules in the soil is high to secure a rapid host response.

Soil factors that influence AMF

It is widely accepted that the development of mycorrhizas and their effect on plant growth is greater in soils with low or unbalanced nutrient status, in particular when P is in short supply.

Phosphorus effect

The positive effect of mycorrhizas on P nutrition and the ensuing plant growth increase has been widely studied and reviewed. Phosphate is a poorly available ion, because it is fixed in the form of non-soluble aluminium and iron phosphates at low pH, and in the form of calcium phosphates at high pH. The low amounts of P that are available are also poorly accessible to plants due to the low mobility of phosphate (low diffusion constant) in soil. Thus P uptake by the roots rapidly creates depletion zones around the roots, which is compensated by heavy P fertilisation of many crops. It is well known that P fertilisation reduces root colonisation by AMF, but the magnitude of the effect is strongly influenced by the fungi studied, the host plant and other environmental factors (e.g. light, soil water content). In extremely nutrient-poor soils, P fertilisation may even enhance colonisation by improving seedling establishment and pre-symbiotic growth (Suriyapperuma and Koske 1995). The negative effect of P on AMF colonisation is mainly mediated by a high internal P level of the plant which induces lower root exudation, and a higher root growth rate that appears to give the plant resistance against colonisation (Menge et al.

1978; Bruce et al. 1994). Thus, when evaluating a soil for possible AMF inoculation, data on plant available P (e.g. $NaHCO_3$-extractable P (Olsen et al. 1954)) and the P fixing capacity of the soil are indispensable.

pH effect

Soils with high or low pH are particularly interesting situations for the use of AM technology since P availability is reduced under these conditions. The effect of pH on AMF under field situations has received little attention. Laboratory experiments do, however, show that spores have their highest germination rate at the pH of the soil from which the fungi were isolated (Clark 1997; Giovanetti 2000). The BEG (Banque Européenne de Glomales) is an excellent source of information on the origin of the isolates. The genus *Glomus* is by far the more frequent in the bank (83% of the registered isolates), with 30% of the *Glomus* species isolated from neutral-alkaline soils, and 27% from acid environments. In contrast, for *Scutellospora*, with 8 isolates representing 5% of the registered species, 4 isolates come from soils with pH below 7 and only one, *Scutellospora castanea*, was isolated from an alkaline soil. Similarly for *Acaulospora* (9 isolates), only one *A. morrowiae* was isolated from an alkaline soil. There are no registered *Gigaspora* spp. originating from an alkaline soil, although an isolate of *Gigaspora rosea* has been subcultured in a potting mix at pH 7.5.

From our experience in the Mediterranean area of Spain, *Glomus* species appear to be more infective and might be able to displace other genera during the isolation process. The pH is therefore a factor that needs to be taken into account, especially when the values are low, and when, in revegetation processes, the aim is to maintain the fungal biodiversity.

Nitrogen effect

The role of AMF in N nutrition of plants has been overshadowed by the important role of AMF in phosphorus uptake. It is now established that AMF benefit N uptake by plants not only in the *Leguminosae* through an enhanced N_2 fixation (Barea et al. 1987), but also in non-legumes mediated by an increased uptake of N (Subramanian and Charest 1999). Onguene and Habte (1995) showed a decrease of AMF colonisation with the addition of large amounts of N-fertilisers. Brundett et al. (1999) demonstrated how different nutrient levels affected root colonisation by AMF. Low to medium levels increased colonisation and sporulation by AMF, because the plants grew better and produced more roots susceptible to colonisation; at high nutrient levels root colonisation was reduced. The level and the balance between N and P fertilisation regimes are crucial for the establishment of an effective symbiosis. The use of appropriate AMF would allow less fertiliser inputs, especially a reduction in NO_3

amendments, which, in turn, will reduce the contamination of the groundwater, thus reducing a problem in many agricultural areas.

Effect of other mineral elements

The effect of AMF in the absorption of non-mobile nutrients, other than P, by hyphal uptake and translocation towards the plant, is well documented (e.g. for NH_4, Zn and Cu).

Another consideration are the phenomena linked to metal toxicity in the soil as a result of mining activities, municipal wastes, sewage sludge amendments or acid rain pollution. The evidence gathered so far suggest that AMF might sequester heavy metals in the roots and thus reduce exposure to heavy metals (Joner and Leyval 1997; Joner and Leyval 2001). Differences between fungi regarding root colonisation, extracellular metal sequestration, metal uptake and translocation have been found (Weissenhorn and Leyval 1995; Vosatka and Dodd 1998; Tonin et al. 2001; Joner et al. 2000). When considering inoculation with AMF in a heavy metal contaminated site it is important to consider the intended strategy: phytoextraction or increased plant tolerance to allow growth and production (Leyval et al., Turnau at al. in this book). Then the selection of the most appropriate association of plant and fungal inoculum, and the corresponding managing practices can be optimised.

Effect of other soil characteristics

AMF-mediated improvement of soil structure is an effect exerted directly on the soil (Bethlenfalvay et al. 1999; Andrade et al. 1997; Thomas et al. 1986), and a phenomenon that justifies inoculation with AMF on sites exposed to erosion. The formation of soil aggregates has been related to the production of glomalin by AMF (Wright and Anderson 2000) and the production of this glycoprotein differs between fungal isolates (Wright and Upadhyaya 1999), which reinforces the importance of selecting the appropriate fungi for a given situation.

The climatic region is another important consideration when selecting inocula. Isolates of the same species can evolve different patterns of root colonisation within a range of temperatures (Schenck and Smith 1982; Borges and Chaney 1989), and this may be influenced by the host plant (Estaún et al. 1996).

Parameters associated to plant characteristics

Mycorrhizal formation involves the compatible growth of specific fungi within the roots of a host plant. Interactions between these partners at the molecular level will determine whether or not the symbiosis will be mutualistic and to what extent the mycorrhiza will improve plant growth. Trappe (1987) reviewed a major part of the work done on the mycorrhizal status of the angiosperms. In this work he established that in certain families like *Rosaceae* (which includes *Prunus* and *Malus* fruit trees) only 0.6% of the species examined are non-mycotrophs (i.e. have never been found forming mycorrhiza in the field), whilst in families like the *Cariophyllidae* (includes many ruderals) 19% of the species examined were non-mycotrophs. The majority of the agricultural species examined were either obligate mycotrophs (i.e. have never been found without the symbiosis) or facultative mycotrophs.

Once the mycotrophy of the plant is established the next factor to take into account is its dependency on the symbiosis. For practical purposes we can define mycorrhizal dependency as the relative increase in growth of mycorrhizal plants over non-mycorrhizal plants under defined environmental conditions. Nevertheless, as plant species can differ in response to particular isolates of AMF, mycorrhizal dependency should not be used as an absolute value.

The effect of the soil characteristics on the results of the symbiosis has been discussed previously, but the role of the plant genotype in the symbiosis is also of consequence, although the underlying molecular basis is only just starting to be studied.

Breeding programmes of agricultural plants have produced varieties or cultivars with a range of genetic differences. Camprubí and Calvet (1996) showed the varying dependence of several *Citrus* rootstocks on the AM symbiosis with *G. intraradices*. In certain cases, the absence of symbiosis has been able to be traced to a gene (Gianinazzi-Pearson et al. 1991; Morandi et al. 2000); in others, differences in the AM dependency have been related to changes in the patterns of root/shoot C allocation (Graham et al. 1997). AM dependency of the plants to be grown needs to be evaluated in real field conditions. The use of selected inocula should be studied and evaluated against non-mycorrhizal plants but also against plants grown with the native endophyte colonisation.

Selection and production of fungal inoculants

To select and produce fungal inoculum, the first basic requirement is to have a large collection of characterised AMF. To obtain pure and rich inoculum of a single isolate requires several successive cultures, and therefore good inoculum cannot be improvised. The selection of AMF inocula to be used at a specific field site needs to take into account the factors discussed above. Isolates that are effective in a given situation might not be effective in other circumstances, particularly when stress factors are present such as drought, ion toxicity, high or low soil temperatures, etc. (Vosatka and Dodd 1998).

AMF are obligate biotrophs, and need to be cultured in association with the roots of living plants. Pot culture is the most frequently used system both for commercial and experimental inoculum production. Related techniques such as hydroponic/aeroponic culture, and *in vitro* culture on transformed roots, have been developed to improve the production of AMF inoculum. The use of a solid substrate, however, has advantages over these more sophisticated techniques with respect to handling and storage. The optimisation of substrate and fertilisation regimes in traditional pot culture can shorten the time necessary for the production of inoculum and will allow the differentiation of particular growing conditions that help maintain specific traits from the fungal isolates (tolerance to high soil ion content, adaptation to high or low pH). The choice of the host plant used to reproduce the inoculum is also of importance, as it should share a minimum of potential pests and diseases with the targeted plant. Rigorous quality controls need to be implemented to check the inoculum for contamination by pests or diseases. The problem of potential contamination by unwanted pests and/or diseases is entirely avoided with the reproduction of the fungi in axenic root organ cultures, nevertheless this system is currently not usable as a source for commercial inoculum for field inoculation due to its high cost and low yield of spores. The application of inoculum depends largely on the targeted plant/plants. In an agricultural situation several techniques have been used with good results.

For example, the inoculum can be cast on the field in the planting row directly under the seed or the inoculation can be done at each planting hole. The inoculation can also be done at the nursery stage, with the consequent savings in inoculum. Extreme care should be taken, however, in assessing the pH, the nutrient content and the fertiliser and pesticide applications to ensure that the AMF inoculation achieves good root colonisation. Pre-cropping with an inoculated plant (Camprubí and Calvet 1996) has also proved successful in re-establishing a good level of infective propagules in eroded soils, both in agriculture and in revegetation scenarios.

The inoculum used in revegetation programmes should contain several species and isolates to enhance the chance of successful colonisation with a wide range of plants. The inoculum of each fungus used has to be produced separately, as in container-grown plants differences in infectivity tend to displace some isolates in favour of others.

Conclusions

The ecological consequences of field inoculation with mycorrhizal fungi may not be predicted with certainty based on the knowledge that the scientific community currently possesses. However, the spread of these largely mutualistic biotrophs may have very few undesirable effects. The correct use of AMF will help decrease fertiliser inputs yet maintain the same level of growth and production. In revegetation of severely disturbed sites, field inoculation must aim to introduce a variety of fungal species and strains, adapted to the soil characteristics, with the objective of improving the soil structure and re-establishing a stable plant community that will not only thrive but also evolve in the given conditions of the site.

The scenarios that may profit from AMF inoculation with the inoculation technology that exists today are limited by the cost of inoculum production, and may be categorised according to whether it is the crop or the soil that justifies such an investment. Crops that may merit inoculation include high value nursery plants where the value is proportional to the space and time that plants occupy in green houses etc., and where growth rates and survival during transplantation are critical factors. Thus production of fruit trees, ornamentals and horticultural plants have a large potential for inoculation, as also does grass turf for golf courses where both production and maintenance are costly.

Soils that may be subject to inoculation may range from agricultural land to polluted areas. The former may be appropriate when undisturbed soil is cultivated for the first time, as indigenous fungi in such soils may propagate through entities that do not support tillage (mechanical disturbance will break up intact hyphal networks), whereas sporulating species may be rare. Farms converting from conventional to ecological agricultural practices may also benefit from inoculation to maximise the utilisation of organic nutrient amendments (Joner et al. 1995; Joner et al. 2000), as may plantations on erosion-exposed road slopes due to the aggregate-stabilising effects of AMF (Bethlenfalvay et al. 1997). Barren polluted sites also require AMF for successful establishment of a plant cover, whether the rationale for planting is erosion control or phytoremediation (see chapter by Leyval et al. in this book), and restoration of such sites is likely to receive both increasing attention as well as funding in the near future.

References

Abbott L K, Robson A D (1982) The role of vesicular arbuscular mycorrhizal fungi in agriculture and the selection of fungi for inoculation. Australian Journal of Agricultural Research 33:389-408

Andrade G, Mihara K L, Linderman R G and Bethlenfalvay G J (1997) Bacteria from rhizosphere and hyphosphere soils of different arbuscular-mycorrhizal fungi. Plant and Soil 192:71-79.

Barea J M, Azcón-Aguilar C, Azcón R (1987) VA mycorrhiza improve both symbiotic N_2 fixation and N uptake from soil as assessed with a ^{15}N technique under field conditions. New Phytologist 106:717-725

Bethlenfalvay G J, Andrade G, Azcon-Aguilar C (1997) Plant and soil responses to mycorrhizal fungi and rhizobacteria in nodulated or nitrate-fertilized peas (*Pisum sativum* L). Biology and Fertility of Soils 24:164-168

Bethlenfalvay G J, Cantrell I C, Mihara K L, Schreiner R P (1999) Relationships between soil aggregation and mycorrhizae as influenced by soil biota and nitrogen nutrition. Biology and Fertility of Soils 28:356-363

Borges R G, Chaney W R (1989) Root temperature affects mycorrhizal efficacy in *Fraxinus pensylvanica*. New Phytologist 112:411-417

Bruce A, Smith S E, Tester M (1994) The development of mycorrhizal infection in cucumber:effects of P supply on root growth, formation of entry points and growth of infection units. New Phytologist 127:507-514

Brundrett M C and Abbott L K (1994) Mycorrhizal fungus propagules in the jarrah forest. I. Seasonal study of inoculum levels. New Phytologist 127:539-546.

Brundrett M C and Abbott L K (1995) Mycorrhizal fungus propagules in the jarrah forest. 2. Spatial variability in inoculum levels. New Phytologist 131:461-469.

Brundrett M C, Jasper D A, Ashwath N (1999) Glomalean mycorrhizal fungi from tropical Australia.

II. The effect of nutrient levels and host species on the isolation of fungi. Mycorrhiza 8:315-321

Camprubí A, Calvet C (1996) A field inoculation system for citrus nurseries using pre-cropping with mycorrhizal aromatic plants. Fruits 51:133-137

Clapp J P, Young J P W, Merryweather J, Fitter A H (1995) Diversity of fungal symbionts in arbuscular mycorrhizas from a natural community. New Phytologist 130:259-265

Clark R B (1997) Arbuscular mycorrhizal adaptation, spore germination, root colonization, and host plant growth and mineral acquisition at low pH. Plant and Soil 192:15-22.

Chelius M K, Triplett EW (1999) Rapid detection of arbuscular mycorrhizae in roots and soil of an intensively managed turfgrass system by PCR amplification of small subunit rDNA. Mycorrhiza 9:61-64

Estaun V, Camprubi A, Calvet C (1996) Effects of temperature on root colonisation and development of two isolates of *Glomus mosseae.* In: Azcon-Aguilar C and Barea J M (eds) Mycorrhizas in integrated systems. European Commission, Brussels Luxembourg pp 428-431

Gerdeman J W, Nicolson T H (1963) Spores of mycorrhizal *Endogone* species extracted from soil by wet sieving and decanting. Transactions of the British Mycological Society 46:235-244

Gianinazzi-Pearson V, Gianinazzi S, Trouvelot A (1985) Evaluation of the infectivity and effectiveness of indigenous vesicular-arbuscular fungal populations in some agricultural soils in Burgundy. Canadian Journal of Botany 63:1521-1524

Gianinazzi-Pearson V, Gianinazzi S, Guillemein J P, Trouvelot A, Duc G (1991) Genetic and cellular analysis of resistance to vesicular arbuscular (VA) mycorrhizal fungi in pea mutants. In: Hennecke H, Verma D P S (eds) Advances in molecular genetics of plant-microbe interactions. Kluwer Academic Publishers, Dordrecht, The Netherlands.

Giovanetti M (1985) Seasonal variations of vesicular-arbuscular mycorrhizas and endogonaceous spores in maritime sand dune. Transactions of the British Mycological Society 84:679-684

Giovanetti M (2000) Spore germination and pre-symbiotic mycelial growth. In: Kalpunick Y, Douds D D (eds) Arbuscular mycorrhizas: physiology and function. Kluwer Academic Press.

Graham J H, Duncan L W, Eissenstat D M (1997) Carbohydrate allocation patterns in citrus genotypes as affected by phosphorous nutrition, mycorrhizal colonization and mycorrhizal dependency. New Phytologist 135:335-343

Joner E J, Jakobsen I. (1995) Uptake of ^{32}P from labelled organic matter by mycorrhizal and non-mycorrhizal subterranean clover (*Trifolium subterraneum* L.). Plant and Soil 172:221-227.

Joner E J and Leyval C (1997) Uptake of ^{109}Cd by roots and hyphae of a *Glomus mosseae/Trifolium subterraneum* mycorrhiza from soil amended with high and low concentrations of cadmium. New Phytologist 135:353-360.

Joner E J, Briones R and Leyval C (2000) Metal binding capacity of arbuscular mycorrhizal mycelium. Plant and Soil 226:227-234.

Joner E J and Leyval C (2001) Time-course of heavy metal uptake in maize and clover as affected by different mycorrhizal inoculation regimes. Biology & Fertility of Soils (in press)

Louis I, Lim G (1987) Spore density and root colonisation of vesicular-arbuscular mycorrhizas in tropical soil. Transactions of the British Mycological Society 88:207-212

McGee P A, Torrisi V, Pattinson G S (1999) The relatioship between density of *Glomus mosseae* propagules and the initiation and spread of arbuscular mycorrhizas in cotton roots. Mycorrhiza 9:221-225

Menge J A, Steirle D, Bagyaraj D J, Johnson E L V, Leonard R T (1978) Phosphorus concentration in plant responsible for inhibition of mycorrhizal infection. New Phytologist 80:575-578

Merryweather J, Fitter A (1998) The arbuscular mycorrhizal fungi of *Hyacinthoides non-scripta*: I. Diversity of fungal taxa. New Phytologist 138:117-129

Morandi D, Sagan M, Prado-Vivant E, Duc G (2000) Influence of genes determining supernodulation on root colonisation by the mycorrhizal fungus *Glomus mosseae* in *Pisum sativum* and *Medicago truncatula* mutants. Mycorrhiza 10:37-43

Onguene N A, Habte M (1995) Nitrogen and phosphorus requirements for raising mycorrhizal seedlings of *Leucaena leucocephala* in containers. Mycorrhiza 5:347-356

Porter W M (1979) The 'most probable number' method for enumerating infective propagules of vesicular-arbuscular mycorrhizal fungi in soil. Australian Journal of Soil Research 17:515-519

Redecker D, Thierfelder H, Walker C, Werner D (1997) Restriction analysis of PCR-amplified internal transcribed spacers of ribosomal DNA as a tool for species identification in different genera of the order Glomales. Applied and Environmental Microbiology 63:1756- 1761

Requena N, Jeffries P, Barea J M (1996) Assessment of natural mycorrhizal potential in a desertified semiarid ecosystem. Applied and Environmental Microbiology 62:842-847

Sanders F E Tinker P B Black R L B Palmerley S M (1977) The development of endomycorrhizal root systems. I. Spread of infection and growth promoting effects with four species of vesicular-arbuscular mycorrhizas. New Phytologist 78:257-268

Sanders I R, Alt M, Groppe K, Boller T, Wiemken A (1995) Identification of ribosomal DNA polymorphisms among and within spores of the Glomales: application to studies on the genetic diversity of arbuscular mycorrhizal fungal communities. New Phytologist 130:419-427

Schenck N, Smith G S (1982) Response of six species of vesicular-arbuscular fungi and their effects on soybean at four soil temperatures. New Phytologist 92:193-201

Schweiger P F and Jakobsen I (1999) The role of mycorrhizas in plant P nutrition: Fungal uptake kinetics and genotype variation. In: Gissel-Nielsen G, Jensen A (eds) Plant nutrition - Molecular biology and genetics Kluwer Academic Publishers, Dordrecht pp 277-289

Sieverding, E (1991) Vesicular-arbuscular mycorrhiza management in tropical agrosystems Published by GTZ Rossdorf Germany

Smith S E, Walker N A (1981) A quantitative study of mycorrhizal infection in *Trifolium*: separate determination of the rates of infection and of mycelial growth. New Phytologist 89:225-240

Subramanian K S, Charest C (1999) Acquisition of N by external hyphae of an arbuscular mycorrhizal fungus and its impact on physiological responses in maize under drought stressed and well watered conditions. Mycorrhiza 9:69-75

Suriyapperuma S P and Koske R E (1995) Attraction of germ tubes and germination of spores of the arbuscular mycorrhizal fungus *Gigaspora gigantea* in the presence of roots of maize exposed to different concentrations of phosphorus. Mycologia 87:772-778.

Thomas R S, Dakessian S, Ames R N, Brown M S and Bethlenfalvay G J (1986) Aggregation of a silty clay loam soil by mycorrhizal onion roots. Soil Science Society American Journal 50:1494-1499

Tonin C, Vandenkoornhuyse P, Joner E J, Straczek J, Leyval C (2001) Assessment of arbuscular mycorrhizal fungi diversity in the rhizosphere of *Viola calaminaria* and effect of these fungi on heavy metal uptake by clover. Mycorrhiza 10:161-168

Trappe J M (1987) Phylogenetic and ecological aspects of mycotrophy in the angiosperms from an evolutionary standpoint. In: Safir G R (ed) Ecophysiology of VA mycorrhizal plants. CRC Press Boca Raton Florida USA pp 6-27

Van der Heijden M G A, Boller T W, Wiemken A, Sanders I R (1998) Different arbuscular mycorrhizal fungal species are potential determinants of plant community structure. Ecology 79:2082-2091

Vosatka M, Dodd J C (1998) The role of different arbuscular mycorrhizal fungi in the growth of *Calamagrotis villosa* and *Dechampsia flexuosa* in experiments with simulated acid rain. Plant and Soil 200:251-263

Weissenhorn I, Leyval C. (1995) Root colonization of maize by a Cd-sensitive and a Cd-tolerant *Glomus mosseae* and cadmium uptake in sand culture. Plant and Soil 175:233-238.

Wilson J M, Trinick M J (1982) Factors affecting the numbers of infective propagules of vesicular-arbuscular mycorrhizal fungi by the most probable number method. Australian Journal of Soil Research 21:73-81

Wilson G W T, Hartnett D C (1997) Effects of mycorrhizae on plant growth and dynamics in experimental tallgrass praire microcosms. American Journal of Botany 84:478-482

Wright S F, Upadhyaya A (1999) Quantification of arbuscular mycorrhizal fungi activity by the glomalin concentration on hyphal traps. Mycorrhiza 8:283-285

Wright S F, Anderson R L (2000) Aggregate stability and glomalin in alternative crop rotations for the central Great Plains. Biology and Fertility of Soils 31:249-253

Mycorrhizal Technology in Agriculture
ed. by S. Gianinazzi, H. Schüepp, J.M. Barea and K. Haselwandter

Directed inoculum production – shall we be able to design populations of arbuscular mycorrhizal fungi to achieve predictable symbiotic effectiveness?

F. Feldmann and C. Grotkass

Institut für Pflanzenkultur, Solkau 2, D-29465 Schnega, Germany

Keywords: Inoculum production, population dynamics, symbiotic effectiveness, commercialisation

Introduction

Arbuscular mycorrhizal fungi (AMF) form mutualistic associations with most terrestrial plants. They provide several important benefits including enhanced nutrition, drought tolerance, biocontrol of pathogens and tolerance of heavy metals and other pollutants. Maximising these benefits for agriculture to produce high quality plants more sustainably has in recent years been an important aim. Progress has been made in several areas. Notably, benefits have been achieved by inoculating *in vitro* cultivated plants during the weaning stage (Varma and Schuepp 1994) and seeds, seedlings, cuttings, or mature plants Chang 1994). Moreover, the introduction of AMF to target plants has been carried out successfully under greenhouse conditions (Miller et al. 1986), in nurseries (Nemec 1987) and in the field (Thompson 1994). From an inoculum formulation perspective it is advantageous that AMF do not generally exhibit a high degree of host specificity. Consequently, one single AMF species can often be used to inoculate dicotyledons, monocotyledons and ferns (Feldmann 1998a). Furthermore, the same AMF species can often be used in the humid tropics (Sieverding 1991) and in temperate climates (Baltruschat 1993).

In spite of such a spectrum of different environmental and cultivation conditions there is one unique expectation of an AMF inoculation: the developing symbioses have to work successfully and must provide advantages to the target plant. The consequent benefits can be considered as the 'Symbiotic effectiveness', a multifactorial phenomenon with host and fungal genotype and abiotic and biotic environmental conditions all interacting to express the phenotype of the specific symbiosis. 'Positive effectiveness', in an agricultural or horticultural sense, is considered as a positive response of the host in terms of plant growth, yield or stress tolerance. In practice, there are several opportunities to influence the expression of the symbioses, e.g. changing the time of inoculation with respect to the developmental stage of the host, the inoculum potential or the culture condition. Nevertheless, it is difficult to predict accurately the effectiveness of any symbioses resulting from inoculation. In fact, the effectiveness following inoculation ranges from positive to negative (Varma and Schüepp 1994) in an mutualism-parasitism continuum (Johnson et al. 1997).

To deal with this problem screening processes have been developed for AMF strains (Dodd and Thomson 1994) with the aim of identifying the 'best' (i.e. most effective) strain (Baltruschat 1993) or mixtures (Sieverding 1991). However, the outcome of these efforts have been disappointing. AMF effectiveness remained too unpredictable for the commercial application of AMF in horticultural and agricultural practice, especially in moderate climates. Because of these problems commercial interest in AMF inocula and their use in plant production processes have never been sustained (Feldmann 1998a). At present, there are two fundamental questions to be answered for the understanding of the basis of mycorrhizal effectiveness:

a) The 'mycorrhizal dependency' of a host is genetically fixed (Azcon and Ocampo 1981) and the degree of mycorrhizal dependency is expressed on the level of an individual, as a gradient within the host's ecological niche and relevant environmental conditions (Feldmann 1998a). But are we able to predict mycorrhizal dependency under specific conditions? Predicted success of the symbiosis is still based on practical experience and not on any knowledge of the basic mechanisms for host dependency.
b) AMF inoculum was considered to be genetically homogenous because spores reproduce by mitosis (although the initial inoculum is usually derived from several spores). The assumed genetic homogeneity of AMF inoculum was thus the basis for all screening projects on AMF strains. However, it is now known that AMF isolates are not genetically homogenous, and recent experiments on the variability of mycorrhizal phenotypes demonstrated that the mutualism-parasitism-continuum of mycorrhizal effectiveness is even found within one single strain of an AMF containing only single spore descendants Feldmann et al. 1998; Feldmann 1998b).

In those experiments it remained unclear whether the mutualism-parasitism-continuum is based on the function of different AMF genotypes or shows genetic differences between individual host seedlings, i.e. the reaction norm of the host population to a genetically homogenous AMF strain. It is essential to clarify whether different genotypes occur within an AMF strain and, if so, whether their function results in changes of mycorrhizal effectiveness.

In this report we focus on the second question. We assumed that spores or AMF infection units are able to colonize a host root-system independent of their later effectiveness (Feldmann 1998b) and that more than one infection unit of the AMF population will be successful in infecting the roots. To test the hypothesis that different genotypes exist within an AMF population we worked with single spores as distinct fungal units.

Our definition of AMF 'genotype' is a functional one, the AMF isolate reacting to a given environment in a reproducible, predictable way for a minimum of one spore propagation cycle. That means that the phenotypic characteristic of a symbiosis derived from inoculation with single spores will be reproduced when descendants of these spores are used to inoculate homogenous plant material in a subsequent experiment. A functional description of a genotype does, of course, not describe the actual genetic differences between AMF units at the DNA level, but alternatively is focussed on active functional genes for specific interactions. Nevertheless, the selected approach, should reflect genotypes and provides a functional focus to quality control of mycorrhizal inoculation technology.

We present a procedure to handle potential genetic differences in an inoculum by decreasing variability in effectiveness via the directed, technical use of abiotic and biotic selection factors during the inoculum production process. This procedure, that we have called the 'Directed Inoculum Production Process' (DIPP), increases the predictability of the outcome of inoculation. DIPP might serve as prototype for process optimisations that will ultimately lead to the development of AMF inocula with consistently predictable characteristics.

Material and methods

AMF genotype frequencies

For our studies we chose the test plants *Anagallis arvensis* and *Plantago lanceolata* because they occupy a broad ecological niche and are easy to cultivate. The selected plant species occur on arable land, on open, sandy or rocky habitats or on wasteland. They can even be found in polluted areas and on soils with pH between 4.5 and 8.0. Soils poor or rich in nutrients and variable temperature and light are tolerated. Both are highly colonised by AMF (Weissenhorn and Feldmann 1999). The microsymbiont is represented by single spore descendants of a taxonomically not identified *Glomus sp.* strain GK12 (Feldmann 1998b). The spores were produced on *Petroselinum crispum* in sand and used for inoculation after extraction from the soil by wet sieving and decanting techniques (Schenck 1984). Sand was used as the growth substrate and the plants were kept in 25 ml-plastic tubes under controlled greenhouse conditions according to Feldmann et al. (1998a). Plants were illuminated by SON-T AGRO 400 Philipps lamps (360μE. $m^{-2.}$ s^{-1}, 14 h day) and maintained at 60–80% relative humidity and 18–20 °C night, 22–26 °C daytime.

Plants were irrigated daily to below field capacity and fertilised once per week with 1% pot volume of a fertiliser solution (1 g fertiliser/l solution), pH 5.5. The fertiliser contained 15% N (10% nitrate, 5% ammonium), 7% P_2O_5, 22% K_2O, 6% MgO, 0.03% B, 0.05% Mn, 0.01 Zn. The experiments were carried out between March and July 1996 at the Institute for Applied Botany, University of Hamburg, and the Institute for Microbiology (Technical University, Braunschweig, Germany), and between 1997 and 1999 in the Institute for Plant Cultivation, Solkau, Germany). Plants were inoculated by placing single spores (separated with micropipettes) near the root surface of the host plant when cuttings (*Anagallis arvensis*) or seedlings (*Plantago lanceolata*) had a root system of approximately 6–7 cm length and the upper plant parts were at the same developmental stage (i.e. the variation of shoot length, leaf number and leaf size was not larger than 5%).

In order to describe the effectiveness of single spore descendants 20 plants were inoculated with single spores. Plant fresh weight was measured after eight weeks of culture. The fungal population resulting from this first propagation cycle was called C1 and represents 20 sub-strains of specific effectiveness. Each time ten single spores of C1 were then isolated from three colonised host plants of significantly different fresh weight. These single spores were used to inoculate new host individuals. After a further two months their fresh weights were measured. The single spore descendants of C1 were called C2. They were mixed and 15 single spores each isolated from this mixed population to inoculate new host individuals. The third propagation cycle was carried out within the next two months.

In a second experiment (variation of soil pH and P-content) we used *Glomus sp.* strain GK12, too. Ten cuttings of *Anagallis arvensis* per treatment in three parallel repetitions were grown until they developed a root system of minimally 35 cm length (conditions as above). Before inoculation the soil was drenched with nutrient solutions of increasing pH (pH 4.5, 5.0, 5.5, 6.0, 6.5, 7.0, 7.5, 8.0) until the run-off had the same pH as the applied solution.

For inoculation, approximately 100 spores were transferred into the substrate, close to the roots. Twenty-one days later the plants were carefully extracted from the substrate, the roots washed to remove old spores and plants transferred to a larger pot (50 ml) with fresh substrate. They remained in that pot for another four weeks when sporulation of the fungus took place. The plants were then harvested, the shoot fresh weight determined and the mycorrhizal status of the roots analysed.

The spores produced in the substrates of all different treatments were extracted by wet sieving, mixed and called C1. The spores of propagation cycle C1 were inoculated to new host plants and tested under the conditions cited above. The procedure was repeated three times (C2, C3, C4) with all pH treatments. The colonization of the test plants were analysed 21 days after inoculation. An analogous experiment with different phosphate concentrations (5ppm, 15ppm, 30ppm, 60ppm, 90ppm, and 120 ppm) in the substrate was carried out. at pH 5.5.

Mycorrhizal analysis

Mycorrhizal colonisation of complete root systems was estimated after clearing the roots in 10% KOH for 15 min, neutralising with HCl, washing three times with distilled water and then staining for 25 min in 0.05% trypan blue in lactic acid: glycerol (10:1 vol/vol). The mycorrhizal efficiency index (MEI) was estimated according to Bagyaraj (1994).

$$\text{MEI} = \frac{\text{weight of inoculated plant} - \text{weight of uninoculated plant} \times 100}{\text{weight of inoculated plant}}$$

Statistical evaluation of the data was carried out by the one-way analysis of variance (ANOVA) with a critical significance level of 5%.

Natural and experimental adaptation of AMF inocula to environmental stress

Plantago lanceolata was identified at saline and heavy metal polluted areas and was found to be sensitive to both stresses in preliminary tests. It therefore was chosen as host plant for this experiment.

Mycorrhizal populations were isolated from the North Sea salt marshes near Neufelder Koog, Germany. They contained at least three morphologically different spore types. After isolation the spores were stored in sand at room temperature for nine months. In tests we demonstrated that 0,5% NaCl content in the irrigation water permitted the best mycorrhizal effectiveness on the host plant *Plantago lanceolata* (unpublished data). AMF populations from mine spoils near Oker, Germany, were extracted in the same way. Soil from the same site was suspended in water (1 kg in 2 l) for 24 h, decanted into a sieve of 45 µm mesh width, collected and used as heavy metal stock solution in the irrigation water without previous determination of the heavy metal content. The pure heavy metal solution strongly inhibited the growth of the non-mycorrhizal test plant *Plantago lanceolata.* Effectiveness was best when the stock solution was diluted to 30%.

Immediately before the start of the experiment the most probable number of propagules (Feldmann and Idczak 1994) in the test inoculum was assessed. The effectivness of the natural AMF populations was compared with that of *Glomus sp.* GK 12 before and after the selection process shown in experiment 2.

Stability of effectiveness after selection by stressors

In the fourth experiment we inoculated *P. lanceolata* seedlings with approximately 100 propagules of *Glomus* sp. GK12. This inoculum had been preselected as shown above and then propagated on *P. lanceolata* in three subsequent propagation cycles without stressors. The MEI was then measured again under stress conditions.

Mass production of inoculum

In contrast to the single spore isolates used above, for mass production a commercial inoculum was chosen, which already had been used by several companies and in research projects (e.g. Feldmann et al. 1999). The inoculum contained morphologically different spore types and showed variable banding patterns after PCR analysis of single spores (Hildebrandt and Hutter personal communication). The method of the inoculum production followed the detailed description of Feldmann et al. (1999). Mass production here means the 160,000 fold multiplication of the initially inoculated number of AMF propagules.

Predictability of AMF effectiveness with and without DIPP

To demonstrate the impact of a Directed Inoculum Production Process (DIPP) we designed many experiments with the same AMF strain but different host species, different inoculum quantities, environmental conditions, scales and effects. In practice a threshold value of MEI >30 must be exceeded to create interest by a potential customer in the mycorrhizal technology. The positive outcome of an inoculation was 'predicted' if that MEI value was clearly passed under conditions plant producers use.

Results and discussion

Genotype differences in an AMF inoculum

Inoculation with single AMF spores from a commercial inoculum showed a variability in effectiveness from slightly effective to highly effective (Fig. 1, C1). The multiplication of single spores from sub-populations with distinct effectiveness conserved the characteristics in the next propagation cycle (C2), though, the variability of effectiveness increased after a further propagation cycle (C3). Distinct characteristics of the sub-populations did not exist after C3.

Nevertheless, the reproducible response of the clonal host under standard conditions caused by AMF descendants of single spore isolates verified the existence of genotypic differences in the initial spore population. The slight variability of effectiveness during the first propagation process reflects the still existing variability of the plant material. If the variability of effectiveness observed in C1 was the result of phenotypic plasticity of only one fungal genotype, the same variability would have had to occur in C2 also.

After the second propagation cycle the distinct characteristics of the genotypes were modified, with an increase of variability of effectiveness observed in C3 (Fig. 1). The basic mechanisms for this enhanced variability in effectiveness of genotypes still remains unclear. Host gene/AMF gene adaptations are possible, as are high mutation rates of the fungus.

For practical application, the findings are of special importance. If genetically fixed characteristics of AMF spores are stable for only one or two propagation cycles, AMF inoculum production should not be based on the last inoculum but on spore material from stock cultures. This complicates the scaling-up of inoculum production, because slight differences in effectiveness of

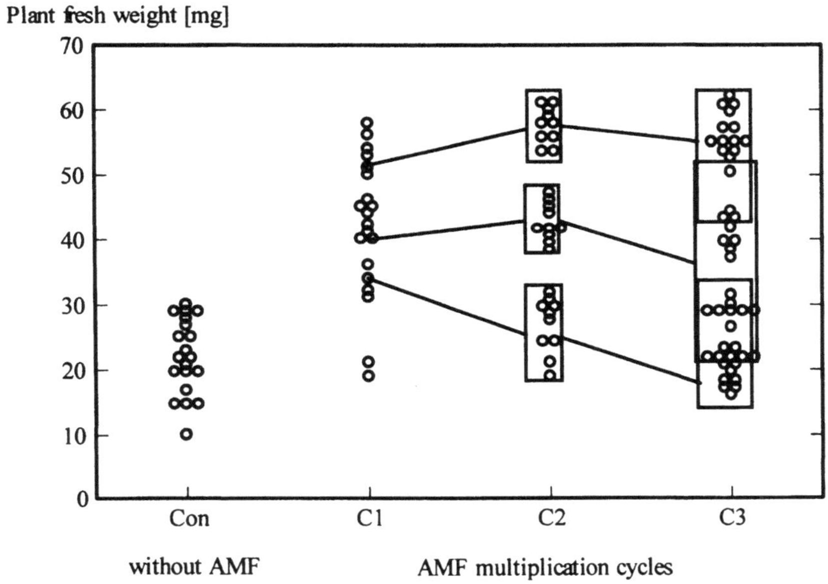

Figure 1. Mycorrhizal effectiveness of AMF single spore descendants (*Glomus sp.* GK 12) on the biomass of *Anagallis arvensis*. Note distinct sub-population characteristics in C2 and overlapping effectiveness in C3.

inocula e.g. C3 (Fig. 1) can create considerable changes in effectiveness of a subsequent inoculum produced (Feldmann et al. 1999).

The relationship between AMF population composition and effectiveness

The initial AMF spore population of *Glomus sp.* GK 12 (Fig. 1) contained spores with different effectiveness (based on biomass production of *Anagallis arvensis* under standard conditions). In the next test we inoculated *Anagallis arvensis* cuttings with approximately 100 spores of that initial inoculum and varied the pH and phosphate concentration of soil solution in independent test systems. Because we did not select spores of a characteristic genotype for the test, the differences in colonisation of the sub-populations under different environmental conditions did not reflect the reaction norm of one genotype; in contrast, the colonisation pattern of the population under changed environmental conditions proves the existence of different genotypes of the initial population.

At extreme soil pH the colonisation of host plants was initially low (Fig. 2a). However, the proportion of spores within the inoculum, able to colonise under extreme conditions was enhanced by separate propagation. Consequently, the effectiveness of the inoculum was enhanced after maintenance under extreme conditions. This provides further evidence for the exis-

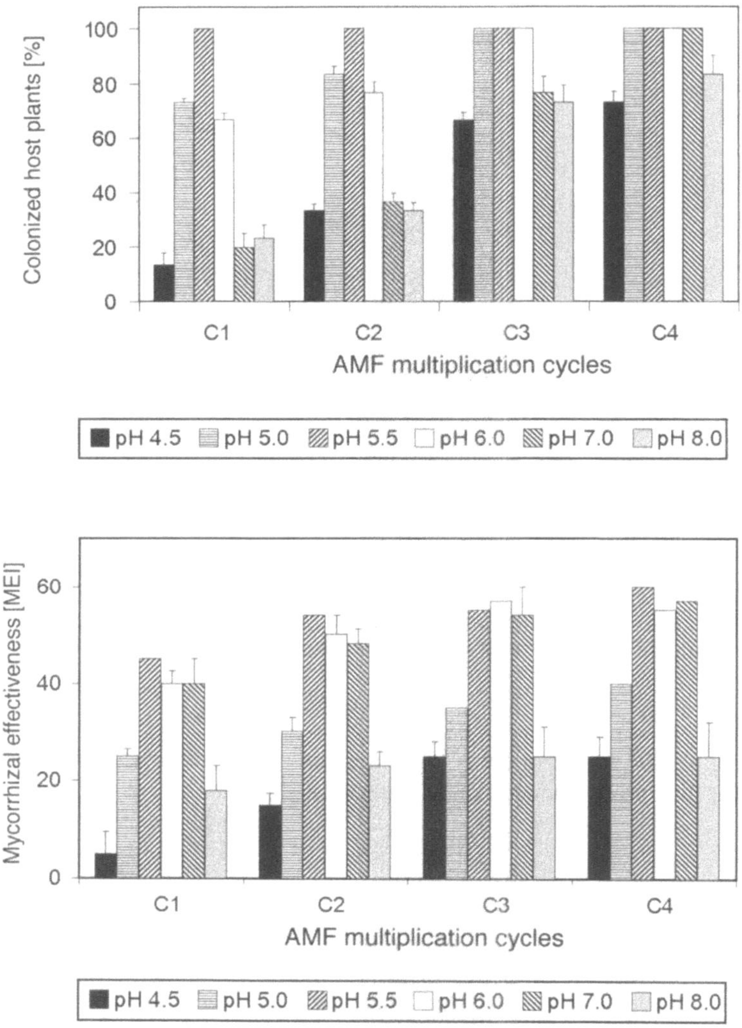

Figure. 2a. Root colonization ability and mycorrhizal effectiveness of AMF populations (*Glomus spec.* GK 12 on *Anagallis arvensis*) with technically modified genotype composition (Selection factor 'soil-pH', details see text). Bars: SD

tence of different genotypes within a strain and is an important finding in the search to develop an effective inoculum production process.

A comparable adaptation process of the strain characteristics could be demonstrated in the case of changes in the phosphate concentration in the soil (Fig. 2b). Under low and medium P-supply (5–60ppm P) the percentage of colonized plants increased. This was not found under optimal supply of phosphorus (90–120ppm P).

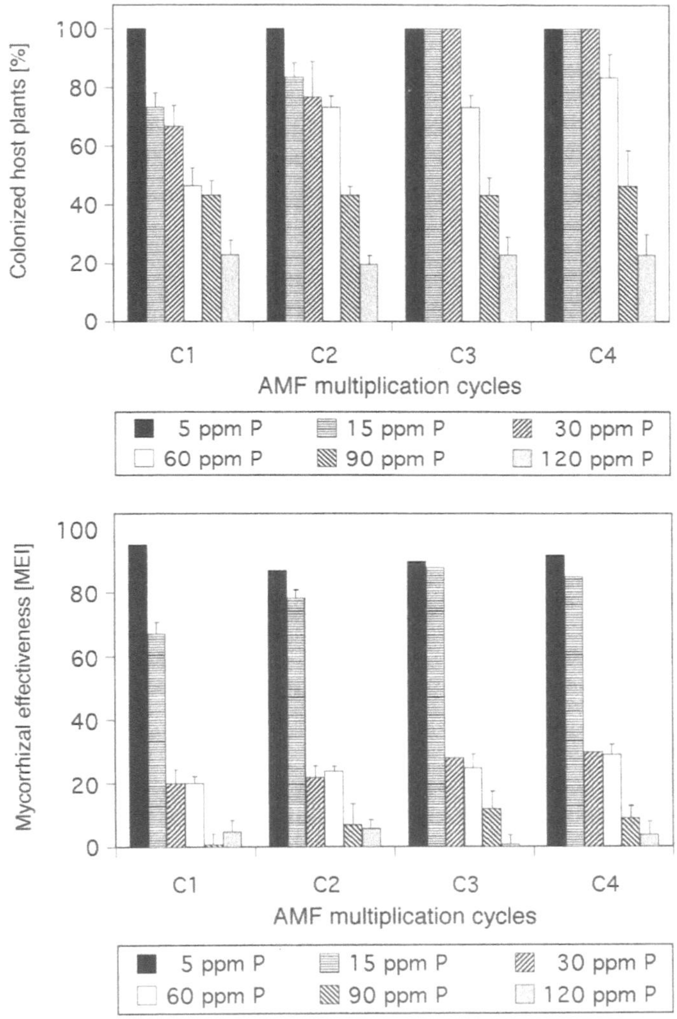

Figure 2b. Root colonization and mycorrhizal effectiveness of AMF populations (*Glomus sp.* GK 12 on *Anagallis arvensis*) with technically modified genotype composition (Selection factor 'soil-P-concentration', see text). Bars: SD

In contrast to the experiment concerning the soil pH, modifications of the percentage of colonized plants led not to changes of the inoculum effectiveness. The reason for this finding might be found in the genetically fixed low dependence of the host under favourable conditions. Under variable environmental conditions the physiological status of the host is probably the main factor that expresses dependency or independency on AMF. Consequently the directed inoculum production process is likely to be particularly successful, if the relationships between later target plants and desired target mycorrhizal effect are clearly defined before the inoculum production starts.

In summary it is possible to influence the genotype composition of an AMF population by directed processing of inoculum production. Abiotic environmental factors can be used to select and better focus AMF genotypes, but the chosen plant species with its specific mycorrhizal dependency also seems to have an important influence on the outcome.

Natural and experimental adaptation of AMF inocula to environmental stresses

After these first experiments there remained the question as to whether selected sub-populations have an effectiveness comparable to ecologically highly adapted ecotypes of AMF. Important questions concerning inoculum production strategies thus remained e.g. does the isolation of ecologically adapted AMF for a specific production process has to be performed or whether technical adaptation (a much cheaper alternative) can be carried out?

In order to describe the relative effectiveness of a standard inoculum we compared *Glomus sp.* GK 12 before and after DIPP with AMF either naturally adapted to salt stress or another population adapted to heavy metal stress. In both stress situations the presence of mycorrhiza is favourable for the host (e.g. Rosendahl and Rosendahl 1991; Hildebrandt et al. 1999) and is intensively studied in applied research projects of the EU (e.g. MYCOREM, EU No. QLRT-1999-0009).

In both cases the effectiveness of naturally adapted AMF populations (without previous technical propagation) was higher than the experimentally non-selected or preselected sub-populations of *Glomus sp.* GK 12 (Tab. 1), with the preselection process resulting in AMF populations with 85% of the effectiveness of naturally occurring salt tolerant AMF and 87% of the effectiveness of heavy metal tolerant AMF. We know from applying the mycorrhizal technology, differences of less than 10% effectiveness are of low significance for a

Table 1. Comparison of effectiveness (MEI) of naturally and experimentally adapted AMF populations on the growth of *Plantago lanceolata* under abiotic stresses (salt and heavy metals, see "Material and methods")

	Glomus sp. GK12		Native AMF
	without	with	without
	phenotype preselection [MEI]		
Salt stress	25.6 ± 3.5	45.4 ± 2.7a	53.4 ± 3.5a
Heavy metal stress	18.8 ± 1.3	65.9 ± 13.6a	75.8 ± 7.8b

The test was repeated three times with 25 individuals per treatment. Values of one row marked with the same letter are not significantly different.

plant grower. Technical pre-selection of sub-populations with favourable characteristics and results in an inoculum with a similar effectivity to an inoculum consisting of naturally adapted AMF from natural sites. Therefore the most rapid and economic strategy for inoculum supply can be chosen by the inoculum producer, without loss of effectiveness.

As demonstrated above the replication of inoculum can change its characteristics. Thus strategies that involve isolating adapted fungi and then multiplying inoculum for several generations are risky. Moreover, the adoption of such a strategy, that necessitates maintenance of a 'gene bank', is resource intensive and thus expensive. Consequently, our approach is to leave AMF ecotypes at sites subject to abiotic stress i.e. maintain them in their natural habitat for further use. The sites themselves serve as '*in situ* conserved gene banks' of specially adapted AMF. For our location, several natural sites are listed (Tab. 2). In the past we have used unclassified AMF, but will identify them in the future using PCR-techniques.

Stability of AMF effectiveness after selection by stressors

Selection of sub-populations by applying stressors is achieved by replication of the strain under conditions of stress (e.g. salt) and separation of the most effective sub-populations afterwards. Since a commercial inoculum must be free of contaminants, e.g. salt or heavy metals originating from the first inoculum production step, the second step, therefore, must involve production without stressors. Despite this the inoculum must still retain the desired characteristics.

In order to ascertain the stability of effectiveness after stress selection we tested *Glomus sp.* GK 12 before and after genotype selection under salt stress

Table 2. Ecologically different areas with AMF populations in Lower Saxony, Germany ('*in situ* conserved gene bank' of IFP)

Vegetation type	Assumed differentiating abiotic eco-factor
Mesophyllic forest	pH > 5.5, moderate water and nutrient availability
Forest on acid soil	pH < 5.0, low water and nutrient availability
Forest on calcareous soil	pH > 6.5, low-moderate water availability
Bog forest	wet areas, flooded over months
Moist green lands	high ground water, high host diversity
Violo-Nardion	dry and sandy soils, low pH.
Rock debris vegetation	especially grasses and mosses and ferns as hosts, partially extreme conditions
Heavy metal vegetation	variable heavy metal content
Primary and secondary salt vegetation	salt stress
Long term agriculture	frequent disturbance, high fertilizer input
Sustainable agriculture	frequent crop rotation, bio-fertilization
Wall vegetation	dryness

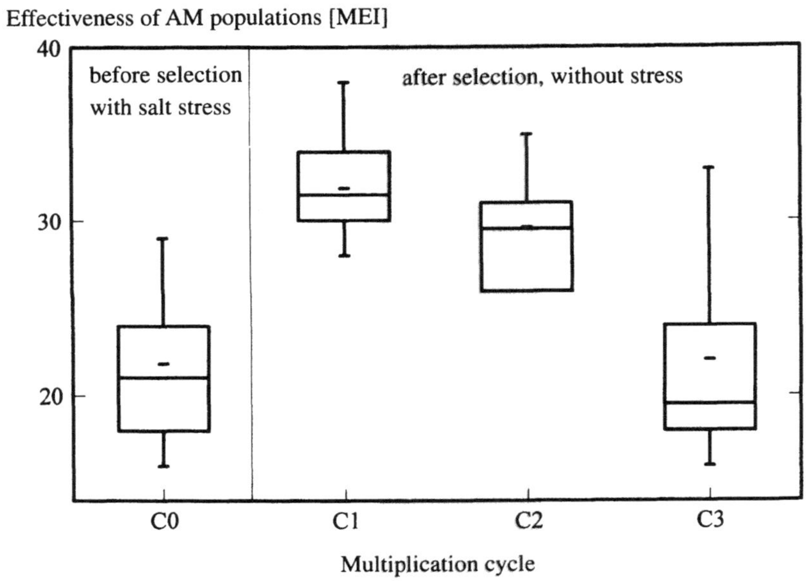

Figure 3. Stability of strain characteristics (effectiveness) before and after sub-population selection under the influence of salt stress. The mycorrhizal effectiveness index (MEI) was calculated according to Bagyaraj (1994)

(Fig. 3). We found an increase in the effectiveness in C1 due to the selection of only very effective sub-populations from the initial population C0. In the subsequent second reproduction step (production without stressor, C2) similar results were obtained, although the variability of effectiveness increased in the third propagation cycle without a stressor (C3).

For the commercial inoculum production these data suggest that the first multiplication step after selection can be carried out without applying a stressor without loosing the desired characteristic. These results are consistent with the findings shown in Figure 1.

Preference of AMF genotypes by host plants

For several years it has been well known that different varieties of host plants react specifically to the same AMF inoculum (Azcon and Ocampo 1981). On one hand, this reflects the mycorrhizal dependency of the host under specific environmental and cultural conditions. But on the other hand, our data suggest that such specific interactions could also be due to selection for AMF populations.

In order to prove whether preferences (Dhillion 1992) of host plant species for AMF genotypes are able to influence the inoculum characteristics we tested, in 1998, 52 separate inocula, i.e. sub-strains of *Glomus sp.* (morphologically similar to *Glomus etunicatum*) on different host combinations (Tab. 3) and measured the mycorrhizal effectiveness index (MEI). Of these 52 sub-strains nine of different effectiveness were chosen and stored for a subsequent test in 1999. A mixture of all 52 sub-strains was included in the experiment. The inocula were used to inoculate a target plant (*Baptisia tinctoria*) that had not been used as host plant during inoculum production.

In the first year an effectiveness spectrum from neutral to positive was expressed in treatments with *Zea mays* and from negative to positive effectiveness with *Tagetes erecta* when the host plant was cultivated alone.

During the next inoculum production cycle the effectiveness of the sub-strains on *Zea mays* changed, with a tendency for each to reach all the same value of MEI. The best sub-strain was still the best but with decreased effectiveness while the others enhanced their effectiveness.

When produced on *Tagetes* the effectiveness of the sub-strains decreased when *Tagetes* was also used as a target plant. Decreases were smaller in sub-strains with higher effectiveness before the second multiplication. When inoculum was produced on *Tagetes* and then used to inoculate maize effects were positive, but showed a positive effectiveness on *Tagetes* itself only when it had been very effective during inoculum production (MEI > 40). In case of lower MEI inoculation of *Tagetes* resulted in neutral or even negative effectiveness.

Table 3. Influence of mass production on the effectiveness of AMF inoculum (strain *Glomus sp. GK12)* inoculated to 52 plots

Inoculum production 1998			Inoculation 1999 [MEI]				
host plants	sub-strain	MEI	Z	T	Z/T		*Baptisia tinctoria*
Z. mays	9	43.1	31.6	28.6	44.1	12.1	31.0
	40	19.8	29.6	12.2	28.7	−7.1	25.3
	41	0.4	23.5	−36.4	24.8	−18.5	16.7
	Mix	-	22.6	0	25.4	−8.2	44.4
T. erecta	24	43.0	36.3	29.4	44.4	4.8	44.7
	32	14.6	21.5	−20	28.5	−34.2	−5.2
	45	12.1	33.0	−36.5	29.4	−23.4	56.5
Z. mays/ T. erecta	52	39.3	40.4		31.8		48.7
		14.0		−1.7		3.1	
	44	32.5	20.7		60.5		25.4
		−52.5		−57.9		−72.2	
	22	36.3	−31.4		67.4		32.4
		−22.8		−18.2		−34.8	

Nine sub-strains of 52 selected with respect to host and target plant species (fresh weight of host shoots). The MEI was calculated on the basis of 50 plants per plot. 'T' *Tagetes erecta*, 'Z' *Zea mays*, 'T/Z' mixed cultivation. *Tagetes* or maize cultivated alone. 'MIX' sub-strain mix.

During co-cultivation of *Zea mays* and *Tagetes erecta* the AMF effectiveness was host specific (MEI of sub-strain 44 positive on maize, negative on *Tagetes*, Table 3). The sub-strain characteristics could be reproduced when the sub-strains were afterwards tested on each host separately: For instance, sub-strain 44 remained negatively effective on *Tagetes erecta* and positively effective on *Zea mays*.

The reasons for the drastic specificity may be partially due to the high degree of competitiveness, with *Zea* reducing the growth of *Tagetes* (Feldmann et al. 1999). Moreover, inoculation with AMF enhanced this negative effect of maize on *Tagetes*. But the persistence of the sub-strain characteristics in case of separately grown and inoculated host species indicates the existence of specific host/fungus genotype combinations resulting in specific interrelationships between the symbiotic partners.

Similar interaction effects appear to exist for *Baptisia tinctoria*. However, a classification of *B. tinctoria* as '*Zea*-type' host or '*Tagetes*-type' host was not possible as there were no strong correlations between effectiveness on host plants and target plants.

Our results can be explained by the following hypothesis: host plant species are preferentially colonised by specific AMF genotypes within the inoculum. The physiological basis for a postulated preference may not be connected with effectiveness of the developing symbiosis i.e. AMF exhibit a colonisation behaviour (Feldmann 1998b) that is transmitted to the next spore 'generation'.

If such AMF genotypes are able to colonise hosts without respect to later effectiveness, as shown for *Petroselinum crispum* (Feldmann 1998b) the response of the host in a symbiotic interaction would be a consequence of the composition of AMF populations. For example, sub-strain 41/*Zea mays* would include more genotypes with negative impact on further symbiotic interactions than, for example, sub-strain 9 (Table.3). If plants, e.g. maize, can control AMF genotypes with negative potential, or favour AMF that promote mutualism, and other plants, e.g. *Tagetes*, are not able to do so, their specific influence on AMF genotype composition would lead to different genotype frequencies in the next spore 'generation'. Consequently, inoculation of *Tagetes erecta* with sub-strains containing a larger number of negatively effective AMF genotypes can lead to negative growth responses of the plant. Furthermore, the postulated inability of plants like *Tagetes* to control the further replication of non-mutulistic AMF genotypes leads to inocula which are still less effective than before. With respect to the 'intermediate' response of *B. tinctoria* to the inoculated sub-strains the question arises whether there could be a continuum in the ability of host plants to control and direct the mycorrhizal influence to their physiology.

The consequence of these findings for the practice of inoculum production is that it is preferable to base predictions of inoculum effectiveness only on well characterised host plant species. In our studies a positive response of inocula on target plants (*Tagetes* and *Baptisia*) was observed only when the previous MEI exceeded 18.6 on *Zea mays* as an inoculum host.

For the mass production of a widely homogenous inoculum with positive effectiveness on the desired target plant species it will be necessary to use only one host with well-characterised selectivity.

The constancy of AMF effectiveness within a mass production process

Mass production of AMF commonly means the production of up to several hundred litres of inoculum containing ca. 80.000 infection units per litre. Inoculum is normally produced in pots of different sizes with up to four host individuals. In the absence of nutrient limitation the growth of the host plants in pots is generally homogenous, due to limited space for root development. Therefore, differences in AMF effectiveness of sub-populations were rarely observed or interpreted as being due to genetic differences between host individuals. If inocula are produced in larger plots limits on root growth are generally removed and up to 50 host plants may be raised (Feldmann et al. 1999).

For our company commercial AMF mass production now means producing more than 25,000 litres inoculum per year (with 100,000 infection units/l). Such amounts of inoculum require a starter inoculum of at least 125,000 infection units. Such quantities of a genetically homogenous inoculum can now be generated using the methods outlined. However, if the process were scaled-up it is possible that this homogeneity would not persist because of the heterogeneity that remains in the AMF. A further segregation of a preselected substrain with high effectiveness into new sub-strains with neutral to high effectiveness occurred (Fig. 4). Nevertheless, more than 90% of the inoculum caused positive growth response in the host (*Zea mays*) during inoculum mass production. If this quota is reproducible, it would be economically feasible to select sub-strains with special effectiveness here a second time and to discard sub-populations of lower effectiveness after mass production.

The Directed Inoculum Production Process (DIPP) and its influence on predictability of inoculum effectiveness

Using our data as a basis we are able to provide a protocol for the direction of later AMF genotype composition included in the inoculum production process which should result in better predictability of inoculum effectiveness. This is as follows:

1. The agricultural or horticultural problem needs to be defined clearly. Based on this assessment a decision is then made as to whether the inoculum can be based on established generalist strains or requires specially adapted AMF from natural areas.
2. Tests with standard plants, e.g. *Zea mays,* are carried out and, if possible, the later target plant species to show the potential effectiveness of the inocula. These small scale tests should be carried out under controlled environ-

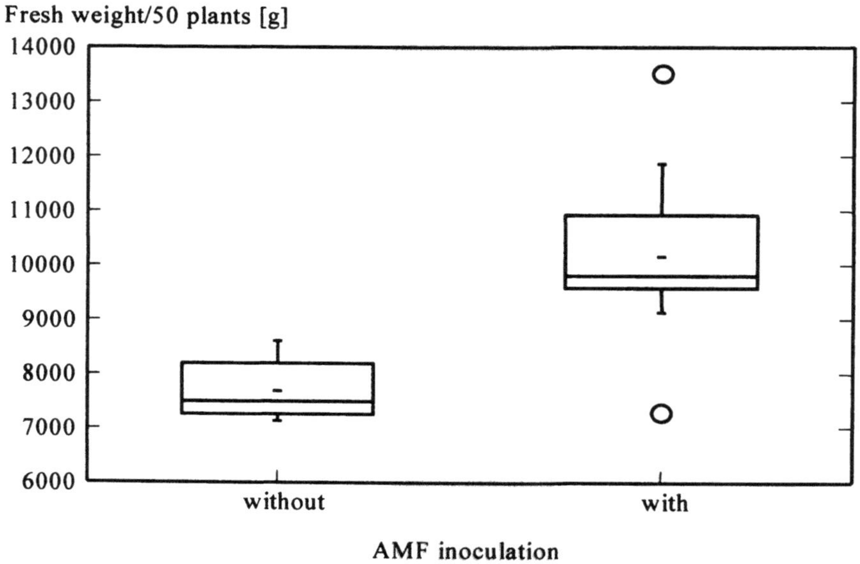

Figure 4. Variability of effectiveness of an AMF strain *Glomus* sp) during inoculum mass production in plots with 50 host individuals (*Zea mays*).

mental conditions in growth chambers and include stressors. The analysis of the variability of AMF effectiveness in a given test system allows the selection of the best sub-strains.

3. The third step includes the mass production. How much inoculum can be produced in this step without any loss of stability is not well known.

The DIPP was introduced to our company in 1996 and has since been optimised. Defining 'predictability of AMF effectiveness' as the frequency of expected host growth response to symbiosis we can compare experiments before and after the introduction of DIPP. The results (Tab. 4) shows that pre-

Table 4. Increase of predictability of mycorrhizal effectiveness with the Directed Inoculum Production Process (DIPP)

Inoculum production	Experiments [n] environment		Predicted success [% experiments] environment	
	constant	variable	constant	variable
with DIPP	16	35	87.5%	68.6%
without DIPP	59	41	52.5%	36.6%

'Constant environments' are greenhouse or growth chamber conditions. Field or garden experiments were carried out under 'variable environments'

dictability could be significantly increased. Nevetheless, we still cannot be certain of the consequences of inoculation. Despite this uncertainty DIPP seems to be the best currently available method to provide effective inocula and represents a major advancement of knowledge.

Perspectives

Producing AMF inoculum is still a 'grey' box process. In defining an AMF 'genotype' we focussed on phenotypic effects which were pronounced in the hosts by single spore inoculation and could be reproduced after replication of single spore descendants. Tommerup (1988) defined AMF at the species level as AMF genotype). Nevertheless, the stability of the characteristics was very low, suggesting that there might be processes ongoing that can change the strain characteristics rapidly (within certain limits). We propose that such changes do not occur spontaneously but are triggered by abiotic or biotic ecofactors including the host itself. If we assume gene/gene interactions of host and fungus to establish and perform a symbiosis (Krishna et al. 1985; Lackie et al. 1988; Gollotte et al. 1993) and if we accept that the quantitative effects of the symbiosis depend on polygenic characters of the host, any increasing or decreasing variability of the host phenotype may be due to a large amount of mycorrhiza induced changes in host physiology.

Of special importance is the multinuclear character of AMF spores (Peterson and Bonfante 1994; Genre and Bonfante 1997; Lingua et al. 1999). We still do not know how much and which nuclei of an AMF spore are active, how they are activated and what influence the heterocaryosis within a spore would have on the effects observed. Does karyogamy exist? Does a population biological process exist that favours the selection of specially adapted nuclei within the population of single spore descendants of an AMF strain? Are strain characteristics under the control of the host? We propose that the relative stability of AMF effectiveness after one propagation cycle suggests there is no arbitrary exchange of information between spores colonising a host but there is host-regulated competition between genotypes. This hypothesis suggests that a 100% predictability of mycorrhizal effectiveness cannot be achieved.

The DIPP presented here integrates many aspects resulting from the practical extrapolation of this theoretical hypothesis and is already leading to more than 85% predictablity under commercial conditions. This means that we have solved a universal problem to an extent which probably approaches the biological limitations of the system. In future we will turn to technical applications of our DIPP, e.g. in bioreactors and *in vitro* techniques. However, clarification of the basis of mycorrhizal dependency of host plant species (Tewari et al. 1993; Boyetchko and Tewari 1995) will be of special importance for the economically successful application of the mycorrhizal technology in agriculture and horticulture in the future.

Acknowledgements
We thank Dr. C. Boyle and Dr. B. Schulz, Institute for Microbiology, Technical University, Braunschweig for intensive discussions on the manuscript. The project was partially financed by the German Bundesstiftung Umwelt.

References

Azcon R, Ocampo J A (1981) Factors affecting the vesicular arbuscular infection and mycorrhizal dependency of thirteen wheat cultivars. New Phytologist 87: 677-685

Bagyaraj D J (1994) Vesicular-arbuscular mycorrhiza: Application in Agriculture. In: Norris J R, Read D J, Varma A K (eds) Techniques for the study of mycorrhiza. Academic Press, London pp 819-833

Baltruschat H (1993) Zur ökosystemaren Bedeutung der VA-Mykorrhiza in Agrarökosystemen und Möglichkeiten der praktischen Nutzung dieser Symbiose. Habilitation in the FB 17 of the University Giessen, Germany

Boyetchko S M, Tewari J P (1995) Susceptibility of barley cultivars to vesicular-arbuscular mycorrhizal fungi. Canadian Journal of Plant Science 75: 269-275

Chang D C N (1994) What is the potential for management of VAM in horticulture? In: Robson A D, Abott L K, Malajczuk N (eds) Management of mycorrhizas in agriculture, horticulture and forestry pp 187-190

Dhillion S S (1992) Evidence for host-mycorrhizal preference in native grassland species. Mycological Research 96:359-362

Dodd J C, Thomson B D (1994) The screening and selection of inoculant arbuscular mycorrhizal and ectomycorrhizal fungi. In: Robson A D, Abott L K, Malajczuk N (eds) Management of mycorrhizas in agriculture, horticulture and forestry: pp 149-158

Feldmann F (1998a) Symbiontentechnologie in der Praxis: Arbuskuläre Mykorrhiza im Gartenbau. Thalacker-Medien, Braunschweig, Germany. ISBN 3-87815-109-8

Feldmann F (1998b): The strain - inherent variability of arbuscular mycorrhizal effectiveness: II. Effectiveness of single spores. Symbiosis 25:131-143

Feldmann F, Hutter I, Niemann P, Weritz J, Grotkass C, Boyle C (1999) Einbindung der Mykorrhizatechnologie in die Heil- und Zierpflanzen-produktion sowie den Endverkauf. Mitteilungen der Biologischen Bundesanstalt 363:6-38.

Feldmann F, Idczak E (1994) Inoculum production of VA-mycorrhizal fungi. In: Norris J R, Read D J, Varma A K (eds): Techniques for mycorrhizal research. Academic Press, San Diego, pp 799-817

Feldmann F, Kruse W, Boyle C, Lieberei R (1998) The Strain - Inherent Variability of Arbuscular Mycorrhizal Effectiveness: I. Development of a Test System Using *Petroselinum crispum* as Host. Symbiosis 25:115-129

Genre A, Bonfante P (1997) A mycorrhizal fungus changes microtubule orientation in tobacco root cells. Protoplasma 199:30-38

Gollotte A, Gianinazzi Pearson V, Giovanetti M, Sbrana C, Avio L, Gianinazzi S (1993) Cellular localization and cytochemical probing of resistance reactions to arbuscular mycorrhizal fungi a 'locus a' myc⁻ mutant of *Pisum sativum* L. Planta 191:112-122

Hildebrandt U, Kaldorf M, Bothe H (1999) The zinc violet and its colonization by arbuscular mycorrhizal fungi. Journal of Plant Physiology 154:709-717

Johnson N C, Graham J H, Smith F A (1997) Functioning and mycorrhizal associations along the mutualism-parasitism continuum. New Phytologist 135:575-586

Krishna K R, Shetty K G, Dart P J, Andrews D J (1985) Genotype-dependent variation in mycorrhizal colonization and response to inoculation of pearl millet (*Pennisetum americanum*). Plant and Soil 86:113-126

Lackie S M, Bowley S R, Peterson R L (1988) Comparison of colonization among half-sib families of *Medicago sativa* L. by *Glomus versiforme* (Daniels and Trappe) Berch. New Phytologist 108:477-482

Lingua G, Sgorbati S, Citterio A, Fusconi A, Trotta A, Gnavi E, Berta G (1999) Arbuscular mycorrhizal colonization delays nucleus senescence in leek root cortical cells. New-Phytologist 141:161-169

Miller J C, Rajzpakse S, Garber R K (1986) VAM in vegetable crops. Hortic Sci. 21:974-984

Nemec S (1987) VAM in horticultural systems. In: Safir S R (ed) Ecophysiology of VAM plants. CRC Press Boca Raton, USA pp 193-221

Peterson R L, Bonfante P (1994) Comparative structure of vesicular-arbuscular mycorrhizas and ectomycorrhizas. Plant and Soil 159:79-88

Rosendahl C N, Rosendahl S (1991) Influence of vesicular-arbuscular mycorrhizal fungi (*Glomus* spp.) on the response of cucumber (*Cucumis sativus* L.) to salt stress. Environmental and Experimental Botany 31:313-318

Schenck N C (1984) Methods and principles of mycorrhizal research. The American Phytopathological Society, St. Paul, Minnesota, USA p 244

Sieverding E (1991) Vesicular-arbuscular mycorrhiza management in tropical ecosystems. Deutsche Gesellschaft für Techn. Zusammenarbeit, Eschborn.

Tewari L, Johri B N, Tandon S M (1993) Host genotype dependency and growth enhancing ability of VA-mycorrhizal fungi for *Eleusine coracana* (finger millet). World Journal of Microbiology & Biotechnology 9:191-195

Thompson J P (1994) What is the potential for management of mycorrhizas in agriculture? In: Robson A D, Abott L K, Malajczuk N (eds) Management of mycorrhizas in agriculture, horticulture and forestry: pp 191-200

Tommerup I C (1988) Long-term preservation by L-drying and storage of vesicular arbuscular mycorrhizal fungi. Transactions Brit. Mycol. Soc. 90:585-592

Varma A, Schuepp H (1994) Infectivity and effectiveness of *Glomus intraradices* on micropropagated plants. Mycorrhiza 5:29-37.

Weissenhorn I, Feldmann F (1999) Perspektiven der Nutzung der arbuskulären Mykorrhiza im niederländischen Gartenbau unter Glas. Mitteilungen der Biologischen Bundesanstalt 363:65-73.

Mycorrhizal Technology in Agriculture
ed. by S. Gianinazzi, H. Schüepp, J.M. Barea and K. Haselwandter

Quality control of arbuscular mycorrhizal fungi inoculum in Europe

H. von Alten[1], B. Blal[4], J.C. Dodd[2], F. Feldmann[3], M. Vosatka[2]

[1] *Mycotec GbR, Oldendorfer Str. 34, D-31840 Hessisch Oldendorf, Germany*
[2] *Plantworks Ltd., 1/19 Innovation Building 1000, Sittingbourne Research Centre, Sittingbourne, Kent ME9 8HL, UK*
[3] *Institut für Pflanzenkultur (IFP), Solkau 2, D-29465 Schnega, Germany*
[4] *Biorize, 8 Rue sainte Anne, F-21000 Dijon, France*

Keywords: Small to medium sized companies (SMEs), production of inocula, commercial use, regulations, code of best practice, quality control, specific quality procedure, physical and chemical properties of inoculum, propagule density, most probable number of propagules (MPN), infection units, guaranteed efficacy, 'preferential selection', microbial contaminants, plant pathogens

Introduction

Most published papers by scientists working in the area of mycorrhizal fungi will mention in the first few lines of their introduction the potential importance of these natural fungi for biotechnology in agriculture and of potential applications of their findings. The growing number of new small to medium sized companies (SMEs) around the world (Sylvia 2001) producing inocula of mycorrhizal fungi indicates that many scientists have seen market opportunities for the commercial use of these fungi to increase in the last decade. Many companies have therefore 'spun-out' of the academic and research world into the business world. It is at this point that, in recent years at least, their products have come under increasing scrutiny by fellow scientists and the end-users alike. Many find that the promises made about their product and the results seen by the end-users are often world's apart. This has led to sweeping generalisations, positive and negative, about the efficacy of mycorrhiza products currently available. As natural biological agents and, for arbuscular mycorrhizal fungi at least, non-axenically culturable fungi, there are great problems in presenting the product in the best state for the target markets. Some have taken the approach of single formulations for every market whilst others produce a range of products for their target buyers. Whatever the approach, it is increasingly likely that greater regulation and controls over the production and selling of such inocula will be introduced in the coming years. This is the reality of the business world but scientists and businesses alike need to begin to look at how a series of 'best practices' can be adopted by these SMEs to allow the market to develop. At present regulation of these products varies between countries in Europe with some having very tight regulations e.g. France while others are less demanding. Over-regulation will prevent the development of SMEs and could destroy the market for what is potentially one of a few biotechnologies using natural microbes available for plant production.

In this article a group of European SMEs have come together to compare notes and make suggestions for a way forward. This may lead to a loose voluntary code of best practice, even a Federation of Mycorrhizal Fungi Inoculum Producers in Europe. Nevertheless this is an attempt to address the complicated problems producers face in their honest attempts at marketing an ethical and useful biotechnology. The article will concentrate on AMF as the production of ectomycorrhizal fungi (ECMF) has its own unique problems. In spite of the fact that each company still has its specific quality procedure adapted to its specific demands and target markets the article will demonstrate the principal problems of the quality assurance, control and efficacy issues of mycorrhizal products.

The issues in quality control

There are many issues associated with 'hyping' the potential benefits of using AMF but these are not different from the marketing of other biological products in an attempt to sell a product, which can (according to the literature):

1. increase plant P uptake and reduce demands for fertilizers
2. potentially increase plant growth and crop uniformity
3. reduce plant mortality
4. reduce root diseases
5. increase plant tolerance of pollutants
6. allow earlier and better flowering
7. increase soil aggregation (soil structure)
8. increases tolerance of water stress
9. act as a mechanism for ecological land restoration

Amongst other claims, all that means that making use of AMF can have a broad range of possible benefits to the end-user. Many of these can justifiably be used to support the use of AMF as a natural 'plant health insurance'. There are criteria that should be fulfilled by the producer of inoculum; one is that by recommending use of the product the plants treated should form mycorrhizas as a result (given proper use of the product); secondly that the producer has taken every care in producing a product free of potential agents which could negatively affect normal plant growth and development; and last that while the production costs have to be kept to a minimum the shelf-life of the product should be sufficient to suit the end-user markets. This should mean that the producer also has a degree of responsibility to educate the end-user and supply back up support. We will look at several aspects of quality control and product declaration, knowing that not all of them will be listed in a final agreement between SMEs at the European level.

Declaration of physical and chemical properties of AMF inoculum

Nutrient content of the inoculum

Customers who intend to introduce mycorrhizal fungal inoculum to target plants must be provided with some basic information concerning the chemical and physical characteristics of the inoculum. The nutrient content and the pH of the substrate solution can be of special importance if high doses of inoculum must be used under controlled fertilizer regimes. Besides this it must be possible to differentiate between the effects of additives like fertilizers or gels and the AMF themselves. Because of the processing a certain amount of nutrients will unavoidably remain in the substrate of conventionally produced AMF inoculum. Nevertheless the processing can reduce the phosphate (P) content of the substrate (Tab. 1). Conventional soil analytic examinations in independent official laboratories can provide inoculum producers with the necessary information about the macro-nutrient content of the inoculum at no great expense. Additionally the independent laboratory can provide a certificate, which guarantees the customers the validity of the data. This enables AMF effects and available nutrient influences to be separated when using the inoculum. Excessive amounts of added nutrients or organic compounds can inhibit seed germination even if they are useful for improving plant growth and mycorrhization in later stages of plant development. Therefore if inocula are combined with biofertilizers or slow release fertilizers, careful tests should be undertaken to check the inhibition of seed germination. Amendment of inocula by additives should be primarily aimed to support mycorrhiza development and therefore the components of these additives should not be general fertilizers but they have to be specifically tuned to be compatible with AMF.

Table 1. Fertilizer content of a commercial inoculum (IFP 01/99)

Nutrient	mg/l
Salt (KCl)	912
Nitrogen (N)	27
Phosphate (P_2O_5)	7
Potassium (K_2O)	29
Magnesium (Mg)	87

The analysis was carried out in an independent laboratory (LUFA, Germany) with standard methods of soil analysis (data presented by IFP).

The pH of the inoculum substrate

Information about the substrate pH is of little importance for the target plant production system because of the relatively low amounts of inoculum (normally 10% v/v) applied to a normally well-buffered soil system. Nevertheless, the range of adaptability of the AMF population within the inoculum to variable pH conditions can be estimated from the basic information provided by the inoculum producer. To our knowledge (IFP), the majority of AMF propagules of an AMF inoculum colonise best under the previous production conditions (Feldmann and Grotkass in this volume). Inoculation under differing pH regimes may often require a longer time to show the desired mycorrhizal effects.

Carrier material

AMF can be bound to a wide range of carrier materials (Backhaus and Feldmann 1996; Jarstfer and Sylvia 1994). Information about the carrier material is important for the subsequent inoculation procedure. For instance, the inoculation of plants on 'roof tops' requires a carrier, which effectively protects the AMF propagules during the 'blowing-up' procedure of the substrate. In our experience (IFP), turf substrates or expanded clays did not survive this rough procedure but other materials like lava were resistant enough and guaranteed effective AMF colonization after a short time. Similarly attapulgite clays, expanded clays or other inert carriers have been very successful carriers in horticulture (Biorize, Mycotec) and landscaping (Plantworks). It, therefore, depends on the target use for the inoculum as to which carrier is best for mycorrhizal fungal propagules.

Testing propagule density

Quantification of mycorrhizal infection units ('MPN'-estimation)

The number of infection units in an inoculum of AMF depends on the number of spores, colonised root fragments and mycelial fragments, which can actually lead to root colonization under favourable conditions. The relevant number of propagules can be determined with various published techniques; the MPN standardized 'most probable number' estimations (An et al. 1990; Daniels et al. 1981; Porter 1979), IP inoculum potential assay (Liu and Luo 1994) and spore counts. In practice, however, the MPN is not a constant and stable measure (Tab. 2). Within the first year after harvest the MPN in a commercial inoculum could vary significantly, probably due to inherent changes in spore maturity, spore dormancy phenomena and the activity of extraradical mycelium (ERM) or colonised root fragments as propagules for later root colonization. After four years of storage the MPN and the number of living spores was nearly identical indicating that the infectivity of the inoculum was mostly based on the spore content at this time. After eight years the remaining infectivity was still 17% of the maximum MPN and the content of living spores still reached

Table 2. Most probable number (MPN) of propagules in a commercial inoculum (IFP 03/89)

Months after harvest	2	6	8	12	24	36	44	82	98
MPN [n]	73 ±9	54 ±5	94 ±7	116 ±12	91 ±4	68 ±3	28 ±6	24 ±4	23 ±1
Spore numbers [n]	48 ±3	51 ±2	42 ±0	47 ±1	37 ±1	24 ±3	22 ±2	24 ±1	20 ±1

The analysis was carried out according to Feldmann and Idczak (1994). The inoculum was stored at room temperature; the carrier material was standard soil (Einheitserde). Spores were counted after the method described by Daniels and Skipper (1984), the most probable number estimation followed the procedure of Porter (1979) and Feldmann and Idczak (1994). (Data provided by IFP)

38% of the maximum count. These data maybe of special importance for inoculum producers and their customers calculating infectivity losses in the course of storage.

In company tests the storage temperature (4, 10 or 20°C) did not affect the colonization potential of individual AMF after various periods of storage nor were there any big differences between isolates of 3 species of *Glomus*. Inocula of these *Glomus* species can therefore be stored in dry substrata at room temperature. Figure 1 (data for *Glomus mosseae* BEG12) shows that this fungus can be stored in sand or attapulgite clay substrates for up to 22 months without real loss of colonization potential. Normally inoculum producers have as a policy of producing to order when possible and inoculum is recycled or discarded after 2–4 years.

The MPN estimation method, however, has its own limitations not the least the inference that one spore leads to one infection unit which underestimates

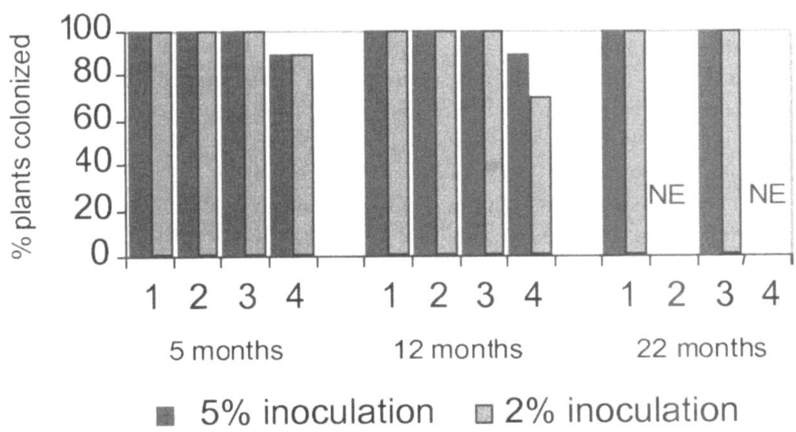

Figure 1. Infectivity assay data for *Glomus mosseae* BEG12 for inocula stored for different lengths of time in four different growth media (1–4; NE = Not Estimated). 5% and 2% inoculation is v/v (data provided by Biorize)

the potential of spores of species of AMF, particularly those in the Gigasporaceae (Dodd et al. 2000). Comparisons between inoculum sources can only occur under single testing conditions due to specific host/AMF interactions, differences in substrates used and the prevailing environmental conditions. In a MPN estimation of a commercial inoculum six host varieties of four plant species were tested in the same substrate. The MPN estimate for a specific host/AMF combination can be up to 4.77 times higher than in another case (Tab. 3). Such a variation makes it impossible to compare estimates made by different laboratories using the MPN procedure, unless attempts of standardisation are being made (identical abiotic and biotic factors).

Table 3. Most probable number estimation (MPN) of propagules in an inoculum of *Glomus etunicatum* with different test plant species

Test plant	*Zea mays*			*Petroselinum crispum*	*Linum usitatissimum*	*Anagallis arvensis*
Variety	Felix	Bad. LM	Blizzard	Mooskrause		
MPN	13 ± 7	20 ± 8	28 ± 6	32 ± 4	37 ± 12	62 ± 9

Before the test a spore content of 50/ml was counted in the substrate after wet sieving. The experiment was repeated three times (data provided by IFP).

This raises the possibility of creating an independent testing service, which could be used by producers in Europe to check that batches of inocula meet the baseline standards established and agreed by individual companies: a voluntary code of best practice. Such a code has been offered to US producers at a price of US$925 (Sylvia 2001); but a more inclusive system could be coordinated, for example, by the BEG (La Banque Européenne des Glomales) if agreement would be reached.

Standard tests for the colonisation potential of commercial inoculum are urgently needed to describe quality differences on the market as shown by recent studies of Plantworks: Estimates of colonisation levels of commercial inocula on different plant species (*Petroselinum crispum, Allium porrum*, and *Zea mays*) grown in the same pot and using an addition of 10% product to 90% attapulgite clay gave evidence for a lack of mycorrhization by certain products under the assay conditions; only one of three tested inocula revealed colonisation of the test plants after 6 weeks. This reveals the problem of variability of inocula being marketed in Europe at present (data provided by Plantworks). The inherent problems of different legal requirements within EU countries is also highlighted by the Biorize's experience where, before marketing Biorize products the company must guarantee a minimum of 10,000 infective propagules per litre of inoculum. The aim of producing EU guidelines, therefore, remains difficult.

Spore numbers

As mentioned earlier spore numbers present can be used as a measure that is more stable than the MPN when single species AMF (*Glomus* or *Gigaspora* spp.) products are used. Unfortunately the method of spore extraction from the substrate and the type of carrier influences the number of spores counted. Usually the mycorrhizal spores of an inoculum are extracted by wet sieving procedures (Pacioni 1994; http://wwwbio.ukc.ac.uk/beg/Protocols/extraction.htm). Variations in the procedure like preparation of substrate samples, centrifugation times, sieve mesh width and so on can, of course, influence the extraction results. But even a standardized extraction procedure leads to misestimations if inoculum carriers are tested which retain the spores and do not allow the extraction (Tab. 4).

Table 4. Spore yield from an AMF inoculum on different carrier materials after wet sieving (data from IFP)

Carrier type	spore number [n/g][a]	extractable spores [n]	extractable spores [%]	living spores [%]
Expanded clay	115 ± 39	23 ± 19	23 ± 6	78 ± 5
Standard soil	76 ± 28	48 ± 23	59 ± 4	87 ± 4
Quartz sand	57 ± 36	49 ± 29	84 ± 7	88 ± 4

[a] The spore number was counted microscopically after diluting 1 g of substrate in water. The same substrate sample was wet sieved afterwards and the extractable spores counted. The portion of living spores was counted according to Glenner (1977). The experiment was repeated three times.

It could be argued that inoculum producers should provide information about the number of extractable spores in their product as many have shown the infectivity of colonised root fragments is greater than that of spores (Sieverding 1991). It is also important to distinguish end use of inoculum in that certain inoculum formulations will have to function in the field situation (landscape use) not in greenhouse assays or production (horticulture) so the adaptability of spores of AMF alone to alien soil conditions may mean that they do not germinate at such high rates (Dodd and Krikun 1984; Tommerup 1983). It is clear, therefore, that only spore counts along with the MPN would serve as a basis for dilution recommendations.

There are other sources of inoculum e.g. aeroponically produced AMF (Sylvia and Jarstfer 1992; Jarstfer and Sylvia 1994) or *in vitro* colonised transformed roots (Declerck et al. 1996); these are potential alternative inoculum sources but only have been successful with a limited number of AMF and cannot be provided in commercial quantities nor at commercially acceptable rates as yet.

Guaranteed efficacy of AMF inoculum

This again finds considerable debate between companies involved in the commercial production of AMF. It has been noted that there are specific host/AMF strain relationships or perhaps 'preferential selection' of AMF by plants (Dodd et al. 1990a, b) and the idea of functional compatibility has arisen as a consequence. This maybe particularly important where single species AMF inocula are produced on monocultures of plants. The same is true in case of qualitative estimates such as predicted effectiveness of the inoculation process. No prediction of the future mycorrhizal effectiveness (as 'strength of desired effect') can be given by the producer if the plant the inoculum is produced on is not the target plant because the outcome of a symbiosis depends on environmental factors, AMF characteristics and plant variables (Tab. 5). Subsequent multiplication on the same host genotype could lead to a decrease of mycorrhizal efficacy. This is probably due to AMF population dynamical processes because it can be reversed by use of other host varieties of the same species (Feldmann 1997).

Furthermore, the type of desired effect can be decisive for the selection of a fungal strain. Strains that enhance biomass in some plants may not increase the stress tolerance of their host and *vice versa*. In different end-user markets it may be more appropriate that the selected consortia of AMF to be used e.g. on trees guarantee long-term ecological benefits rather than biomass increases. Similarly increased flowering by inoculated grasses maybe more beneficial than greater vegetative biomass (Dodd et al. 2001. Plantworks uses multiple host plants in the production of their mixed AMF consortia to increase the potential range of benefits from inoculation. The directed inoculum production

Table 5. Effectiveness (MEI) of three AMF strains on the biomass accumulation of different host plants inoculated with subsequently produced inocula (data provided by IFP)

	Glomus etunicatum HH6				*Glomus etunicatum* HH13				*Glomus intraradices* HH267			
Test year	1.	2.	3.	4.	1.	2.	3.	4.	1.	2.	3.	4.
Zea mays	**31**	**20**	**15**	7	**43**	**37**	**14**	**14**	**27**	**24**	4	5
Pelargonium zonale	**26**	-	**49**	**30**	**28**	-	**56**	**25**	**20**	-	**23**	**28**
Trifolium repens	-	**12**	**20**	-	-	**30**	−1	-	-	**13**	−1	-
Petroselinum crispum	-	9	**13**	**21**	-	**11**	−8	−4	-	**17**	−5	**21**
Baptisia tinctoria	-	7	-	18	-	5	-	**20**	-	−2	-	5
Helianthus annuus	1	−3	-	-	−4	2	-	-	5	5	-	-
Triticum aestivum	1	4	−2	-	**−16**	−9	**−15**	-	9	**19**	−1	-

Values printed in bold are significantly different (t-test, $p < 0{,}05$) from values for control plants. The mycorrhizal effectiveness index was calculated according to Plenchette et al. (1983). Positive values indicate an increase of fresh weight; negative values demonstrate lower fresh weight than in control plants. The number of plants was $n = 30$ per year; the maximum standard deviation was 9.5% of the cited average. The inoculum for the tests was produced each year in a subsequent process on *Zea mays* cv. Felix.

process (Feldmann and Grotkass, in this volume) appears to cover most phenomena and helps to enhance the predictability of AMF inoculum effectiveness because of a technologically adaptation between fungal partner and target plant species with regard to desired effects and explicit environments.

Nevertheless, a quality control of commercial inoculum must deal with the cited problems and find procedures which give the customer some basic information he needs for the decision to introduce effective mycorrhizal fungi. A reference system of information concerning AMF effectiveness could include results from standard tests (Tab. 6) and furthermore a list of examples where the relevant inoculum had already successfully been used before (Feldmann 1998). This could be presented in associated company literature or on the company's website.

Table 6. Mycorrhizal effectiveness index of two commercial inocula on three test plant varieties of different mycorrhizal dependency under standardised conditions (data from IFP)

Commercial inoculum	*Zea mays* cv. Blizzard	*Tagetes erecta* cv. Orange Prince	*Phaseolus vulgaris* cv. Saxa
IFP	48 ± 12	27 ± 9	14 ± 2
Mycotec	44 ± 7	31 ± 7	7 ± 4

Note: the mycorrhizal dependency of a test plant is a variable of the chosen environmental conditions! The standard test was carried out in climate chambers with light (480µE. $m^{-2.}$ s^{-1}) of e.g. SON-T AGRO 400 Philipps bulbs, 14 h/d; minimal temperature 15 °C, maximum temperature 25 °C; 40–60% relative air humidity; irrigation below field capacity; substrate 30% loam, 50% quartz sand, 10% peat and 10% expanded clay; fertilization 2 times per week with a quantity calculated as 10% of the pot volume. The fertilizer used is Flory 9 (1%), pH 5.5; content: 15% N, 7% P_2O_5, 22% K_2O, 6% MgO, 0.03% B, 0.05% Mn, 0.01 Zn. Three test plant species were chosen which were known to be hosts but a) highly dependent on AMF to reach maximum growth under the relevant conditions, e.g. *Zea mays* cv. Blizzard, b) intermediately dependent, e.g. *Tagetes erecta* cv. Orange Prince, and c) of low dependence, e.g. *Phaseolus vulgaris* cv. Saxa

If these data are provided for customers they can be sure that the offered inoculum has a principal ability to act mutualistically with the target host. The inocula, however, cannot be compared on the basis of the standard test since the target plant has to be included in the efficacy tests for an appropriate validation.

Detection of microbial contaminants

As long as the production of large (m^3 or tons) amounts of AMF inocula is connected to the necessity for open culturing of plants in non-sterile greenhouses or open-air systems, these inocula will not be free from other associated microorganisms. Table 7 and 8 show bacterial and fungal presence on par-

Table 7. Amount of different types of bacteria on inoculum of AMF (particles of expanded clay; cfu per g inoculum; data provided by Mycotec)

	Inoculum: isolate number (Mycotec) and host plant during production			
Bacterial col. Type	510 maize	510 tagetes	92 tagetes	114 tagetes
1	1,900,000	1,100,0000	3300	0
2	5000	710	27,000	15,0000
3	0	1000	2400	25,000
4	11,000	0	46,000	11,000
5	1700	0	25,0000	18,0000
6	0	0	39,000	13,000
7	0	0	6400	33,0000
others	3100	840	100	43,0000
Total	1.9×10^6	1.1×10^7	3.7×10^5	1.1×10^6

Table 8. Amount of fungal propagules of different genera on inoculum of AMF

	Inoculum: isolate number (Mycotec) and host plant during production			
Genera	510 maize	510 tagetes	92 tagetes	114 tagetes
Aspergillus	200	530	220	0
Colletotrichum	400	0	0	0
Fusarium	1100	1500	6700	8600
Mucor	0	20	0	400
Penicillium	14,0000	220	220	22,0000
Rhinocladiella	0	1100	3000	0
Rhizopus	1000	0	0	0
Total	1.4×10^5	3.4×10^3	10^4	2.3×10^5

(particles of expanded clay; cfu per g inoculum; data provided by Mycotec)

ticles of expanded clay used as carrier material for propagules of AMF. The composition of the microbes found varied with the host plant and in particular with the AMF isolate used. These microorganisms may include saprophytes that live in the rhizosphere of the host plants; however, there is the possibility that phytopathogenic organisms could be transferred via the inoculum from one host plant to the next albeit in very low propagule numbers. In contrast there are microorganisms that can accelerate the development of the symbiosis or improve plant health. They are called MHB (mycorrhiza helper bacteria) or PGPR (plant growth promoting rhizobacteria).

To avoid unwanted microorganisms producers have in principle two possibilities, the selection of a host plant for inoculum production that is more or less resistant to root diseases or the control of root health during production as often as possible, and using all measures to keep away soil-borne pathogens from the host plants. The selection of a suitable host plant is of primary importance. In general the host plant should never be the same as the plant to be inoculated by the user of the inoculum during production.

This is a basic rule of plant pathology and does not contradict with the idea of directing inoculum production (Feldman and Grotkass in this volume). The efficacy of the mycorrhizal fungi is improved and transfers of specific pathogens e.g. the wilt pathogens within the genus *Fusarium* is avoided. It should be noted that the detection of the genus *Fusarium* does not infer pathogenic presence, as many species are saprophytic. Non-specific pathogens causing root-rots, like species of *Pythium*, *Rhizoctonia*, and *Thielaviopsis* are potentially much more dangerous. If present they can threaten many possible target host plants. It is best if these fungal pathogens can be excluded from the inoculum production totally.

Contamination and spread of plant pathogens can be avoided by Good Horticultural Practice (GHP, Feldmann et al. 1999). If there is any risk of infection suitable fungicides can be used to control pathogens of the host plant (e.g. Plantworks and Mycotec) where the product is not used for the organic market. A removal of pathogens from contaminated inocula at present is not possible because of the recently changing legal situation of plant production products. However, due to the fact that fungicides have to be chosen for the host plant/pathogen combination without consideration of AMF they can be severely influenced by the fungicide application itself. In older studies oomycetes having a special metabolism, could be controlled easily using fungicides like prothiocarb, which can selectively kill the pathogen but leave the AMF unaffected (Tab. 9). Given the background of new legal regulations (directive 91/414/EEC) there is an urgent need to find and also to test registered plant protection products, which are useful in the inoculum production process. Preliminary studies are on the way and will be presented soon.

For the producer of AMF inoculum it is surely impossible to guarantee zero presence of pathogens for his material if the inoculum is produced in open pot

Table 9. Decontamination of AMF inoculum (expanded clay carrier) contaminated with *Pythium* spp. during the production by weekly applications of fungicides to the substratum

Treatment	% root pieces with mycorrhizal structures	*Pythium*-infected root pieces
Untreated	48	19
metalaxyl (0,05%)	15	15
furalaxyl (0,05%)	7	1
prothiocarb (0,2%)	51	1

(host plant *Zea mays* cv Blizzard, Dehne 1984 personal communication)
Note that only metalaxyl is officially registered in Germany

culture, even if they took all precautions. In consequence a quality control of AMF inoculum must cover this aspect of unwanted presence of microorganisms. When selling his material the producer needs to know whether they might have to face any economic risk arising from claims for indemnity. Some companies take out a liability insurance of some sort. However, the strict European product liability legislation has the consequence that insurance policies will not protect pathogen contamination. So producers cannot relax about undertaking appropriate checking procedures for potential pathogens.

During the inoculum production root samples are microscopically checked for the presence of potential pathogenic fungi. Of course it will not be possible to check each plant or each pot; however, root samples are taken in a representative way to provide a maximum of safety. These microscopic checks could become a standard operating procedure for companies and scientific institutions working with the commercial production of AMF and plant pathogens. For SME's, however, there can be an economic problem, not only a good (often expensive) microscope is needed but also personnel with the ability to recognise and identify pathogens as well as to judge their relative significance. In one case a blue staining fungus has developed around the roots but has not colonised them. On closer inspection this was found to be a *Rhizopus* spp. which can commonly be found in partially sterilised soils. This may be mistaken for AMF but no roots in these samples had typical vesicles, arbuscules or any intraradical mycelium development (data provided by Plantworks).

Additional tests can also be used to back-up microscopic examination, for example, a simple possibility is the use of trap plants for inoculation with the inoculum to be tested. The plants used must be highly susceptible to root pathogens. It might even be appropriate to use assortments of plant species covering susceptibilities to all possible pathogens but this may not be feasible. The ideal trap plant should germinate and grow fast, be extremely susceptible and show clear symptoms easily recognisable even for unskilled personnel. One example for such a plant is cress because it is very susceptible to root rots caused e.g. by *Pythium* spp. Seeds of *Lepidium sativum* (cress) can be sown in a Petri-dish containing a sterilized attapulgite clay control as well as pure inocula and dilutions (recommended application rates) of AMF. The percentage germination can be assessed after 8 days of culture in the dark.

Results showed that neither *Glomus mosseae* BEG12 produced in France nor *Glomus mosseae* BEG29 produced in Finland, contained pathogens or an excess of nutrients that would have significantly reduced germination of the cress seed (Tab. 10). The main reason for inhibition of germination of cress seeds were the high nutrient levels or excessive amounts of organic compounds present in the two other inocula tested, although some other non-pathogenic fungi were present. This assay can therefore be useful but care must be taken not to infer pathogen presence; if inhibition of germination occurs it may be related to other factors such as the levels of added nutrients within the inoculum.

Table 10. Germination of cress seeds after 8 days growth on inoculum of two species of AMF

	% of inoculum mixed in the sterile media (v/v)			
	0	5	10	20
AMF	% germination			
Glomus mosseae BEG12	97	98	100	93
Glomus mosseae BEG29	97	98	100	98

(data provided by Biorize)

AMF can colonise an enormous variety of plants and inoculum of these fungi can be used for the majority of cultured plants. A producer of commercial inoculum can offer its product for use in only one plant species or as a more non-specific promoter of many plants. Even if the inoculum is used only for one species different cultivars can show significant differences regarding susceptibility to root pathogens. In consequence the best way of testing for contaminating pathogens is to use the same plant cultivar, under more or less the same growing conditions, as the buyer of the inoculum would do. A small screening for the best AMF isolate using the customer's plants should be part of the service while selling the inoculum.

Conclusions

The quality control of AMF inoculum is still an obligation of the inoculum producing companies and is not yet under control of independent institutions. The producer declares, e.g. on a control sheet which is accompanying the product, what he proved, which methods he used and what results he found. As an example the control sheet of the Committee of Mycorrhizal Application Germany (CMAG 1997) is demonstrated in Table 11.

During consultation (or on the package) the producer should in addition give a recommendation of maximum and/or minimum dilution factors for the inoculation process but use of the optimum amount can be specified to ensure mycorrhization. The producer can also state any relevant adaptations of their fungi to target conditions, which will help the end-user choice. The aim of this is to ensure that the buyer, paying a premium price for a mycorrhiza product, is receiving a product that should, if used properly, ensure mycorrhization of the plants treated. It should not merely be a support carrier improving the activity of other additives in the mix. The issues involved with ectomycorrhizal fungi and their infectivity characters and shelf-life have not been discussed here but it should be noted that some products do combine AMF and ECMF in their mixes to be used on trees.

The companies unified under the cover of this article agree that independent institutions should carry out the quality control of mycorrhizal products,

Table 11. Quality control parameters of AMF inoculum according to the agreement of the Committee of Mycorrhiza Application Germany (CMAG, 12/1997)

Test Parameters
pH
Content of fertilizer of the substrate [mg/l]
Salt (KCl)
Nitrogen (N)
Phosphate (P_2O_5)
Potassium (K_2O)
Magnesium (Mg)
AMF species/strain
Most probable number of propagules (on host plant variety [n/cm^3])
Effectiveness (on *Zea mays, Tagetes erecta, Phaseolus vulgaris*) [MEI]
Germination inhibition (on Lactuca sativa, *Lolium perenne, Phaseolus vulgaris, Lepidium sativum*)
Fungal contaminants
Potential phytopathogens
Hyperparasitic fungi
other saprophytic fungi
Pathogenity of contaminants (on *Tagetes erecta, Zea mays, target plants*)
Potential phytophageous faunistic contaminants
Diptera
Coleoptera, -larva
Collembola
Acari
Nematoda
Gastropoda
Botanical contaminants
Algae (Diatomeae, Cyanophyceae, Chlorophyceae)
'Weeds'

possibly on the national level or even on the European level. It is not clear who could do that job, but first contacts to independent laboratories do exist in different countries. The criteria, which have to be fulfilled, will be defined by the market for mycorrhizal fungi inoculum itself and will hopefully be standardized on the European level after discussion between the related companies. We think that the foundation of a specific certificate for quality of mycorrhizal fungi inoculum would lead to the spread of the idea to sell only high quality products on the European market.

Each company, of course, defines the marketing strategy. But we would like to stress a point of special importance: no producer of mycorrhizal fungi inoculum should declare, as an intended use of the product, the phytosanitary effect of the inoculum. If the marketing strategy points out any effect against phytopathogens the mycorrhizal fungal inoculum would have to be considered as biological plant protection product with the consequence of the necessity to reach authorised registration under the directive 91/414/EEC. The costs of that process would eliminate our attempts to introduce the mycorrhizal technology to plant production systems. This problem already was intensively discussed in the COST Action 838 and formulated in a position paper (Annual report of COST Action 838, 1999). This paper deals in addition with data requirements for a registration process that seems to be of low importance if no one will register mycorrhizal fungi inoculum as a plant protection product. We, as inoculum producers, would appreciate support by the COST Action 838 to find viable and low cost ways of quality control and proposals for the establishment of that quality control in independent institutions. Furthermore rapid and accurate methods, e.g. PCR-techniques, have to be adapted to the demands of quality control. Not only AMF species have to be distinguished in the quality control, but even strains and sub-strains. This problem remains unsolved. Research activities of European partners of the COST Action 838 are needed on another field of quality control: we urgently need rapid assessment protocols for the recognition of the degree of host dependency on AMF.

The aims of controlling unwanted microorganisms in inocula have been discussed above and approaches to limiting their impact made. The use of mycorrhizal fungi for natural plant production is still in its infancy and will require added value by companies to reassure end-users of its great potential. This means more information and guidance to growers and not a hard-sell approach. There are few biotechnologies using natural microbes available to aid sustainable plant production. It is therefore imperative that scientists and business collaborate more regularly to develop this market, as much further research will be needed to tune the products for the markets. We hope that this article is taken as a first attempt to bridge the mycorrhizal science-business gap.

Acknowledgements
Some of the work presented was partly supported by funds contributed from the BEGNET project EU Framework IV, contract number BIO4-CT97-2225.

References

An Z-Q, Hendrix J W, Hershman D E, Henson G T (1990) Evaluation of the 'Most Probable Number' (MPN) and wet-sieving methods for determining soil-borne populations of endogonaceous mycorrhizal fungi. Mycologia, 82:576-581

Annual report of COST Action 838 (1999): http://www.dijon.inra.fr/cost/cost838/rapports/rapport1999.html

Backhaus G F, Feldmann F (1996) Mykorrhiza in gärtnerischen Substraten - endlich einsatzreif? Taspo Gartenbaumagazin 4:12-14

Daniels B A, Skipper H D (1984) Methods for the recovery and quantitative estimation of propagules from soil. In: Schenck N C (ed) Methods and principles of mycorrhizal research, The American Phytopathological Society, St. Paul, Minnesota, USA pp 29-38

Declerck S, Strullu D G, Plenchette C (1996) In vitro mass-production of the arbuscular mycorrhizal fungus, *Glomus versiforme*, associated with Ri T-DNA transformed carrot roots. Mycol. Res. 100:1237-1242

Dehne H-W, Backhaus G F (1986) The use of vesicular-arbuscular mycorrhizal fungi in plant production. I. Inoculum production. J. Plant Dis. a. Prot. 93 (4): 415-424
Dodd J C, Dougall T A G, Clapp J P, Jeffries P (2001) The role of arbuscular mycorrhizal fungi in plant community establishment at Samphire Hoe, Kent, UK –The reclamation platform created during the building of the Channel Tunnel. Biodiversity and Conservation (In Press)
Dodd J C, Krikun J (1984) Observations on endogonaceous spores in the Negev desert (Israel). Transactions of the British Mycological Society 82:536-540
Dodd J C, Boddington C L, Rodriguez A, Gonzalez-Chavez C, Mansur I (2000) Mycelium of Arbuscular Mycorrhizal fungi (AMF) from different genera: form, function and detection. Plant and Soil 226:131-151
Dodd J C, Arias I, Koomen I, Hayman D S (1990a) The management of populations of vesicular-arbuscular mycorrhizal fungi in acid-infertile soils of a savanna ecosystem. I. The effect of pre-cropping and inoculation with VAM-fungi on plant growth and nutrition in the field. Plant Soil 122:229-240
Dodd J C, Arias I, Koomen I and Hayman D S (1990b) The management of populations of vesicular-arbuscular mycorrhizal fungi in acid-infertile soils of a savanna ecosystem. II. The effect of pre-crops on the spore populations of native and introduced VAM fungi. Plant Soil 122:241-247
Feldmann F (1997) Stabilität und Reproduzierbarkeit der Wirksamkeit von Inokulum arbuskulärer Mykorrhizapilze. Mitt. Biol. Bundesanstalt 332, 54-65
Feldmann F (1998) Symbiontentechnologie in der Praxis: Arbuskuläre Mykorrhiza im Gartenbau. Thalacker-Medien, Braunschweig, Germany. ISBN 3-87815-109-8
Feldmann F, Idczak E (1994) Inoculum production of VA-mycorrhizal fungi. In: Norris J R, Read D J, Varma A K (eds) Techniques for mycorrhizal research. Academic Press, San Diego pp 799-817
Feldmann F, Hutter I, Niemann P, Weritz J, Grotkass C, Boyle C (1999) Einbindung der Mykorrhizatechnologie in die Heil- und Zierpflanzenproduktion sowie den Endverkauf. Mitteilungen der Biologischen Bundesanstalt 363:6-38
Glenner G G (1977) Formazans and tetrazolium salts. In: Lillie R D (ed) Biological stains, Williams and Wilkins Co., Baltimore pp 225-235
Jarstfer A G, Sylvia D M (1994) Aeroponic culture of VAM fungi. In: Varma A K, Hock B (eds) Mycorrhiza: structure, function, molecular biology and bio-technology. Springer-Verlag, Berlin pp 427-441
Liu R-J, Luo X-S (1994) A new method to quantify the inoculum potential of an arbuscular mycorrhizal fungi. New Phytol. 128:89-92
Pacioni G (1994) Wet-sieving and decanting techniques for the extraction of spores of VAM fungi. In: Norris J R, Read D J, Varma A K (eds) Techniques for mycorrhizal research. Academic Press pp 777-782
Plenchette C, Fortin J A, Furlan V (1983) Growth responses of several plant species to mycorrhizae in a soil of moderate P-fertility. I. Mycorrhizal dependency under field conditions. Plant Soil 70:199-209
Porter W M (1979) The most probable number method for enumerating infective propagules of vesicular arbuscular mycorrhizal fungi in soil. Australian Journal of Soil Research, 17:515-519
Sieverding E (1991) Vesicular-arbuscular mycorrhiza management in tropical agro-systems. Deutsche Gesellschaft für Technische Zusannenarbeit (GTZ), Eschborn, Germany.
Sylvia D M, Jarstfer A G (1992) Sheared-root inocula of vesicular-arbuscular mycor-rhizal fungi. Appl. Environ. Microbiol. 58:229-232
Sylvia D M, Jarstfer A G (1994) Production of inoculum and inoculation with arbuscular mycorrhizal fungi. In: Robson A D, Abbott L K, Malajczuk N. (eds) Management of mycorrhizas in agriculture, horticulture and forestry pp 231-238
Sylvia D M (2001) http://dmsylvia.ifas.ufl.edu/Commercial.htm
Tommerup I C (1983) Spore dormancy in vesicular-arbuscular mycorrhizal fungi. Trans. Brit. Mycol. Soc. 81:37-45

MIX
Papier aus verantwortungsvollen Quellen
Paper from responsible sources
FSC® C105338

If you have any concerns about our products,
you can contact us on
ProductSafety@springernature.com

In case Publisher is established outside the EU,
the EU authorized representative is:
Springer Nature Customer Service Center GmbH
Europaplatz 3, 69115 Heidelberg, Germany

Printed by Libri Plureos GmbH
in Hamburg, Germany